Livestock Development
in Subsaharan Africa

Westview Replica Editions

The concept of Westview Replica Editions is a response to the continuing crisis in academic and informational publishing. Library budgets for books have been severely curtailed. Ever larger portions of general library budgets are being diverted from the purchase of books and used for data banks, computers, micromedia, and other methods of information retrieval. Interlibrary loan structures further reduce the edition sizes required to satisfy the needs of the scholarly community. Economic pressures on the university presses and the few private scholarly publishing companies have severely limited the capacity of the industry to properly serve the academic and research communities. As a result, many manuscripts dealing with important subjects, often representing the highest level of scholarship, are no longer economically viable publishing projects--or, if accepted for publication, are typically subject to lead times ranging from one to three years.

Westview Replica Editions are our practical solution to the problem. We accept a manuscript in camera-ready form, typed according to our specifications, and move it immediately into the production process. As always, the selection criteria include the importance of the subject, the work's contribution to scholarship, and its insight, originality of thought, and excellence of exposition. The responsibility for editing and proofreading lies with the author or sponsoring institution. We prepare chapter headings and display pages, file for copyright, and obtain Library of Congress Cataloging in Publication Data. A detailed manual contains simple instructions for preparing the final typescript, and our editorial staff is always available to answer questions.

The end result is a book printed on acid-free paper and bound in sturdy library-quality soft covers. We manufacture these books ourselves using equipment that does not require a lengthy make-ready process and that allows us to publish first editions of 300 to 600 copies and to reprint even smaller quantities as needed. Thus, we can produce Replica Editions quickly and can keep even very specialized books in print as long as there is a demand for them.

About the Book and Editors

Livestock Development in Subsaharan Africa:
Constraints, Prospects, Policy
edited by James R. Simpson
and Phylo Evangelou

The nations of Subsaharan Africa experienced declining levels of food production per capita throughout the 1970s and early 1980s, particularly in the area of livestock production. Addressing that problem, the authors of this book assess in a systems context the environmental, biological, and social constraints on future African livestock development and consider prospects for improving productivity. They focus especially on changes needed in production and marketing systems, pointing to important policy considerations.

The book is divided into four parts containing twenty-one chapters, each authored by one or more respective authorities in his or her field. Each section in its own way addresses the entire set of questions; topics include aspects of animal breeding and nutrition, anthropology, economics, ecology, farming systems, governmental policy, land tenure, marketing, modelling, and veterinary medicine.

James R. Simpson is professor and livestock marketing economist in the Food and Resource Economics Department, University of Florida. Phylo Evangelou recently received a Ph.D. from the University of Florida in agricultural economics.

To the livestock producers of Subsaharan
Africa, and to individuals and organizations committed
to that sector's growth and development.

Livestock Development in Subsaharan Africa
Constraints, Prospects, Policy

edited by
James R. Simpson
and Phylo Evangelou

Routledge
Taylor & Francis Group

LONDON AND NEW YORK

First published 1984 by Westview Press

Published 2018 by Routledge
52 Vanderbilt Avenue, New York, NY 10017
2 Park Square, Milton Park, Abingdon, Oxon OX14 4RN

Routledge is an imprint of the Taylor & Francis Group, an informa business

Library of Congress Cataloging in Publication Data
Main entry under title:
Livestock development in Subsaharan Africa.
     (A Westview replica edition)
     1. Livestock--Africa, Sub-Saharan. 2. Animal industry--Africa,
Sub-Saharan. I. Simpson, James R.
II. Evangelou, Phylo.
SF55.A38L58 1984   636'.00967   83-19816

ISBN 13: 978-0-367-01714-9 (hbk)
ISBN 13: 978-0-367-16701-1 (pbk)

Contents

Illustrations

Illustrations.

NOTE : Photographs follow p.138

Foreword

Africa is overwhelmingly a rural continent. Depending upon the country, three to four out of every five people live and work in the rural sector. Agriculture is far and away the most important economic activity on the continent, both in terms of personal and national income. Yet, African agriculture is today in a state of crisis. · As Carl Eicher recently noted in Foreign Affairs 61 (Fall, 1982), "The most intractable food problem facing the world in the 1980s is the food and hunger crisis in sub-Saharan Africa--the poorest part of the world" [p.152]. Furthermore the crisis has been building steadily for the past two decades, and it "stems from a seamless web of political, technical, and structural constraints which are a product of colonial surplus extraction strategies, misguided development plans and priorities of African states since independence, and faulty advice from many expatriate planning advisers" [p.157].

In the same way that the crisis has deep historical roots and has been building over a long period of time, so too will the efforts to resolve the crisis be long-term in nature. A failure to undertake such efforts, or undertaking them half-heartedly and without a sufficient sense of urgency and commitment, will result in an ever-intensifying crisis. Of course, much of the work in addressing these issues must take place through institutions in Africa such as the International Livestock Centre for Africa (ILCA). Yet, the responsibility is great for institutions and organizations around the world that are concerned with Africa to grapple with the food and hunger crisis. Hence, the University of Florida's Center for Africa Studies and the Institute of Food and Agricultural Sciences sponsorship brought together the authors of this book to discuss their differences and personal perspectives.

No single book, conference or even series of conferences can, in and of themselves, alleviate the food production crisis and economic stagnation in

Africa. Yet, drawing together specialists from a number
of disciplines who have been engaged in both basic and
applied research serves the important purpose of sum-
marizing the current state of knowledge and pointing out
the direction for future research and applied work.
This objective has been achieved in this book by interaction between authors and careful editing to insure a
tightly focused, well-integrated product rather than
simply a collection of papers. Consequently, by examining livestock production, and more specifically issues
related to its stagnation in the countries of Subsaharan
Africa, this collection makes a meaningful contribution
toward resolution of the food and hunger crisis.

Three sets of questions are at the heart of this
book. First, what are the constraints to livestock production in Subsaharan Africa? Why is it languishing?
Secondly, what are the prospects for overcoming these
constraints and increasing livestock production?
Finally, in light of the constraints and the prospects,
what policies will be most effective for promoting realistic livestock development options? The editors of
this volume have been addressing these questions in
their own research and writing. James R. Simpson, Professor of Food and Resource Economics at the University
of Florida, has long been involved in the economics of
livestock systems in tropical areas through his research
and teaching. Phylo Evangelou has worked in Botswana and
more recently conducted research on livestock marketing
in Kenya under ILCA auspices. Both have individual
contributions to this book that illustrate their own
approaches to grappling with the three basic sets of
questions that underlie this volume.

To answer the basic questions, the co-editors
have organized the separate chapters, each by one or
more respected authorities in his or her own field, into
four sections to form a coherent whole. Each section
in its own way addresses the entire set of questions.
The first group of chapters constitutes an overview
of the trends that portend ever growing shortages in
livestock production for much of Subsaharan Africa.
Individually, the authors address themselves to the
overall issue of problems and constraints, to ecological
and nutritional factors, and to the appropriateness of
systems analysis for understanding livestock production.
The second group of chapters examines specific aspects
of production and marketing, such as disease deterrents,
small ruminant productivity, animal breeding, cattle
marketing, and the impact of governmental policies.

Insights into past efforts provide valuable lessons
for the present and future. Thus, the third section of
of the book takes up this topic with an examination of
development projects and experiences. The individual
contributors focus on evolving donor assistance strat-

egies, land use development, a historical perspective on the role of livestock, identification of critical socio-economic variables, range management, and a review of livestock projects in the Sahel. The final section looks to the years ahead. Important issues that need to be resolved in terms of future strategies and approaches involve land tenure policy, the question of fenced versus open-range ranching, ways to improve the effectiveness of veterinary services, the economic roles of cattle, and sources of meat and milk other than cattle.

Just as the food and hunger crisis in Subsaharan Africa, and more directly the stagnation of livestock production, must be understood in terms of process and not as a single, isolated event, so too must the resolution of this crisis be viewed in terms of process. That this book stands as an element in this process, and not merely as a single event, adds to its significance. Individually and collectively the contributors to this volume and the editors have furthered our understanding of the nature of the seamless web of political, technical, and structural constraints that have served to create the stagnation of livestock production as part of the larger intractable food problem confronting Africa and its people. The book thus merits the careful attention of those individuals, organizations, and governments who are serious in their efforts to implement positive changes in existing livestock production-marketing systems for the purpose of better meeting the food needs of Africans.

R. Hunt Davis Jr.
Director, Center for African Studies
University of Florida
Gainesville, Florida
December 1983

Acknowledgments

Grateful recognition is extended by the editors to the University of Florida's Center for Tropical Agriculture of the Institute of Food and Agricultural Sciences, and Center for African Studies, for their generous financial and moral support in preparation of this book. The Center for Tropical Agriculture, directed by Dr. Hugh Popenoe, is involved in graduate and undergraduate training, foreign student coordination, and technical assistance and cooperation with other internationally oriented organizations and centers at the university. The Center for African Studies, under the direction of Dr. R. Hunt Davis, Jr., coordinates interdisciplinary instruction and research related to Africa, and provides an active outreach program for schools and the community at large. The editors would also like to thank International Minerals and Chemical Corporation (IMC) for their generous financial support. A well deserved note of appreciation is due Richard Harris and Terri Parsons for the attentive care taken in word processing and typing.

James R. Simpson
and Phylo Evangelou

PART 1

Overview

Subsaharan Africa encompasses a rich diversity of resources put to equally multifarious uses. Nevertheless, areas of considerable homogeneity exist among the countries, such as major economic dependence on the export of two or three primary commodities, principally rural populations, dominance of labor-intensive agricultural systems and political fragility. Livestock have traditionally been an important component in the economies of most subsaharan states, but little progress has been made toward either using resources more efficiently or improving productivity.

There are about 347 million people in Subsaharan Africa, up from 261 million in 1970. Population is projected to double by 2000 to 639 million people. James Simpson, in the first chapter of Part I, calculates that cattle inventory would have to reach 280 million head (from 147 million head in 1980) by the end of this century to meet total consumption requirements if there were no change in per capita consumption levels. Likewise, small stock (sheep and goats) would have to increase to 410 million head from the 1980 level of 230 million head. Increased levels of consumption could, of course, be met solely by productivity increases. This would mean raising the annual production of meat from cattle and small stock from 13.2 and 3.7 kg per head of inventory, respectively, to 25.0 and 6.0 kg—still very low levels when compared with the developed economies.

Expanded total meat and milk production will, undoubtedly, need to take place through both inventory expansion and productivity improvements. The accomplishment of these twin objectives by technicians and policy makers is an incredibly complex task due to the multitude of goals involved, lack of agreement on tradeoffs between policies (let alone the political weighting of tradeoffs), and the interrelatedness of variables. For example, livestock as a source of milk

1

is an extremely important aspect of the production decision making processes of many small holder and pastoral households. Yet, national goals, to the extent they are formulated to provide low-cost meat to urban consumers, may lead to a different set of policy recommendations and orientation of technical assistance programs.

The focus of this book is on improving cattle and small stock production, and yet other sources of protein for human consumption in Subsaharan Africa cannot be ignored. Game ranching, for example, in which wild ungulates are domesticated and herded with livestock capitalizes on the biological efficiency and greater tolerance of disease of the former animals [Surujbally 1977]. Many questions remain unanswered, but it is clear that wildlife, even when unmanaged, are presently a major source of meat for many areas of the continent (Table I.1). In Botswana and Zaire for example, most of the meat consumed is game [Krostitz 1979]. Rodents may well be of greater importance in many rural areas than bigger game animals, since they are numerous in relatively densely populated areas and have high rates of reproduction, allowing continuous harvest [de Vos 1977]. The role of game--not to mention poultry, pig meat, and other domestic meat sources--cannot be overlooked in solving Africa's food problems. Expansion of such enterprises, while beyond the scope of this book, may be a means of more effectively exploiting the productive potential of particular environments, especially in arid and semiarid regions.

The vast ecological differences in Subsaharan Africa are described by David Pratt in Chapter 2 using a zonal analysis. The interaction between ecology and livestock is described in terms of the distinct set of animal environments vis-a-vis climatic stress, feed and water supply, and disease hazard. He concludes that care must be taken to avoid pursuing an ecological perspective too far or too vigorously. Rather, a middle course should be followed in which ecological priorities are considered in terms of people, their needs, and their perspectives. In effect, planning must be based on both economic development opportunities, ecological potential, and conservation demands.

The very important aspect of feed and animal nutrition is followed up by Robert McDowell who provides an overview of different producton situations and ways in which nutrition research and policy can be developed.

Virtually all chapters in this book are based on a systems approach to livestock industry development. Ahmed Sidahmed and L.J. Koong define the term system in this section's last chapter as a set of components linked together for some common purpose or function.

Table I.1. Estimated annual game meat production for
selected Subsaharan African countries, 1977

Country	Total Production	Per capita Production
	--1,000 MT--	--kg--
Angola	6	0.9
Botswana	6	7.5
Cameroon	4	0.6
Chad	3	0.7
Congo	6	4.0
Benin	6	1.9
Ethiopia	7	0.2
Gambia	1	1.8
Ghana	28	2.7
Guinea	4	0.8
Ivory Coast	13	2.5
Kenya	7	0.5
Lesotho	4	2.9
Liberia	5	3.0
Nigeria	95	1.2
Rwanda	6	1.3
Sudan	7	0.3
Tanzania	8	0.5
Togo	4	1.7
Uganda	14	1.2
Zaire	68	2.6
Zambia	20	2.7

Source: Adapted from Krostitz [1979], Table 2.

4

An application of systems analysis using Sudan as an
example is presented, from which it is concluded that
certain livestock development projects in that country
have failed where a systems approach would have provided greater likelihood of success. They persuasively
argue that any attempts toward development or intervention in existing systems should be preceded by construction of a model which contains as many of the
production systems' essential elements as possible.

REFERENCES

de Vos, A.   "Game as Food: A Report on its Significance
     in Africa and Latin America."   Unasylva 29(1977),
     no. 116, pp. 2-12.
Krostitz, W.   "The New International Market for Game
     Meat." Unasylva 31(1979), no. 123, pp. 32-36.
Surujbally, R.   "Game Farming is a Reality." Unasylva
     29(1977), no. 116, pp.13-16.

1
Problems and Constraints, Goals and Policy: Conflict Resolution in Development of Subsaharan Africa's Livestock Industry

James R. Simpson

Subsaharan Africa[1] is characterized by extreme contrasts and wide diversity of resources and use of them. Some of the 45 countries in this region are rich in oil and mineral resources, while the economies of others are almost entirely based on agriculture and livestock [Eicher and Baker 1982]. Some countries are large while others are tiny; some have relatively low population densities while others are already very heavily populated. Some have experienced substantial economic growth while others are characterized as dismal economic failures.

Despite the many dissimilarities, areas of considerable homogeneity also exist. For the most part, the economies are all small in terms of total Gross National Product (GNP), as well as GNP per capita. The economies are typically open which means that foreign trade generally accounts for about a quarter of GNP [World Bank 1981]. Most of them are quite specialized and are often dependent on export of two or three primary commodities for generation of foreign exchange as well as stimulation of domestic growth. Another point of similarity is that modern wage employment only accounts for a small proportion of the work force. Many of the similarities between the subsaharan nations also act as constraints to economic development; recent emergence from colonial rule and consequent political fragility, low percentage of educated people, small managerial class, wide ethnic diversity, language differences within countries, dominance of land-extensive agricultural systems, and wide swings in policy are a response to changes in economic problems.

Livestock have traditionally been an important component in the economies of virtually all the subsaharan states, whether household inventory be two pigs and one goat on a one hectare farm in a primarily agricultural area of West Africa, or 40 head of cattle belonging to a pastoral family in East Africa. But, livestock production in the countries of Subsaharan

Africa has remained essentially stagnant while populations have grown at unprecedented rates. These trends
portend severe meat shortages in the near future unless
positive changes in existing production and marketing
systems are implemented.

Development alternatives can only be cogently
evaluated by delineating the continuum of constraints
which presently limit output. Then, the constraints
must be assessed according to their relative significance after which various possibilities for overcoming
them can be presented in technical, economic, and
socio-political terms. Options must be assessed from
the producer's as well as society's viewpoint if programs are to be successful and beneficial. The objectives of this book are emphasizing the wide variety of
factors influencing development of Africa's livestock
industry, underscoring the dynamic interrelationships
between the constraining factors, presenting techniques
and supporting frameworks for the analysis of development alternatives, and illustrating the application of
the techniques in the analysis of livestock problems in
present-day Africa.

HAS THERE BEEN PROGRESS?

Population in the 39 principal countries making up
Subsaharan Africa stood at 261 million in 1970.2
Just 10 years later it had increased to 347 million
people, a growth rate of 2.9 percent annually (Appendix
1). The World Bank reckons it will double by 2000 to
639 million people--and stabilize at 1.8 billion people
in the next century--about 5 times the current level.
Considerable uncertainty exists about the absolute
level of the projections but the message is clear--
considerably more animal products will be needed than
are now being produced.

There are about 21.6 million km^2 in Subsaharan
Africa which meant an average population density of 16
people per square kilometer in 1980 (Appendix 2). By
2000 the density will have increased to 30 people per
km^2 and, by the time population stabilizes, it will
be about 85 people per km^2. In contrast, worldwide
the density was about 33 people in 1980.

Total cattle inventory in the 39 principal countries amounted to 112 million head in 1961-65 (Appendix
3), climbed to 134 million head in 1970 (Appendix 4),
and reached 147 million head by 1980 (Appendix 5).
Total sheep and goat inventory was 143 million head,
170 million head and 230 million head for the three
periods respectively (Appendixes 3, 6 and 7).

One means of calculating the importance of forage
based animals is by placing them on an animal unit (AU)
basis. Considerable controversy exists over the appro-

priate weight to assign the base (usually one mature cow), the makeup of animal classes within national herds, size differences across countries, and the conversion factor for small stock. For purposes of this book each head of cattle has been assigned one AU while each sheep or goat is assigned 0.2 AU. On this basis there were 193 million AUs in the 39 subsaharan countries in 1980 (Appendix 8). Overall, cattle represented 76 percent of total AUs, with Zambia recording cattle at 97 percent of the total in contrast to 7 percent in Gabon. There was an average of 0.56 AUs per person in the subsaharan region with the highest ratio in Botswana (3.84) where cattle make up 95 percent of total AUs, down to 0.07 per person in Gabon.

Subsaharan Africa has quite a large trade in sheep, goats and cattle. For example, in 1980 there were 1.3 million head of goats and sheep exported and 3.5 million head imported by the 39 countries (Appendix 9). Most of the exports were by the low income semi-arid countries while the Ivory Coast, Nigeria and Senegal were the major importers. There were about 1 million head of cattle exported and 800 thousand imported by the subsaharan countries.

Has there been progress made in the subsaharan livestock industry?. Although a number of indicators are available for measuring progress such as reduction in cost of production and increased offtake per hectare, two of the more easily calculated measures using the data provided in Table 1.1 and the appendixes are production per head of livestock, inventory per person and production per person. Clearly, there are the well-recognized statistical limitations to production and inventory data, especially for the longer term, but the data and ratios derived from them nevertheless do provide an indication of relative change.

Indigenous production per head of cattle inventory showed a slight increase over the period 1961-65 to 1980, from 11.9 to 13.2 kg per animal for the 39 countries (Table 1.1). The highest productivities are now found in Mozambique and Liberia (25.7 and 25.6 kg) while the lowest (8.3 and 8.7 kg per head) is in Guinea and Sierra Leone (Appendix 5). Indigenous production per head of sheep and goats averaged 3.3 kg in 1961-65 and 3.7 kg in 1980, a slight increase similiar to that found in cattle.

Inventories per person for both cattle and small stock remained at about the same levels (0.5 and 0.7 head respectively) over the 17 year period 1961-65 to 1980. In effect, inventory of livestock is just keeping pace with population growth. The third measure, indigenous production per person, showed a slight decline for beef and veal, from 6.2 kg in 1961-65, to 5.6 kg in 1980. Indigenous production of goat meat, lamb and

8

Table 1.1 Summary of statistics related to progress in Subsaharan Africa's livestock industry[a]

Item	Units	Low Income Semi-Arid	Low Income Other	Middle Income Oil Importers	Oil Exporters	Total 39 Countries
Population						
1970	Mil.	22.8	125.5	49.0	63.6	260.9
1980	Mil.	30.3	162.6	67.9	82.3	347.1
2000	Mil.	49.0	285.0	128.0	177.0	639.0
Stationary	Mil.	140.0	818.0	368.0	503.0	1,829.0
Population density						
1980	Per km^2	5	17	19	31	16
2000	Per km^2	9	30	35	64	30
Total land, 1980	1,000 km^2	5,687	9,504	3,607	2,757	21,555
Arable land, 1980	Pct.	7	8	9	13	8
Permanent pastures, 1980	Pct.	29	26	30	23	27
Inventory, cattle						
1961-65	Mil. HD	18.7	62.0	18.5	12.4	112.1
1970	Mil. HD	22.7	75.7	21.9	14.1	134.3
1980	Mil. HD	20.3	81.8	29.2	15.5	146.8
Inventory sheep and goats						
1961-65	Mil. HD	37.6	59.0	17.7	29.2	143.6
1970	Mil. HD	43.9	71.6	21.9	32.6	170.1
1980	Mil. HD	66.7	99.9	26.0	37.7	230.4
Indigenous production of beef and veal						
1961-65	1,000 MT	198	667	316	152	1,333
1970	1,000 MT	233	790	245	158	1,426
1980	1,000 MT	297	918	473	255	1,943
Indigenous production of lamb and mutton						
1961-65	1,000 MT	115	212	52	92	471
1970	1,000 MT	177	223	65	102	567
1980	1,000 MT	227	357	92	167	843
Indigenous production per head of cattle						
1961-65	KG	10.6	10.8	17.1	12.2	11.9
1970	KG	10.3	10.4	11.2	11.2	10.6
1980	KG	14.6	11.2	11.2	16.4	13.2
Indigenous production per head of sheep and goats						
1961-65	KG	3.1	3.6	2.9	3.2	3.3
1970	KG	4.0	3.1	3.0	3.1	3.3
1980	KG	3.4	3.6	3.5	4.4	3.7
Inventory per person, cattle						
1961-65	HD	1.0	0.6	0.5	0.2	0.5
1970	HD	1.0	0.6	0.5	0.2	0.5
1980	HD	0.7	0.5	0.4	0.2	0.4
Inventory per person, goats and sheep						
1961-65	HD	2.9	0.6	0.4	0.6	0.7
1970	HD	1.9	0.6	0.5	0.5	0.7
1980	HD	2.2	0.6	0.4	0.4	0.7
Indigenous production per person, beef and veal						
1961-65	KG	10.5	6.5	7.9	2.9	6.2
1970	KG	10.2	6.3	5.0	2.5	5.5
1980	KG	9.8	5.7	7.0	3.0	5.6

-continued-

Indigenous production per person, goat meat, lamb and mutton						
1961-65	KG	6.1	2.1	1.3	1.8	2.2
1970	KG	7.8	1.8	1.3	1.6	2.2
1980	KG	7.9	2.2	1.4	1.9	2.4

[a] See appendices for country listings. Six countries, Cape Verde, Comoros, Djibouti, Equatorial Guinea, Sao Tome and Principe,and Seychelles are not included due to lack of data or because the numbers are too small for effective presentation.

mutton had remained about constant at 2.2 to 2.4 kg. Again, considerable discussion could be generated about accuracy of the data, especially for individual countries, but the essential message is indisputable-- only very minimal progress has been made in increasing productivity of ruminant livestock, and growth in output is derived from expanded herd numbers. In contrast, considerable progress has been made in other regions. Europe expanded output 31 percent from 1961-65 to 1980, up to 80 kg per head of inventory which is about 6 times the level of Subsaharan Africa (Table 1.2). Oceania increased its output 27 percent over that period to 58 kg while the USSR registered a 40 percent increase to also reach that same level [Simpson and Farris 1982]. Regardless of whether 1980 as the last year in the data series is a "typical" or "average" year, a second message is clear--Subsaharan Africa's absolute level of productivity technically could be expanded an enormous amount.

CONSTRAINT IDENTIFICATION

Barriers to development, or constraints as they are commonly termed, can be thought of as a continuum ranging from ones which are extremely difficult to overcome and are thus essentially limits to expanding or developing a certain practice or policy, to ones for which relatively easy solutions do exist. An effective way to analyze constraints is by classifying them as external and internal. If, for example, a Ministry of Agriculture officer is planning a project, it might be useful to divide the constraints into national and international categories. Internal constraints, from this perspective, might be budget restrictions or availability of local technicians. External constraints can be thought of as market outlets for livestock products, availability of foreign credits and producer response.

Ministry officers, or for that matter specialists from donor agencies, frequently only consider projects from their point of view, or at least that of the

Table 1.2   Indigenous production of beef and veal per head of
cattle inventory, 1961-65, 1970 and 1980, selected
regions and countries of the world

Country or Region	Year			Change 1961-65 to 1980
	1961-65	1970	1980	
	----------kg----------			-Percent-
Subsaharan Africa	11.9	10.6	13.2	11
North and Central America (other than USA)	34.1	34.3	35.0	03
USA	77.0	89.1	87.5	14
South America	32.8	30.5	31.7	-03
Asia	8.0	11.5	10.8	35
Europe	61.4	71.7	80.4	31
Oceania	46.1	46.4	58.4	27
USSR	41.1	56.2	58.0	40

Source: Compiled from FAO Production Yearbook 1981 and 1973.

nation. But, to be successful, any project must be evaluated from the potential client's viewpoint. In effect change agents (to borrow a term from community development) should think of constraints from the producer's perspective. Following are some of the more obvious objectives of each participant in the system, set forth in an effort to highlight the contrast which, itself, is a constraint to "improving" the livestock sector and thus meeting the needs of individuals and society as a whole [Simpson and Sullivan 1983].

Government, as a trustholder for society, inherently embodies numerous objectives but, perhaps most important, it is primarily interested in preservation of itself and the nation as a whole. As a consequence, some of the more striking national goals with respect to the livestock sector are:

-Improve the balance of trade through increased livestock product exports or, alternatively, reduce imports of livestock products.

-Improve urban diets by increasing per capita meat consumption.

-Reduce the cost of meat at retail. This allows for more meat consumption and also liberates some personal income for purchase of other items.

-Integrate livestock sector planning with the rest of the economy, recognizing that sustained economic growth implies "improvement" in each and every sector of the economy.

-Improve per capita incomes in society including those of livestock raisers.

-Reduce health problems and, as a corollary, improve sanitation at both the production and marketing levels.

-"Modernize" society subject to some nationalistic view.

-Promote a stable political situation while improving equity.

-Predominantly, reduce cost per kilo of meat and associated products, and increase total production.

Livestock owners and herders, in contrast to government, are much more diverse in their points of view. Whereas some individuals are quite traditional in the sense of desiring that no (or very little) change take place in their life-styles, others have an explicit (or perhaps subliminal) desire to become "modernized". This leads to postulation of the following objectives for livestock producers:

-Improve quality of life, which may reflect very little change in life-style, or may involve deep institutional change.

-Improve security, which can vary from owning more livestock to having money in the bank and assurance that current productive efforts will have a later reward.

-Expand esteem or prestige in their peer group.

-Decrease daily "inconveniences" such as personal health problems or societal difficulties.

Many more objectives could be postulated, but these are sufficient to demonstrate that while government has a vested interest in increasing productivity and reducing cost of meat and associated products at the retail level, individual producers do not hold these objectives or, if they do, only marginally, as their major interest lies in the welfare of themselves, their families and, to an ever lesser degree, their community.

There are a number of commonly held axioms or beliefs among development planners, researchers and government officials about land tenure in the African livestock sector. A valuable exercise is delineation of these attitudes as they provide guidelines in setting forth philosophies about sectoral, regional and national programs [Simpson 1978]. The following list is not prioritized nor is any attempt made to quantify or evaluate the extent to which the philosophies are held. In other words they are tendencies. Some typical beliefs are:

-Land has a certain mystique, i.e., it is more than just a factor of production.

-Grazing is a free good, i.e., an inviolate "right".

-Government or tribes should control grazing land rather than private individuals.

-Livestock raising is a way of life which should be continued.

-Cattle or other livestock are a better store of wealth than cash or savings in modern financial institutions.

-Rural life is "good". It is "better" to be poor in a rural area than in an urban area.

THE NEED FOR A VISION

A major constraining factor to development of Africa's livestock industry is lack of a vision about what is desirable as viewed by government, religious authorities and others in a leadership role. Naturally, what is desired by elements of society may be different from that of the leaders. The problem, of course, is the nebulous nature of both the vision and the dominant forces in setting it. The lack of a vision is exemplified by violent shifts in economic systems in some countries, radical price policy changes and, most important, failure to establish a clear-cut realistic land tenure policy [Simpson 1983]. The point is that the appropriate production technology exists for Africa to easily double, and perhaps triple output per animal unit of livestock if a decision were made to reach that goal, and appropriate measures taken to provide legislation which would pull development rather than attempting to push it in ways which are counter to economic and social realities.

Vision is based on philosophy--what is desired in terms of societal makeup or organization at some long-term point in the future. If the destination is known, then a type of roadmap, or series of short-term development goals can be set forth. In some cases circuitous routes must be chosen due to obstacles. If the impediments are sufficiently large, they will form absolute constraints which may prevent the destination from ever being reached. It is clear that lack of a vision, and failure to "make a decision to make a decision" are preventing Subsaharan Africa's livestock industry from being improved in terms of output per animal unit.

The need for a vision is underscored by recognition that livestock development in Africa is primarily a people--not a resource problem. As an example, rural pastoral populations in the widespread semi-arid areas of Subsaharan Africa are growing quite rapidly despite continued migration to urban areas. As a result, herd size per family is inexorably declining and will continue to decline--and each family will steadily become poorer since, with a general policy of communal grazing, there is almost no room left for technological advance. Families will generally not adopt herd control strategies which, from a social viewpoint, are optimum due to the local institutional and cultural framework surrounding them. In effect, if substantial increases in output per animal unit are the desired goal, then the barriers must be removed. The basic question in the vision is: how much modernization or, alternatively, how fast a move from traditionalism? The problem: rapid population growth precludes perpetuation of traditional ways of life.

Small farmer mixed crop/livestock holdings clearly reflect the need for a development vision, and there is no question about the need for accelerated research and extension efforts on small farm systems. But, in addition, if the objective is providing lower cost meat and animal products to an increasingly urbanized society, then attention will also have to be placed on stimulation of operations sufficiently large to effectively adopt known technologies. The implication is there must be substantial privatization of land, even greater migration from livestock producing areas to urban centers, and attendant ethical considerations about loss of traditional cultures [Pratt, n.d.; Tweeten 1982].

Examples abound of simple known technologies which can and do yield great economic returns provided they are used in a commercial manner. Some of the relatively simple ones are implanting cattle with growth stimulants, worming for internal parasites, control of external parasites, controlled crossbreeding, appropriate culling, testing herd sires, upgrading by improved bloodlines, proper nutrition, confinement or intensive forage finishing to slaughter weights, use of records in decision making, price strategy development, introduction of improved forages, and proper forage management, just to mention a few. The main drawback is most of these technologies, while carrying very high benefit-cost ratios, are not appropriate for smallholders because that group generally does not maintain animals primarily as an income generating mechanism.

A secondary reason for the low adoption rates of known technology and one reason for including it in a "vision" is lack of an effective extension service mechanism. Third, and even more constraining, is that intensive practices require management skills that the typical smallholder has not learned. The bottom line, to borrow a phrase frequently used by businessmen, is that substantial increases in terms of output per head of inventory can only be expected to take place under a commercial system. But, fomenting such an orientation carries with it very important implications in terms of a larger vision about societal organization as well as policy on research, teaching, extension and project development efforts.

PROJECTIONS TO THE YEAR 2000

One of the more difficult aspects in setting forth a long-term vision, apart from its nebulous and ethical nature, is visualizing what a country would be like at some distant point in the future under alternative scenarios. Just thinking five years in the future at which time a specific project may be completed is a difficult task; projecting out 16 years to the year 2000 (from a base of 1984) is a heroic undertaking, and 130 years (close to the time Africa's population is expected to stabilize at 1.8 billion people) is virtually beyond our ability. Nevertheless, let us look at the year 2000 from a policy making standpoint by first projecting total consumption of beef and veal, and goatmeat, lamb and mutton, and then calculating the inventory of cattle, goat and sheep needed to satisfy that level of consumption.

There are two means of measuring meat production; that derived from indigenous production, i.e. from animals raised within a country, and production based on animals slaughtered in the country. The latter, adjusted for imports and exports, provides an estimate of consumption and, when divided by population, yields an estimate of consumption per capita. Following this approach, it is estimated that total net production or, alternatively, total consumption of beef and veal in the 39 principal subsaharan countries, was 2.0 million metric tons in 1980 which, on a per capita basis, was 4.9 kg (Appendix 10).

The highest per capita consumption of beef in Subsaharan Africa during 1980 was in Swaziland (25.8 kg) while the lowest was 0.8 kg in Zaire. Total mutton, lamb and goat meat production was 0.8 million tons, while the average was 2.3 kg per capita, which is about 40 percent that of beef (Appendix 11). The highest mutton, lamb and goat meat consumption per capita was in Somalia (14.1 kg) while the lowest was in Gabon and Mozambique (0.2 kg). In contrast, about 50 kg of beef were consumed in Australia during 1980, 40 kg in the United States and 21.6 kg in Europe. Mutton, lamb and goat meat consumption was 24.8, 0.7 and 3.0 kg for these same countries and region.

The approach taken for the consumption projections is utilization of two scenarios for the year 2000, one in which there is no change in per capita consumption, and one in which per capita consumption increases a total of 20 percent, which is a compound annual growth rate of 0.9 percent annually.[3] The projections, which are really a simulation exercise, are carried out individually on the four regions under which the 39 principal subsaharan countries are grouped in the appendixes, as well as on a total basis.

Multiplying the 1980 base per capita consumption of beef, (5.6 kg for the 39 countries), times a projected population of 639 million in 2000 indicates that about 3.7 million tons of beef and veal will be required to meet total consumption (Table 1.3). Likewise, multiplying the base of 2.3 kg of mutton, lamb and goat meat by the projected population in 2000 means that total consumption would reach 1.4 million tons. If per capita consumption were to increase 20 percent, then total consumption would be 4.5 million tons of beef and veal, and 1.7 million tons of meat from small stock.

The next step taken in Table 1.3 is calculating the livestock inventory required to meet consumption under the assumption that no productivity changes take place, i.e. that all additional production is derived from increased numbers of livestock, and that the ratio of indigenous production to total consumption remains unchanged. If effect, it is assumed that the same amount of trade would take place in 2000 as in 1980. This means that current exporters would ship out proportionately more livestock and meat, and importers would bring in a higher total quantity. Results of the calculations show that under the "no change in per capita meat consumption scenario" cattle inventory would have to about double to 280 million head from the current 147 million head. Likewise, total small stock inventory would also have to almost double, from 230 million head to 410 million head.

If per capita consumption were to increase just 20 percent (which for the bases of 5.6 kg and 2.3 kg is really a very minor amount) total inventories would have to more than double, to 337 million head of cattle, and 492 million head of sheep and goats. A better understanding and appreciation of what doubling in 20 years means can be obtained by referring to the 17 year historical period 1961-65 to 1980 for which data are provided in the appendixes during which cattle inventory only increased 30 percent and small stock 61 percent.

Production increases can, of course, come about through animal productivity improvements. Continuation of the projection exercise in Table 1.3 reveals that without change in inventories, indigenous production of beef and veal would have to be 25 kg per head of inventory, and 6.6 kg for small stock in order to meet the projected consumption requirements (no change in per capita consumption). The outputs per head are now 13.2 and 3.7 kg so productivity would roughly have to double. This represents an average annual (compounded) growth rate of 3.3 percent for cattle and 2.9 percent for small stock. The rates for cattle and small stock

Table 1.3. Projections of beef and veal, and mutton and goat consumption and cattle, goat and sheep inventory requirements, Subsaharan Africa, 2000

Item	Units	Low Income Semi-Arid	Low Income Other	Middle Income Oil Emporters	Oil Exporters	Total
Per capita consumption, 1980						
Beef and veal	KG	6.5	5.6	7.5	4.9	5.6
Mutton, lamb and goat meat	KG	6.1	2.1	1.5	2.0	2.3
Total population						
1980	Mill	30	163	68	86	347
2000	Mill	49	285	128	177	639
Total consumption, 1980						
Beef and veal	1,000 MT	197.7	919.4	506.6	332.6	1,956.3
Mutton, lamb and goat meat	1,000 MT	185.0	341.6	104.4	170.3	801.3
Total consumption, 2000 No change in per capita						
Beef and veal	1,000 MT	318.5	1,596.0	960.0	867.3	3,741.8
Mutton, lamb and goat meat	1,000 MT	298.8	598.5	192.0	354.0	1,443.4
20 percent increase in per capita (0.9 percent annually)						
Beef and veal	1,000 MT	382.2	1,909.5	1,152.0	1,044.3	4,488.0
Mutton, lamb and goat meat	1,000 MT	358.7	718.2	230.4	424.8	1,732.1
Indigenous production, 1980						
Beef and veal	1,000 MT	297.0	918.0	473.0	255.0	1,943.0
Mutton, lamb and goat meat	1,000 MT	227.0	357.0	92.0	167.0	843.0
Ratio of indigenous production as a proportion of total consumption, 1980						
Beef and veal	Ratio	1.50	1.00	0.93	0.77	0.99
Mutton, lamb and goat meat	Ratio	1.23	1.05	0.88	0.98	1.05
Indigenous production per head of inventory; 1980						
Beef and veal	KG	14.6	11.2	16.2	16.4	13.2
Mutton, lamb and goat meat	KG	3.4	3.6	3.5	4.4	3.7
Inventory, 1980						
Cattle	1,000 HD	20,347	81,757	29,189	15,527	146,820
Sheep and goats	1,000 HD	66,665	99,900	26,045	37,746	230,356

Economic Group[a]

-continued-

Inventory required to keep 1980 ratio of indigenous production to total consumption, 2000[b] No change per capita cons.						
Cattle	1,000 HD	32,730	142,500	55,110	40,720	280,640
Sheep and goats	1,000 HD	108,130	174,560	48,280	78,800	409,620
20 percent increase in per capita cons.						
Cattle	1,000 HD	39,270	170,490	66,130	49,030	336,600
Sheep and goats	1,000 HD	129,770	209,480	57,930	94,620	491,550
Indigenous production per head of inventory required for no increase in inventory to meet consumption projections, 2000[c] No change per capita cons.						
Cattle	KG	23.5	19.5	30.6	43.0	25.2
Sheep and goats	KG	5.5	6.3	6.5	9.2	6.6
20 percent increase in per capita cons.						
Cattle	KG	28.2	23.4	36.7	51.8	30.3
Sheep and goats	KG	6.6	7.6	7.8	11.0	7.9
Annual growth rate in production per head of inventory required for no increase in inventory to meet consumptions, 2000 No change per capita cons.						
Cattle	Pct.	2.4	2.8	3.2	4.9	3.3
Sheep and goats	Pct.	2.5	2.8	3.2	3.8	2.9
20 percent increase in per capita cons.						
Cattle	Pct.	3.4	3.8	4.2	5.9	4.3
Sheep and goats	Pct.	3.4	3.8	4.1	4.7	3.9

Source: From Appendix Tables.

[a] See Appendixes for country listings.

[b] Total consumption in 2000 divided by indigenous production per head and the result multiplied by the ratio of indigenous production to total consumption. Eg for low income semi-arid, 318.5 ÷ 14.6 = 26.18 times 1.5 = 39.3 million head.

[c] Total consumption in 2000 times ratio of consumption to indigenous production divided by 1980 inventory. For example, for low income semi-arid countries, 318,500 thousand metric tons of beef and veal consumption times 1.5 = 477,750,000 tons divided by 20,347,000 head = 23.48 kg.

under the 20 percent increase in per capita consumption would be 4.3 and 3.9 percent, respectively.

Unquestionably, total inventory will continue to increase and, quite likely, will make up a substantial portion of increased total production. Nevertheless, the scenarios are useful for they show the dramatic impact that improving productivity can have on total production. Given the very low levels of productivity in Subsaharan Africa compared with other countries or regions (Table 1.2) it is not unreasonable to expect that the productivity increases of the magnitude described could take place. The problem is identification of constraints and development of viable solutions for overcoming them.

CONCLUSIONS

Population in Subsaharan Africa has grown very rapidly and will continue to increase at an astounding pace in the foreseeable future. This rapid growth carries with it several major implications in the design of livestock programs. First, although there will be a rapid increase in demand for meat due to an increasingly urbanized society, it appears that a dualistic policy will have to be developed to an even greater extent than today. On the one hand, development will be needed in the small farm or subsistence sector to make more efficient use of the resources available to small producers.

The second part of a dualistic policy, and one partially tied to the small-holder subsector, is deliberate development of commercial, private enterprises to meet the major portion of increased demand. Conflict is inherent to such a policy, resolution of which will be difficult, but necessary.

It can be concluded that due to the long-term nature of livestock production a vision about what is wanted and desirable in terms of industry organization is needed. Technically, demands for livestock products can be met by increasing productivity; the problem is one of understanding the constraints and planning to overcome them. This is an awesome and complex undertaking, but is not insurmountable.

NOTES

[1]The definition of Subsaharan Africa follows that by the World Bank [1981] and, as such, does not include Namibia or South Africa.

[2]The country tables in the Appendix embrace 39 countries. The other 6 countries (Cape Verde, Comoros, Djibouti, Equatorial Guinea, Sao Tome and Principe, and

Seychelles) are not included due to lack of data or
because the numbers are too small for effective
presentation. Countries are grouped in the Appendixes
after the World Bank [1981].

[3]As a more sophisticated approach, See M'Pia
and Simpson [forthcoming].

REFERENCES

Eicher, Carl K. and Doyle C. Baker. Research on Agri-
cultural Development in Sub-Saharan Africa: A
Critical Survey. MSU. Int. Dev. Paper 1, Michigan
State University, 1982.

FAO. Production Yearbook. FAO Trade Yearbook. various
issues. Rome

M'Pia, Elengasa and James R. Simpson. "Use of an
Economic Demographic Simulation Model for Planning
in Zaire's Beef Cattle Subsector." J. African
Studies, forthcoming.

Pratt, David J. "The Ecological Management of Arid and
Semi-Arid Rangeland in Africa and The Near East."
United Nations Food and Agriculture Organization.
Report No. AGPC:MISC/25. n.d.

Simpson, James R. "Ethics and Regional Development: A
Meeting of Philosophy, Economics and Anthropology." Staff Paper 80. Food and Resource Economics Department, University of Florida, Gainesville, April, 1978.

Simpson, James R. "Identification of Goals and
Strategies in Designing Technological Change for
Developing Countries." ed. Molnar, Joseph J. and
Howard A. Clonts, Transferring Food Production
Technology to Developing Nations: Economics and
Social Dimensions. Boulder, Colorado: Westview
Press, 1983.

Simpson, James R. and Donald E. Farris. The World's
Beef Business. Ames: Iowa State University
Press, 1982.

Simpson, James R. and Greg Sullivan, "On Institutional
Change in Utilization of African Common Property
Range Resources." Unpublished Paper, April 28,
1983.

Tweeten, Luther. "Food for People and Profit: The
Ethics of a Capitalist Food System." Paper
presented at the Resource Conservation Act
Symposium, Washington D.C., 7 Dec. 1982.

World Bank. Accelerated Development in Sub-Saharan
Africa: An Agenda for Action. Washington, DC,
October, 1981.

2
Ecology and Livestock

David J. Pratt

This chapter examines relationships between ecology and livestock in tropical Africa. It deals first with basic ecological influences and interactions before reviewing their implications for development. Limitations of an ecological perspective are recognized, but these stem mainly from misapplication of the perspective and should not detract from its value. Ecology is undoubtedly the most basic and pervasive of the many determinants of the livestock production systems of Africa. To ignore or oppose ecology is to court disaster. The size of the literature reflects the importance of the subject but is not reviewed here in any detail; rather the aim is simply to set the scene for later discussions.

ECOLOGICAL INFLUENCES AND INTERACTIONS

To set the scene it is necessary to begin by identifying the components of ecology and livestock production which interact most strongly. For this purpose it is convenient to separate ecology into climate, landform/soil and vegetation and to subdivide livestock production according to the population, functions and management of the livestock concerned. The interaction between ecology and livestock also introduces a distinct set of animal environments which are considered here by reference to climatic stress, feed and water supply, and disease hazard.

Ecology

The range of ecological conditions in tropical Africa is remarkedly diverse, extending across gradients from coast to mountain glacier and from desert to forest. In any one locality it is the totality of all environmental factors which characterizes the site, but it is useful to look first at the range and impact of individual factors.

21

Climate. Day length, temperature and rainfall are all influential features of the climate of tropical Africa. However, day length is rather uniform and temperature is a dominant influence only where it is extremely hot or where elevation or latitude produces a cooling effect, particularly in those few areas where frost occurs. The dominant influence overall is therefore rainfall. Where mean annual rainfall is less than 300 mm perennial plants are sparse and the aspect is semidesert; with double that rainfall vegetation is more diverse and agriculture commonplace, and where it is doubled again forest begins to appear.

Although there is a close correlation between mean annual rainfall and land use and potential, the correlation is even closer when water availability is substituted for rainfall. Water availability depends partly on the absorptive capacity and retentivity of the soil in relation to evaporation, but it is determined primarily by the seasonality of rainfall. Thus, 500 mm of rainfall concentrated in a single 3-4 month season (as in the Sahel of western Africa) is adequate for grain production but, because of the 8-9 month dry season, inadequate for perennial grasses and year-round livestock productivity. In contrast, the same total rainfall distributed equally between two wet seasons (as in parts of eastern Africa) provides two unreliable cropping seasons combined with potential for perennial grassland and an extended grazing season.

The lower the annual mean, the more important are small variations in rainfall amount and distribution. The limitations of a low rainfall are further accentuated by the fact that, as mean annual rainfall decreases, so does the reliability of receiving that total. On the other hand, as elevation increases and evaporation decreases so does the usefulness of rainfall increase. Thus, a rainfall regime which is marginal for cropping or intensive grazing at low elevation may reliably permit all of these activities at an elevation of 2,000 m. Highland masses in an arid environment can also attract mists (as well as rainfall) which favor forms of vegetation and land use superior to those which would be expected from the rainfall data.

Not surprisingly there are many schemes for classifying climate. Even those which focus on water availability for agriculture involve several criteria for defining relative aridity or humidity. It is sufficient here to reproduce the comparison of zonations presented by Jahnke [1982] in his analysis of livestock production systems in tropical Africa. The comparison is shown in Figure 2.1, along with a zonal map in Figure 2.2. The zonation preferred by Jahnke, based on length of growing season, has the merit of simplicity

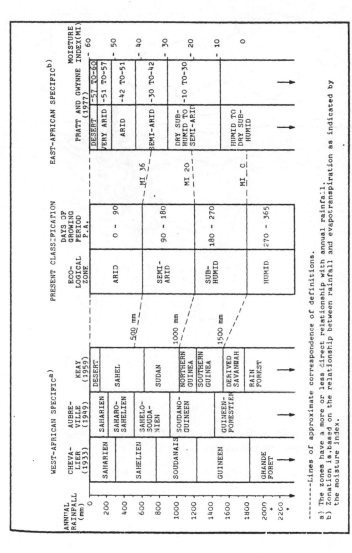

Figure 2.1. Ecological classification scheme used by Jahnke and its correspondence with other classification schemes

Source: Jahnke [1982].

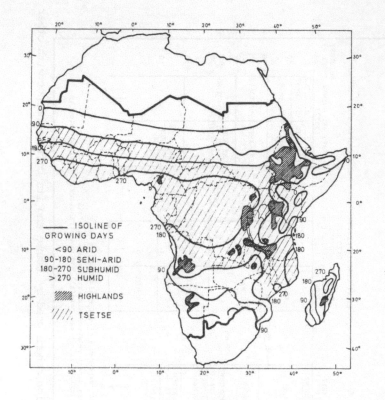

Figure 2.2.   Ecological zones of tropical Africa and the extent
of tsetse infestation

Source: Jahnke [1982], from data by Ford and Katondo, and FAO.

but it does not recognize the different needs of crop and livestock production. The ideal classification for livestock production would give as much emphasis to dry as to wet seasons. However, whichever way zones are defined, the basic distinction emerges of arid and humid zones with more than half of the livestock population of tropical Africa located in the arid zones and a further concentration occurring in the highlands, as illustrated in Table 2.1.

Landform and soil. Reference has already been made to the effect of relief and elevation on climate. These effects are profound -- highland environments have always attracted settlement and diversified agriculture -- but they are also limited in their extent. More widespread are the effects of landform and soil on the shedding and collection of rainfall and on the release of water and nutrients for plant growth. On these depends much of the variation in vegetation and agricultural potential that occurs within climatic zones.

This variation is of critical concern for crop production but has impact too on livestock production. The occurrence of large flood plains, such as the interior delta of the Niger River, provides dry season grazing to support herds which, at other times of the year, utilize much larger tracts of dry grazing land. More locally, depressions in an arid landscape can be all important sites for herbage production, collecting and utilizing the run-off from desert showers which would contribute little if they entered the soil where they fell. On landform and soil also depends the occurrence of stock watering points, without which there would be no livestock production.

Other influences include accessibility. Very rugged terrain obviously limits access by both man and livestock. Seasonal flooding or waterlogging also restricts access, though this is not necessarily an impediment if it preserves grazing for dry season use. Some such influences affect all livestock species equally while others (such as terrain) affect some species more than others. There is also a distinction in that landform, like climate, is almost impossible to change whereas soil constraints are often open to amelioration by tillage or fertilizers. However, it does not follow that such improvements are economic, as is illustrated for the Sahel by Penning de Vries and Djiteye [1982].

Vegetation. An important feature of vegetation is the several values which it can hold concurrently. It can be regarded as a timber resource, a tsetse or disease habitat, an indicator of climatic conditions, an impediment to cultivation, a unit of certain grazing

Table 2.1. Ruminant livestock population by species and ecological zone in tropical Africa, 1979

Ecological Zone		Ruminant Livestock Populations					Grazing Density
Title	Area	Cattle	Sheep	Goats	Total[a]	Proportion	
	—Million km²—	—————————Million hd—————————				—Percent—	—Hd/km²—
Arid	8.3	31.5	37.1	48.3	128.0	33.0	15.4
Semi-arid	4.1	45.4	23.1	33.2	101.7	26.2	24.8
Subhumid	4.9	32.8	14.1	20.3	67.2	17.3	13.7
Humid	4.1	8.8	8.2	11.6	28.6	7.4	7.0
Highlands	1.0	29.0	21.4	11.9	62.3	16.1	62.3
Total	22.4	147.5	103.9	125.3	387.8	100.0	

Source: Jahnke [1982].

[a]Includes 11.1 million camels incorporated in the arid zone total.

potential, a habitat for wild animals or a scenic
attraction. Nor is this list (from Pratt and Gwynne
[1977], p.22) exhaustive. For example, vegetation is an
indicator of more than climatic conditions. In its
broader characteristics (its formation type) it indicates climate but in its species composition and vigor
it also indicates soil status and the history of land
use. More research is needed to elucidate finer
relationships, but extreme soil conditions and overgrazing are relatively easy to detect from species
composition. It is also common to use vegetation to
assess livestock carrying capacity.

Carrying capacity is simple in concept but complicated in use and utility. The fact that different
authorities often produce different figures for carrying capacity has led to considerable confusion, as
illustrated by Sandford [1983]. Yet to expect there to
be one figure for carrying capacity is as absurd as to
expect there to be one figure for how many people can
share a cake. The number of animals that a given area
can reasonably support depends on the animals concerned
(including their feeding preferences and requirements,
their expected productivity and the watering regime
proposed) and on the range management strategy adopted
(including provision for drought and recuperative rest
and the intended frequency of burning). Variation in
one factor can change carrying capacity by 50 percent.
An increase of that magnitude could follow the
substitution of browsing animals for grazers in certain
types of bushland, and a similar decrease could be
associated with provision for rotational resting and
burning. There is also methodological variation,
depending on whether carrying capacity is calculated in
terms of feed supply and requirements, or assessed in
terms of ecological conditions. The latter method
(normal in anglophone Africa) is usually more
conservative than the former (associated more with
francophone Africa). Rarely, however, does accuracy
really matter, for in most of Africa fine tuning of
animal numbers is impracticable and it is sufficient to
know whether an area is radically under or over-stocked
so that the direction in which to move (or try to move)
is clear. Monitoring of range condition and animal
performance will then indicate progress or otherwise,
and the need for further action.

The concepts of range condition and overgrazing
are interrelated with carrying capacity and are subject
to the same sort of variables. Range condition, in
particular, needs to be expressed in terms of a standard of comparison and the animal species to be accommodated. Overgrazing is self-evident if acute, but
still needs care in interpretation, especially in the
arid zone of eastern Africa where the dynamics relating

perennial grasses, annual grasses and woody species are particularly complex. At the lower rainfall limit for the survival of perennial grasses (around 250 mm in Kenya) a cycle of dry years results in annual grassland irrespective of grazing pressure, while grassland under higher rainfall (around 600 mm) can show remarkable powers of recuperation from excessive grazing. Under intermediate rainfall grassland is sometimes one phase in a long term cyclic succession between grassland and bushland, the dynamics of which are apparently determined by rainfall and fire.

Fire is a potent force in shaping natural vegetation and it can also be used as a management tool for bush-control. More so than other environmental components, vegetation is open to manipulation. Fire and grazing management are the two most generally useful tools for improving vegetation but, for more rapid change where appropriate, mechanical and chemical methods of bush-control and a number of seeding techniques are available. The utility, justifiable cost and methods of improvement by these means are basically all determined by land use and potential. Their role in East Africa is reviewed in Pratt and Gwynne ([1977], Chapters 16-18).

Livestock Production

The livestock production systems of tropical Africa range from traditional to modern, and from extensive to intensive, involving up to nine principal livestock species and as many products or functions. The production systems may be multipurpose or specialist, subsistence or market orientated and combined or not with crop production. Livestock production is practiced on ranches, farms and smallholdings, in feedlots and village compounds, under tree crops and, in pastoral areas, according to the differing needs and preferences of about thirty major ethnic groups. In the paragraphs which follow it is possible only to highlight principal components in relation to ecology. For a more comprehensive analysis reference should be made to Jahnke [1982].

Population. Some population data for ruminant livestock have already been given in Table 2.1. The total number of ruminants in tropical Africa approaches 400 million. In addition, there are as many or more chickens, about 7 million pigs and over 10 million horses and donkeys. However, these non-ruminants receive little attention here because (except for horses) their distribution and productivity is not so greatly affected by ecology and because, together, they contribute less than 10 percent to the livestock biomass. In

terms of biomass,the dominant livestock are cattle,
principally Bos indicus with some B. taurus and crossbreeds. Camels, sheep and goats contribute less than
10 percent each to the livestock biomass, though sheep
and goats, as shown in Table 2.1, are almost as abundant as cattle.

The species most restricted in distribution is the
camel. Not only is it confined to the arid zone but 85
percent of the population is concentrated in eastern
Africa. The residual 15 percent occurs in the Sahel.
(Surprisingly, in view of the attention it receives as
a livestock region, the Sahel supports no more than 15
percent of the total ruminant population of tropical
Africa. Western Africa as a whole supports around 30
percent of the total and central and southern Africa
another 15 percent, leaving eastern Africa with the
majority. These geographical distinctions are partly
ecological - related to the distribution of zones - but
they are too intimately related to historical and cultural factors for analysis here.) Relative to camels
other ruminants are highly adaptable and, with the help
of breed variation, extend over most of tropical
Africa. There are, however, distinctions between
species both in zonal distribution and in the causes of
this distribution.

The fact that goats outnumber other species in the
arid and the humid zone is probably ecologically determined. The breeds concerned -- the sleek coated goat of
the arid zone and the black dwarf breed of the forest
-- are very different, but both browse actively and
browsing certainly helps in the utilization of both
environments. Moreover the forest goat must be the
product of very selective ecological pressures for, as
with N'Dama cattle, any breed which is to survive in
the forest environment has to develop adequate tolerance to trypanosomiasis. Cattle, too, predominate in
the zones to which they are best adapted -- in the
transitional zones between arid and humid and in the
highlands -- but they also extend in substantial numbers into certain arid bushlands for which they are ill
equipped, as do sheep. The reasons why animals are
forced into environments to which they are not adapted
relates mainly to the functions which they are required
to perform.

Functions. Livestock in Africa contribute more
than meat, milk and clothing materials. When associated with cropland they often contribute manure and traction. It is also common for livestock to be used for
transport and as a form of capital and security. In
pastoral societies they may contribute blood to the
diet and invariably they are used to build and maintain
social relationships. Hides and skins are used for
housing and water containers as well as for clothing,

and dung may be used (and even sold) as fuel instead of manure. Nor are these functions necessarily secondary to food production. Cattle in the Ethiopian highlands are kept as much for dung and traction as for milk.

Different species perform different functions in different livestock production systems. There are many variations, but typically camels provide milk and transport, cattle provide either milk or milk and traction, or meat for the commercial market, sheep and goats provide meat (in handy family-sized packages) and sometimes milk, and horses and donkeys provide transport and sometimes traction. Camels and cattle are the main species used for investment, security and building social relationships and sheep and goats are used as small change to meet everyday expenses and obligations. All livestock contribute hair, wool, skins or hides according to their kind and all contribute dung, though most of that used as fuel comes from cattle.

The functions which livestock fulfill in a given situation derive from an interaction between species, environment, human need and custom. It should not be assumed that function derives directly from species or that species derives directly from environment or need. The possible permutations are many. Thus a pastoral society which became dependent on cow milk for subsistence because cattle were well-suited to the orginal pastoral environment is quite likely, if the society extends into country unfavorable for cattle, to seek to retain cattle in preference to other, better adapted, species or diets.

Management. Very different management considerations and practices apply across the livestock production spectrum from pastoralism through ranching and mixed farming to feedlots. A full review is beyond the scope of this chapter but is should be appreciated that ecology both influences the occurrence or choice of production system and the manner in which the system is managed. This is especially true of subsistence pastoralism, which is both a response to environment and, in its execution, a living exercise in applied ecology. Its central concern is to maintain animal numbers and milk yield by seeking good grazing and adequate water and avoiding predators and know foci of disease. As often as not its movement patterns are a direct response to the distribution of grazing and water. At the other end of the scale, feedlot management centers on price ratios and markets and is virtually independent of ecology.

Animal Environments

It is appropriate to conclude this section by returning to the environment to look briefly at those aspects of overriding importance to the animal.

Climatic stress. In view of the extremes of temperature and humidity experienced in parts of tropical Africa, climatic stress might be expected to be a major constraint to livestock production. In practice, however, the indigenous breeds in their natural environments have little difficulty in coping with these stresses. Problems arise with these animals only when they are deprived of water or customary shade, or when they are moved into an alien environment. This happens sometimes when cattle and sheep are pressed into the semidesert, or when cattle are moved into humid coastal areas. The sensitivity of many goats to even smallscale movements may be due partly to climatic change but is better attributed to change in the total environmental complex.

The main impact of climatic stress is on livestock development schemes which call for the stocking of underutilized areas and/or breed improvement with exotic stock. Breeds of European origin are particularly prone to climatic and other stresses and, if introduced outside the highlands, need very careful husbandry.

Feed and water supply. With the few exceptions just described, and others to follow under "disease hazard", the essential feature of any environment for an animal is the feed and water which is on offer.

Usually water is most limiting in the arid zones, where livestock may drink only every second or third day. When watering at this frequency, the livestock will probably be ranging over 500-1,500 km^2 of grazing which, though often sparse and desiccated, should be relatively high in feeding value, having been killed and cured while still physiologically young. In wetter transition zones dry season herbage cures more slowly and is usually subject to leaching by late rain, thereby reducing its feeding value well below maintenance requirements. The best hope in these circumstances is that the livestock have access to browse or crop residues or, in integrated crop/livestock systems, to forage produced within the farming system. Other forms of supplementary feeding may be possible and economic if livestock production has a commercial outlet, but not usually if it is practiced only for subsistence.

Where water is deficient, its provision becomes top priority. The form which water development takes depends on the physical environment (including its hydrogeology), the season for which the water is required and the individual and total needs of the animals concerned. It also depends on cost and it should depend on the grazing potential of the area served. Relatively small surface water supplies are often preferable to permanent supplies because of the limit they impose on over-concentration of livestock. Where evaporation is high control may be sought through reservoir design or the use of surfactants.

Opportunities for increasing feed supply in the arid zones are limited to what is possible through range management and, sometimes, through browse plantations and the incorporation of crops like cowpeas into dryland farming systems. Opportunities are much greater in the subhumid zone and the highlands. The latter usually has high population pressure but, such is its potential, half a hectare of cultivated forage can contribute significantly to livestock production. The potential of the subhumid zone lies in its size and under-utilization in relation to rainfall, which is sufficient for reliable cropping and yet not enough to form forest, which is difficult to exploit for livestock production. The subhumid zone has potential to support a greatly expanded livestock industry based on a combination of natural grazing, planted grasses and legumes and agricultural byproducts.

Disease hazard. Disease exercises a profound influence on livestock distribution and productivity in tropical Africa. The prime example is trypanosomiasis, which excludes all but trypanotolerant breeds from large tracts of country and reduces productivity in peripheral areas either by infection or by denying livestock seasonal access to productive grazing. Because infection is related to the presence of tsetse flies and hence of tsetse habitats, there is a close correlation in this case with ecological zone. The presence of the fly is felt over vitually all of the humid and subhumid zones and extends in riparian vegetation even into the arid zone and the fringes of the highlands. Other diseases are less closely correlated with ecology, though regional and zonal differences do occur. Peste des petits ruminants and dermatophilosis, for example, are more associated with the humid zones of western Africa than with other parts; while some theilerioses are confined to eastern Africa, and the highlands are spared some of the disease problems of lowland Africa. There are also more local associations, as between liver flukes and standing water, and between malignant catarrh and concentrations of wildebeest.

To evaluate environment in terms of disease hazard is important both to assessing risk and deciding an appropriate response. Environment partly determines what can justifiably be spent on veterinary measures and, through factors such as access, the control measures which are most appropriate. It is also possible, in the case of some diseases, to reduce the hazard by changing the environment, whether by bush-control (against tsetse) or by draining standing water (against flukes). The threat of disease transmission by wildlife is not easily overcome (short of eliminating the wildlife) but preventative treatment or rotational grazing may help.

IMPLICATIONS FOR LIVESTOCK DEVELOPMENT

It is not possible, within the confines of this review, to show how the relationships outlined above apply to individual livestock production systems. However, Table 2.2 summarizes the situation by reference to five ecological zones. These are not precisely the zones set out in Figure 2.1 but they are derived from them and are seen as the most basic divisions for planning livestock development in tropical Africa. They separate semidesert from the rest of the arid zone but do not give separate attention to the semi-arid zone on the grounds that the drier parts of this zone are not basically different from the arid rangelands and the wetter parts, of agricultural potential, offer similar though reduced development opportunities to the sub-humid zone. The entry for the highlands refers only to the "normal" range of conditions and omits those parts which are very dry (less than 600 mm rainfall) or very high (over 3,000 m elevation). Emphasis in the table is given to land use and livestock development, which is also the concluding theme of this chapter. Development is considered by reference to its operational components of planning, research and implementation, and the chapter ends with comments on the limitations of an ecological perspective.

Planning

There is no better starting point for development planning than an ecological zonation which focuses on major differences in land potential. It is necessary, however, to adjust the detail and the interpretation of the zonation to the scale of planning concerned.

National planning. The allocation of national resources between areas and sectors is inevitably subject to political and social pressures which have little to do with ecology. However, insofar as economic costbenefit is a consideration, an eco-climatic zonation provides a valid framework for decision making. The zonal differences which have already been discussed translate readily (if not precisely) into figures for output and human support capacity. In terms of livestock development they indicate where individual species and commodities are concentrated, and they provide a generalized measure of the con-

Table 2.2. Principal livestock zones of tropical Africa

	Physical Environment				Vegetation and Land Use				Livestock Development			
Zone	Annual Rainfall (mm)	Temperature and Humidity	Elevation (m) and Relief	Natural Vegetation	Fire	Tsetse	Land Use	Main Live-Stock	Over-Grazing	Pot-tential	Objectives	Priorities
Semi-desert	< 250	Hot and very dry with no cool season except in S	Variable but most <1000 with minor relief	Sparse shrub grassland concentrated in depressions & water courses	Rare	Absent	Seasonal & nomadic use, with some commercial ranching in S (karakuls)	Camels Goats (Sheep)	Local (near water)	Very low	Controlled seasonal use by present users, keeping desert at bay	(1) Control desert creep (2) Local water harvesting (3) Restricted water development
Arid rangeland	> 250 < 750	Relatively hot and dry, often with seasonal fluctuation	Extends over wide range of altitude & relief	Thorn bushland and grassland; grasses mainly in W with more perennials in E & S	Quite common	Very Local (near water)	Pastoralism & localized ranching & cultivation often with wildlife	Zebu cattle Goats Sheep Camels Donkeys	Common	Moderate	Sustained pastoralism & local ranching & cropping with increased off-take	(1) Build local management units (2) Develop forage potential of favored sites (3) Marketing
Semi-arid & sub-humid lowlands of agricultural potential	> 750 < 1250	Relatively hot and seasonally humid with marked dry season	<1500 mostly plains & hills	Savanna woodland & bushland with perennial grasses (tall & woody in sub-humid zone)	Common	Common especially in sub-humid zone[a]	Crop & livestock production with much of sub-humid zone only lightly used (± wildlife)	Zebu & N'Dama cattle Sheep Goats Poultry	Local	High	Integrated crop/livestock production & new enterprises for underused areas	(1) Improve dry season feed (2) Tsetse & bush control (3) Multiply adapted breeds
Humid lowlands	>1250	Warm and oppressive because of high humidity	<1200 mostly flat	Forest and derivatives with patchy grazing except where forest is cleared	Local	Rarely absent	Tree & root crops & forestry with relatively few livestock; more intensive use in E	Goats N'Dama cattle Poultry Pigs	Rare	High (when forest cleared)	Increased meat production from small stock pending longer-term development	(1) Improve small stock husbandry (2) Research trypano-tolerance & new farm systems
Highlands of agricultural potential	> 600	Warm days & cool nights with relatively low evaporation	>1200 <3000 plateaux, scarps & hills	Varies with rainfall from savanna to forest, often with productive grassland	Quite common	Rare	Mostly small-holdings with some forestry; high livestock & human population	Zebu & exotic cattle Sheep Donkeys Poultry	Local	Very high	Increased milk & sheep production by breed & forage crop improvement	(1) Increase on-farm forages (2) Upgrade live-stock (3) Improve animal traction

Source: Author.

[a] The semi-arid and subhumid lowlands can legitimately be subdivided according to tsetse presence or absence or seasonal occurrence since tsetse and trypanosomiasis challenge profoundly influences livestock production strategies and economics in terms of choice of breed, chemotherapy, grazing pattern and inputs into tsetse and bush-control. Seasonal occurrence is important because tsetse expands greatly in distribution during wet seasons, and areas of high challenge at such times may be tsetse-free at other times of the year.

straints, costs, and benefits associated with development. In themselves they do not show appropriate inputs but they do provide guidance on what not to do or
expect in terms of land use. Adoption of them should
avoid gross errors like installing ranch development in
the semidesert or placing imported cattle in a tsetse
zone.

District planning. The process of deciding the
appropriate inputs for a zone or district has to involve field survey. All of the components of ecology
and livestock production which have been described need
to be investigated and their interrelationships established. Particular attention has to be given to functions of livestock and, in pastoral societies, to
social territorial organization. Many of the disap-
pointments in livestock development stem either from
lack of understanding of the multipurpose role of livestock in subsistence economies or, in pastoral areas,
from failure to identify an appropriate form of local
organization to ensure land management and active
participation in the development process.

Topics such as the function of livestock and the
choice of pastoral organization might be thought the
preserve of sociology and not directly relevant to
ecology. There is, in fact, interaction in three
directions. Not only does the environment help determine the choice of function(s) and organization at the
time that development is planned but, once a development program is implemented, the choice that has been
made in these areas will largely determine the ecological impact of the development program. Moreover, the
sociological assessment that is made of current practices at the outset is likely to be much enhanced if a
basis for these practices is sought not only in history
and social values but also in the ecology of pastoralism. Indeed, one of the principal uses of ecology in
development planning is to provide a framework within
which scientists of all disciplines can apply and
coordinate their individual skills.

Site planning. As planning becomes more specific
so does the need for data. The contribution which
ecology can make to the planning of individual sites
and enterprises depends greatly on the nature of the
enterprise. In the case of feedlots an ecological
perspective has little to contribute except possibly
for feedlots which fatten store cattle, when assessment
may be needed of the likely availability and seasonal
flow of store animals from arid rangeland or of the
suitability and productivity of land for home-grown
fodder. The latter application is relevant also to
crop/livestock enterprises, in order to select crops
(and animals) appropriate to the environment and to set

ecological limits on the spread of cultivation. By far
the greatest contribution of ecology, however, arises
in pastoral and ranching situations.

The starting point for planning the development of
a specific pastoral area is the existing form(s) of
social-territorial organization. The end-point similarly is the definition of pastoral units appropriate to
development. In between, an ecological perspective
needs to be applied to the assessment of grazing and
livestock   resources,   movement   patterns   and   water
distribution, and to options for range management and
improvement, including needs for bush-control and/or
seeding. Mapping of ecological land units is needed at
a scale of a least 1:100,000.   (For the purposes
described   at   the   national   and   district   level,
ecological maps at scales, respectively, of 1:1 million
and 1:250,000 are usually sufficient.)

The form(s) of pastoral organization chosen for
development can vary from loosely structured grazing
associations through group ranches and cooperatives to
commercial ranching companies. In practice the less
structured organizations are the best suited to extensive arid zone pastoralism, and company ranches are
usually restricted to commerically minded investors
with access to relatively high potential land [Pratt
1974]. The group ranches of Kenya (now extended to
other parts of Africa, [Oxby 1981]) are a special
case. Although the original concept is often misinter-
preted, group ranches were conceived simply as a framework for pastoral development so that those with
customary right to land could have security of tenure
to mobilize and apply inputs and management. Ownership
of livestock remains in the hands of individual owners
who, through elected group representatives, decide the
use of group lands. Group ranches therefore have more
in common with semi-sedentary pastoralism than with
commercial ranching. Their establishment has to be
preceded by close examination of the distribution of
livestock between families and of the ecological potential and human support capacity of the group lands.

Such studies take time to execute, and first
impressions can be dangerously wrong. Nor is it to be
expected that there is a practical solution for all
situations. If the the pastoral population greatly
exceeds the ecological potential of the land, subsistence for some will have to be found outside the
pastoral economy before pastoral development can be
effective.

Research

The purpose in introducing the subject of research is not to list proposals for ecological research but to show how ecology can influence research priorities and methods.

Priorities. In the same manner as an ecological zonation guides development planning so it can be used to guide the allocation of resources for research. This is not to say that a high potential zone should necessarily receive more resources than one with low potential. Certainly the potential payoff is an important factor in allocating resources but, depending on the nature of the constraints, the immediate returns from research may not be directly correlated with land potential. (Thus research on tsetse-control is likely to be better concentrated at the drier limits of tsetse distribution rather than in areas of higher potential deep within the tsetse habitat.) It is also a fact that arid zones research is relatively long-term and expensive so that, if there is to be any research at all in these zones, it needs disproportionately more funding than its potential would suggest.

An ecological zonation is also useful as a framework for building research teams and encouraging appropriate specialization. A completely different set of production constraints exists between ecological extremes which can (and should) be reflected in team composition and specialization. It is significant in this context that the International Livestock Centre for Africa has organized its research on a zonal basis, with separate teams for the arid zones, the subhumid and the humid zones and for the highlands [ILCA 1980]. Although all major disciplines are represented in each team, the emphasis and focus varies greatly between teams. Whereas emphasis in the arid zones is on the ecology and improvement of pastoralism, it lies with trypanotolerance and small ruminant production in the humid zone, animal nutrition in the subhumid zone and with forages and breed improvement for milk and traction in the highlands.

The relevance of an ecological approach to research is obviously greater in environments which present major constraints to production than in environments favorable for agriculture. In the former case -- as in the arid zones -- a significant part of the research effort is likely to be focused directly on ecological problems, be they concerned with the dynamics of bush encroachment, the role of fire, or the effects on vegetation of different grazing regimes. To ignore ecological constraints when planning research is one step towards irrelevant research.

Methods. The application of an ecological perspective does not greatly influence experimental
design, except perhaps in the choice of sites or number
of replicates. However, the holistic approach which
comes with ecology leads very naturally to the adoption
of a systems approach to research. This is not the
place to elaborate at length on the systems approach,
but it does relate closely to ecology -- combining
social and economic factors with those inherent in the
ecosystem concept -- and it does have particular
relevance to livestock research in Africa. It is, for
example, applied by the International Livestock Centre
for Africa to the selection and study of specific livestock production systems so that, by a better understanding of real production situations, the essential
constraints and the most profitable lines of specialist
research can be pursued. As practiced by ILCA it takes
the major research effort out of a research station
environment onto real farms and herds, with the aim of
increasing the relevance of research to development.
And relevance in research -- in the choice of avenues
of research and the transferability of results -- must
be the prime consideration of anyone concerned about
rural development in Africa.

Implementation

Success in livestock development depends not just
on identifying appropriate development paths and technical inputs. Much also depends on project design, the
sequence of inputs, and project management. A brief
comment on the influence of ecology in these fields is
therefore appropriate.

Project design. Although it is convenient to
write in terms of "projects", one of the current concerns in livestock development in tropical Africa is
whether a project framework, rigidly prescribed in time
and space, is an actual impediment to development, and
better suppressed in favor of a more flexible and open-
ended program approach. Certainly, from an ecological
perspective, there are inherent dangers in rigid and
ambitious prescriptions. In areas of low and erratic
rainfall, flexibility and mobility are cardinal vir-
tues, and in intricate pastoral systems, where people,
livestock and land are interwoven and interdependent on
the chance of seasonal rainfall, an incremental
approach to development seems the only safe course.
Ecology, combined with social anthropology, can show
where a cautious approach is needed, and it can also
show which inputs are least likely to disrupt the
existing system or lead to environmental damage.

Inputs. The choice and sequence of development inputs should be decided by reference to all perspectives; economic, social and technical as well as ecological (and local as well as governmental). Ecology is of most help in prescribing breeds or varieties appropriate to a given situation and in avoiding environmental mistakes. Examples where failure to heed ecology would be particularly catastrophic include (a) the dissemination of European breeds of cattle outside the highlands, (b) breed improvement which in any way reduces the trypanotolerance of ruminants in the tsetse belt or the hardiness of arid zone stock, (c) adding or confining livestock to a grazing area beyond the limits imposed by its critical environmental constraint (be that the spatial or temporal irregularity of rainfall or a soil nutrient deficiency) and (d) allowing water development in a pastoral area to precede the establishment of an effective mechanism for controlling land use.

The last example is related specifically to the sequence in which inputs are applied. This can be just as important to effective development as the actual choice of inputs. In some cases inputs applied out of sequence merely cause inconvenience or waste of money; the particular example of premature water development, however, can cause irreparable damage to arid rangeland and needs constant vigilance.

Management. Project managers have an awesome responsibility. Planning, like research, can be a comfortable sinecure, but poor project management is there for all to see and impinges directly on the lives of those entrained in the development process. More attention needs to be given to the training of present and future managers, and, in this context, to providing them with a sufficient appreciation of ecology to take sensible management decisions. Universities, in turn, need to improve their capability for providing such training.

Limitations of an Ecological Perspective

Being close to the end of this conducted tour through the highways and byways of ecology, one could admit that the passage has not been easy. This is partly because of the breadth of country traversed, and having to move rapidly over areas which need time to explore, but also it results from the need to keep to the ecology pathway when other avenues invite inspection.

In reality, livestock production and livestock development in Africa rest squarely on a combination of ecology, economics, biology and sociology. Ecology happens to be the most fundamental of influences, but to elevate one component above the rest is inherently dangerous. Ecology can provide a useful meeting point for other disciplines but, for most purposes, the more useful framework is that provided by livestock production systems. Ecology has a large part to play in molding production systems as well as in their evaluation, but it is the production system which is the summation of all determining factors. It is both neutral territory in the event of disciplinary conflict and, most important, it is the natural focus of the development process. One seldom develops an ecological zone or a vegetation type -- not at least as one management unit -- whereas the individual farms or pastoral groupings which make up a production system are themselves discrete development units.

This digression is not an indictment against ecology. It is far more dangerous to give free rein in development to an animal scientist or an economist than to an ecologist. For the most part problems arise only when the ecological perspective is pursued too far or too vigorously. To override ecology and cause pastoral settlement in the semidesert is wrong, but to insist than an area of high ecological potential receive priority over one of low potential, irrespective of social benefit, is also wrong. The problem is accentuated when ecological priorities are seen in terms of conservation rather than development. It is good that there are ecologists whose principal concern is conservation, but those with regard for Africa have somehow to tread a middle course which gives pre-eminence to the people and their perspectives while preventing ill-advised and destructive land use.

REFERENCES

The few references cited in this chapter and listed below represent no more than 1 percent of the literature available. Almost each paragraph could have carried a list of equal length. However, such a listing would have been inappropriate to the intended purpose, which is simply to present a scenario and a point of reference for later discussion. Instead, references have been included which include extensive bibliographies of their own. For example, Pratt and Gwynne [1977] cites over 500 references, Sandford [1983] around 300, and Jahnke [1982] and Penning de Vries and Djiteye [1982] over 200 each.

International Livestock Centre for Africa. ILCA The First Years. Addis Ababa: ILCA, 1980.
Jahnke, H.E. Livestock Productions Systems and Livestock Development in Tropical Africa. Kiel:Kieler Wissenschaftsverlag Vauk, 1982.

Oxby, C. Group Ranches in Africa. Rome: Food and Agriculture Organization of the United Nations, 1981.

Penning de Vries, F.W.T. and Djitèye, M.A. ed. La productivité des pâturages sahéliens. Une étude des sols, des vegetations et de l'exploitation de cette ressource naturelle. Wageningen: Centre for Agricultural Publishing and Documentation, 1982.

Pratt, D.J. The Ecological Management of Arid and Semi-Arid Rangeland in Africa and the Near East. The Concept of "Discrete Development Areas" as applied to Range Development. Rome: FAO Paper AGPC:MISC/25, 1974.

Pratt, D.J. and Gwynne, M.D., ed. Rangeland Management and Ecology in East Africa. London: Hodder and Stoughton, 1977.

Sandford, S. Management of Pastoral Development in the Third World. Chichester, England: John Wiley & Sons, 1983 .

3
Livestock Nutrition in Subsaharan Africa: An Overview

Robert E. McDowell

INTRODUCTION

There is general acceptance that the subsistence function of livestock is the principal characteristic of pastoral systems. Based largely on short-term surveillance, both social and animal scientists view the pastoral systems of Subsaharan Africa as creating and maintaining a depressed sector. This image has led to the assumption that a political change, along with concomitant adjustments in policies on food production, will induce change in output from pastoral systems. A facilitating political environment can be an important precondition to increasing productivity, but it is insufficient unless supported by appropriate technology.

Although there is the general impression that pastoralists are irrational in such matters as stocking rate and management of grazing, recent observations indicate that within their economic and national environment, they are as rational and productive as their counterparts elsewhere [de Leeuw and Konandreas 1982] However, even though pastoralists may operate in a practical way, their systems are of concern to governments and international agencies, as governments often view them as a mobile force with disruptive potential and as groups which contribute little to national productivity. To a large extent animal scientists and other technologists view pastoralism as a system with potential for larger outputs of animal products, but these scientists are presently frustrated on how to induce viable changes. There are also certain groups, such as ecologists and conservationists, who frequently consider pastoral systems a destructive force in a fragile environment.

The total animal biomass in Subsaharan Africa is unknown, but estimates are that the indigenous animal life, consisting of game species, rodents, ants, locusts, termites, reptiles, birds and others, accounts for a large proportion of biomass apart from the predominant domestic species; camels, cattle, goats and sheep, along with some equines. The whole of Africa is highly dependent on its animals. The estimated annual output for livestock is $10 billion with 50 percent attributed to meat, milk, fiber and skins and the rest from services or non-food products such as manure, traction and transport [ILCA 1982]. Game animals contribute $3.5 billion from meat, tourism and trophies. Seafood from rivers, lakes and coastal waters add an estimated $3-4 billion to food supplies. Production by animals far exceeds Africa's annual cereal output, valued at $8.5 billion.

To meet a rapid rise in demand for additional livestock products, African governments have invested an average of about $1 billion per year over the past 15 years in attempts to improve pastoral systems [ILCA 1983a]. But, despite this apparent large input, assistance agencies and governments concur, in general, that results have been disappointing, largely due to an inadequate knowledge base on which development interventions were designed [Jahnke 1982]. For example, the rush to relieve or prevent starvation among pastoralists led to attempts to transfer animal production technology without suitable pre-testing. Unless more linkages among biological and social scientists can be developed to bring about visible changes in human welfare in the subsaharan region in the near future, there is a strong possiblity that support for livestock programs will wane. The discussion in this chapter begins with some of the shortcomings of attempts to induce change and, simultaneously, draw attention to possible useful modifications in developing future strategies for livestock production.

MISCONCEPTIONS

A first step in designing a livestock development strategy for Africa is to convince those concerned with environmental deterioration that livestock have not been the major contributor to increased desertification. It is timely to create awareness that an early destructive force was the cutting of trees to fire furnaces for iron making [Haaland 1979]. Cropping has also been a degrading force so that, in much of the Subsahara, domestic livestock have become the "last or residual users". Furthermore, it can be argued that animals have actually slowed desertification because the plants which have survived produce largely hard coated seeds. Passage of these seeds through ruminants scarifies the seeds, thereby making them more responsive to moisture.

A second point is that animal scientists have largely viewed pastoral systems as static. They assumed

that natural environmental pressures were the major
force in developing various livestock phenotypes found
in the region. But, on the contrary, it is pastoralists who, due to their adapative systems have selected
for a large number of animals now found. In addition,
pastoralists have focused management to meet their
needs as, for example, their sheep and goats not only
breed throughout the year, but are species of high
fecundity. Also, cattle, goats and sheep in the Subsahara have a better tolerance to low water intake than
species found in cool climates. Pastoralists have
accentuated development of fatty deposits in the rump
or tail of sheep in order to have a readily available
supply of fat needed for a number of household uses at
the time of slaughter. The pastoralist is also well
aware that low fat coverage on meat permits cooking
over open fires without charring as would be the case
for lambs of western breeds. Short wool or hair covering has been the focus for sheep in the higher rainfall
areas and coarse wool (mixture of hair and wool) in dry
areas. Coarse wool has a high utility for hand weav-
ing, and in the production of clothing and other goods
highly useful in harsh environments.

The cattle of Subsaharan Africa are low milk producers by western standards, but pastoralists have
found that continuity in the yield of a few liters of
milk per day better fits their needs than improved
breeds with high peak yields but ones which show a
rapid decline when feed resources are limited. The
pastoralist also views modest milk yield as associated
with longevity and herds with a life of 10 to 12 years
are common compared to ones with a much shorter life in
intensive systems. Pastoralists are aware of peculiar-
ities in feeding habits of camels, cattle, goats and
sheep, and they generally exploit these differences in
herd composition to make fullest utilization of
grasses, forbs and browse plants in various ecosystems.
Mature cows are preferred to permit more rapid conception than would occur for immature cows when there is a
spurt in the growth of forages.

Studies of African rangelands have resulted in
recommendations for lower stocking rates on the assumption that more residual biomass would result in higher
forage production. Heavy grazing pressure has been
assumed to lead to destruction of some plants and the
invasion of undesirable plant species. More recent
evidence has indicated this may not always be the case,
as areas which are grazed most heavily often have the
greatest populations of the most nutritious grasses
[Mosienyane 1979; Reed 1983]. Pastoralists are aware
that grasses, such as Themada triandra and Eragrostis
rigidor, are low in feeding value unless kept closely
cropped. They are also cognizant that a grass biomass

20 cm or higher will deteriorate rather rapidly and a high proportion will be lost in the absence of grazing or fire through natural decay, termites, ants, locusts, birds, and rodents. In northern Nigeria, for example, up to 50 percent of the sorghum residue is destroyed by termites [Von Kaufman 1982]. Pastoralists know that close grazing minimizes the presence of pests which may prey on adjacent agricultural crops. The major problem of the pastoralist is low flexibility in number of animals for food needs under highly variable rainfall conditions.

Jahnke [1982] takes the position that pastoralists execute logical risk-averting strategies for their livestock. They attempt to maximize herd and flock size to the limits of labor for herding and watering. They have mixed holdings of two or more species; herds are split into separate management units; animals are lent out to relatives; a high proportion of females are kept for rapid rebuilding of herds or flocks following drought; and they take risk in crop production to enhance their food security. Other recent evidence [de Leeuw and Kondandreas 1982] also demonstrates that pastoralists perform rationally in the management and utilization of resources. These authors state that when livestock sales are expressed in terms of biomass, the annual offtake ranges from 16-28 percent which suggests a rather efficient strategy of operation.

TRANSFER OF TECHNOLOGY

Thus far success in the transfer of animal production technology has been limited, as single inputs of technology instead of packages of practices have been recommended most of the time. For instance, males or semen from males of improved breeds have been used in crossing with local cows, does or ewes. The resulting crossbreeds have come under severe nutritional stress from low feed supplies and diseases such as rinderpest, resulting in much lower performance and frequently a shorter life than expected [McDowell 1983]. Considering that pastoralists have been dealing with species of livestock which were not indigenous to Africa, they have done well as animal breeders. As an example, the rinderpest control program waged during the 1960s and 1970s was a single stage or phase input. It resulted in more animals being kept alive, but with a larger animal biomass competing for approximately the same feed. Vaccination programs were, to a large extent, short-term insurance.

High mortality (17 to 45 percent) has been recorded for calves, kids and lambs between birth and three months of age. It is often recommended that highest priority be given to means for reducing mortality

[Eicher and Baker 1982; Oxby 1981; World Bank 1981], but success has been low. Although identifiable health problems may be the ultimate cause of mortality, it is becoming more evident that malnutrition is a major cause of high losses. Calves are poorly fed because of the need to use milk for human consumption, and mortality may be higher when women replace men for the milking process due to differences in objectives. Women see milk more as food or higher returns from sales as opposed to men who are interested in expansion of stock numbers mainly for prestige purposes [Nicholson 1983].

Low nutrition level for does or ewes and the high expenditure of energy by kids and lambs to follow their mothers while feeding contributes to high losses observed in these species [Donaldson 1982]. Also, the death rate among adult cattle is approximately twice that of intensive African systems (5 to 6 versus 2 to 3 percent). The higher rate in pastoral herds is attributed more to the long herd life (greater than 10 years for cows, and greater than 6 years for does and ewes) than to specific health problems [Nsibandze 1982]. Probably the best means of improving viability of young would be to decrease dependence of pastoralists on milk through crop agriculture, or improve food security in other ways such as off-farm work [Nicholson 1983].

It can be hypothesized that reproductive efficiency in pastoral enterprises could be improved through closer attention to their livestock, but implementation of means to improve breeding efficiency by pastoralists has been poor. The usual assumption is that higher reproduction rates are desirable and there are few constraints. Even if a pastoralist deemed more frequent parturitions desirable, that person may not view as acceptable returns for the extra labor required for closer attention to individual animals.

More examples to illustrate some of the shortcomings in attempts at direct transfer of technology could be cited, but the central issue is that the record to date is far less than desired. Obviously, single phase technology and direct transfer do not meet the needs for producing visible and viable changes in traditional systems.

POTENTIAL FOR CHANGE

New Technology

Although the transfer of technology picture has been less than satisfactory, there are numerous new developments in animal science technology which can be beneficial in more effective research, and some that can be directly employed in pastoral herds. Among those applicable to animal nutrition research are a new

methodology for determining the feeding value of rangeland forages [Van Soest 1982]; more precise knowledge
of digestive physiology of herbivores [Demment and Van
Soest 1982]; more exact means for determining the
influence of chemical compounds, such as tannins, on
the utilization of browse plants [Reed et al. 1982];
and improvement in procedures for identification of
areas where mineral deficiencies may be serious for
livestock [Kapu 1982; Mtimuni 1982].

New technology potentially applicable to mixed
farms for improving feed resources include alkali
treatment of crop residues to increase digestibility;
use of ammoniated compounds for direct feeding to
ruminants or applied to forages at the time of harvest
for storage to raise the crude protein content; and the
addition of preservatives to aid in the preservation of
quality of forages with a low dry matter content when
cut in the wet season.

Compounds useful in synchronization of estrus to
improve reproductive efficiency are available. Transfer of fertilized embryos has high potential. This
procedure could be especially useful in rapid multiplication of cattle breeds with a high resistance to the
disease trypanosomiasis. However, these technologies
in nutrition and physiology must await application
until cost is lowered and national agencies in Africa
have sufficient infrastructure for effective implementation.

Several modifications of traditional management
practices are feasible in some situations, such as the
use of a controlled or restricted breeding season which
would lessen neonatal mortality and improve growth
rate. This procedure could be applied on fenced group
or cooperative ranches but it will not be effective for
pastoral herds until the herders have closer control
over their herds or flocks. A number of recent developments in rangeland management could be useful as
well, for instance, closer attention to animal feeding
habits. In traditional pastoral systems the grazing
patterns and vegetation types on which ruminants feed
are controlled by the livestock herders. Keeping
animals in bomas at night severely restricts their
feeding time, but restrictions on feeding time may have
been an important feature in the adaptation of cattle,
goats and sheep to this system. Dentition, muzzle
width and manner of grazing or browsing are important
in the ease with which ruminants can select plants with
different morphology [McCammon-Feldman et al. 1981].
Goats, for example, have prehensile lips and narrow
muzzels which permit easier selection of leaves from
the thorny Acacia species. In contrast, the broad
muzzle and large tongue of cattle allow rapid harvesting of grass.

Goats are the most selective of the region's domestic stock as they prefer grass when protein content and digestibility are high, but shift to browse (twigs and leaves of woody plants) when their nutritive value may be greater. This means the goat has a wide range in ability to alter diet with season. Camels and sheep have been determined as intermediate feeders, prefering grass, but browsing in order to maintain high quality diets. Sheep are able to graze shorter growing plants and, with their wool covering can feed further underneath thorn bushes than cattle or goats.

A high degree of selectivity in feeding is not without disadvantages due to energy expenditures in securing feed. A camel may travel up to 15 km per day, goats 10 km, sheep 6 km and cattle less than 5 km in the same feeding area. Energy expended in walking may reach 50 percent of maintenance requirements. Herd or flock size will influence energy expenditure in grazing. For example, a herd of 100 cattle will need to walk 34 km per day to have the same opportunity for feeding as a herd of 10 head walking 6 km [Bayer and Otchere 1982].

Cattle are the least selective feeders in the region. Their metabolic rate is lower than goats or sheep and retention time longer, resulting in greater ability to digest fiber. The vast bacteria population of the cattle rumen becomes a significant source of protein, hence they can survive on a diet of poor quality grazing. These observations on feeding strategy demonstrate the need for appreciation of the plant ecosystem composition in order to obtain best utilization by grazing animals.

Cattle, goats and sheep will have near equal efficiency in using the forage dry matter of grass pastures in temperate areas, but the interchangeability of species concept is not appropriate to the subtropics and tropics. The problem is that plants of tropical areas tend to be more differentiated i.e. they have highly lignified as well as unlignified parts. The digestive physiology of the species determines feed strategies [Demment and Van Soest 1983], and animals which can choose the most digestible parts have advantages in these situations. Where feed resources are generally less than optimal a mixture of animal species makes the best use of an ecosystem comprised of natural vegetation.

Recent studies on the nutritional value of browse and forb plants in Kenya's rangelands showed that the feeding value of numerous species, especially browse plants, is substantially lower than commonly assumed based on crude protein analysis because of the high tannin content which lowers palatability, protein utilization and digestibility. These results indicate the need for further evaluation of the plant inhibitors in Subsaharan Africa, particularly where browse plants and leguminous trees may be introduced for dry season feeding. Research by personnel at the University of Florida and others show that the mineral deficiencies in the forages could be important, and direct supplementation to animals may be highly viable [Mtimuni 1982].

Some researchers, such as Talbot et al. [1965] and Hopcraft [1975] have attempted to demonstrate that greater returns could be obtained from the use of game for meat production in the semi-arid regions. There are also claims that game will make more efficient use of plant ecosystems than livestock, and problems of erosion and desertification would be reduced. However, evidence has accrued to show these are tenuous hypotheses [McDowell et al 1983]. As an example, the best species for marketable meat, Thomason's gazelle, Grant's gazelle, kongoni and wildebeast, are selective feeders. They tend to prefer high quality grass, but shift to browse when quality declines. When game numbers are high, their feeding strategy tends to make them partially competitive with livestock as game need a low selection feeder like cattle to condition the grass stand. In the large range areas, such as the Serengeti Plains, the zebra largely fills the role of low selector, but in pastoral systems game follow cattle [Reed 1983].

Costs of harvesting and marketing game meat are more than twice that or beef, hence the price of game meat in the market is above the reach of low income groups. But, the use or sale of game meat, skins and trophies can enhance returns to pastoralists. The reproduction rate of game is as subject to variations in environmental stress as domestic livestock, hence their viability has been over estimated. Also, those who are optimistic about game overlook the high value of milk from livestock for humans in pastoral systems. Nevertheless, it can be concluded [MoDowell et al. 1983] that a mix of game and cattle could give higher returns for meat production than either pure game or cattle operations. Furthermore, since game do tend to inhabit pastoral areas the combining of game harvest with livestock in pastoral systems does appear worthy of further investigation.

Feed Resources

It is difficult to derive estimates of the rangeland's feeding value as so much depends upon the ecosystem and the amount of rainfall and stocking rate. For example, with a near pure stand of _Eragrostis rigidior_ cattle would be able to obtain energy needs

for maintenance requirements for less than one month
[Mosienyane 1978]. Under managed grazing, and with
mixtures of grasses, rangelands can be a satisfactory
source of feed for grazing ruminants for several months
per year as illustrated in Table 3.1. With maximum
intake from tropical grass pastures by cattle at 2.2 to
2.4 times the maintenance level [McDowell et al. 1975],
African forages are excellent for about three months.
But, when intake level is just 2.0 times maintenance,
approximately 60 percent of the energy consumed will be
available for growth or milk production. At this level
cattle may gain 0.6 to 1.0 kg per day, but when intake
declines to 1.5 times maintenance level only 30 percent
of the nutrient intake is used for production [Bayer
and Otchere 1982]. On the northern ranges of Nigeria,
cattle will be in a weight declining state for about 5
months, hence gains per year are low (Table 3.1). In
contrast, Botswana rangelands are no better at their
peak than in Nigeria, but quality remains higher due to
a longer rainfall period, greater density of better
quality grasses, and more available browse than in
Nigeria. Much of Kenya's rangeland have a bimodal
rainfall which affords definite advantages. To obtain
the feeding levels indicated in Table 3.1, about 4 to 5
ha per adult cow unit will be required in Kenya, and 7
to 10 ha in Nigeria.

Several reports, e.g. World Bank [1981] and Oxby
[1981], have conveyed the impression that the rangelands are often underutilized. This assumption is
generally based on the presence of residual plant biomass. If there is ungrazed material there is usually
one or more limiting factors, such as low nitrogen
content [Humphreys 1978] or water for animals. In
general, it may be concluded that a managed stocking
rate and periodic burning, along with assuring water
supplies and appropriate mixtures of animals, will make
best utilization of rangelands. It is not economically
feasible at this stage to attempt large substitutions
of grasses. However, there are numerous promising
species of trees which could be useful as fuel and feed
resources [ILCA 1980].

Supplementary Feeding

In the uni-modal rainfall areas pastoralists
expend a great portion of their energies seeking to
supplement rangeland grazing both for quantity and
quality [Waters-Bayer, 1983]. Seeking the additional
feed is a source of antagonism with the agricultural
sector due to crop damage, but it does serve to increase interdependence between herders and crop
farmers. Herders want use of crop residues for milk
production to exchange for grains, and crop farmers

Table 3.1. Estimated energy intake by months of maintenance multiples for 250-300 kg cattle on rangeland grazing[a]

Month	Northern Nigeria[b]	Central District Botswana[c]	Machakos District Kenya[d]
January	.8	2.1	1.4
February	.7	2.0	1.2
March	.6	1.7	1.5
April	.6	1.4	1.9
May	.5	1.2	2.0
June	1.5	1.0	1.6
July	2.3	.8	.9
August	2.2	.7	.8
September	2.0	.6	1.0
October	1.5	.6	1.5
November	1.2	1.6	2.0
December	.9	2.1	1.6
Mean	1.23	1.29	1.45
Mean daily gain for year (kg)	.20	.24	.30

Source: Adapted from McDowell [1978], Mosienyane [1978], and Reed [1983].

[a]An example of interpreting this table is that in January, 250-300 kg cattle in Northern Nigeria only intake 80 percent of the nutrient energy required just to meet maintenance needs.

[b]Annual rainfall 450-500 mm with 97 percent from late May to mid-September.

[c]Annual rainfall 400-500 mm with 95 percent from late November to mid-May.

[d]Annual rainfall 500-550 mm with long rainy season, March-June and short rains, October-December.

desire manure to improve crop yields and milk to augment their diets [Wilson 1982].

Following more than six years of analyzing the linkages and interactions of the major livestock production systems components in Africa, the International Livestock Centre for Africa (ILCA) has given highest priority to improvement in supplementary feed resources [ILCA 1982, 1983a]. The major focus is on improvement in quality and expansion of protein feed sources. There are numerous potentials with one initial finding being that feeding of agro-industrial by-products as a supplement to rangeland forages will increase milk offtake, but at a level insufficient to meet costs except in areas where there is infrastructure to collect and move milk to urban areas [Von Kaufmann 1982]. ILCA has formed the African Research Network of Agricultural By-products (ARNAB) to coordinate research on agricultural by-products in Africa. ARNAB publishes a quarterly newsletter to disseminate findings. The main objectives of the network are to evaluate additives to improve the nutritive value of crop residues and local by-products, and develop methods for preservation of them.

One example of improving by-product use is treatment of cereal grain straws, and sorghum and millet stovers with alkali compounds. e.g. sodium hydroxide to improve digestibility. But, costs at present are not justified unless a high protein feed is available. The addition of ammoniated compounds to increase nitrogen appears beneficial [Von Kaufmann 1982] but, again, cost may restrict use. In Nigeria, the cutting of the upper 1.0 to 1.2 m of sorghum stalks at the time of harvest, and storing in trees to avoid termites has resulted in a satisfactory dry season supplement feed [Bayer and Otchre 1982.]

Another major topic is development and testing of improved forages and crop legumes. The objective is to increase production and nutritive value of rangelands as well as supplementation of the crop residues many animals rely upon [Brumby 1983]. Use of small plantings of the forage legume Stylosanthes in northern Nigeria has proven successful as a source of supplementary protein feed for lactating cows [ILCA 1983b]. In Mali, new varieties of the legume cowpeas are cropped to enhance human food and livestock feed sources [Wilson 1982]. Crop and forage legumes offer great promise for Africa, but potential will not be realized until seed is locally available as importation of seeds is too expensive. Africa has a wealth of legumes, but none are under commercial seed production. ILCA has begun to collect legumes and test their suitability for field use through a network of national institutions [Brumby 1983].

Browse from bushes and shrubs often contributes a significant amount of feed resources especially for goats and camels. A number of these are being tested for content of chemical inhibitors, such as tannins, with the objective of advising herders and range management personnel on the most useful species. A number of browse are good feed resources whereas numerous others have low utility [ILCA 1982].

Hay making offers some promise. However, in all of the Subsahara, it is only a traditional practice in the highlands of Ethiopia. Consequently, the introduction of legumes, such as vetch, in that area as a hay crop has been well accepted. [Gryseels and Anderson 1982] But, in general, hay making from rangelands is not promising for several reasons among which are risk from spoilage if cut during the rainy season when quality is best, low density of grasses requiring high labor inputs, competition with labor demands for weeding crops, and high storage losses from mold or pests, e.g. termites.

Preservation of feed as silage has frequently been advocated, but results have been poor and costs excessive. A major problem is that crops most frequently available in the Subsahara are low in dry matter and water soluble carbohydrates (WSC) and such materials are particularly prone to secondary fermentation [Wilkinson 1983]. An adequate level of WSC for fermentation to a stable low-pH, lactic acid-dominant, well preserved silage is 3 percent of the fresh crop weight, but most tropical forages are below 2 percent in WSC [Van Soest 1982]. McDowell (1978) reported that silages made from Pennisetum purpureum in West Africa underwent secondary fermentation resulting in high butric acid contents. All the silages tested were marginal to submarginal in crude protein content and digestibility was not high enough to provide maintenance needs for cattle. Hence, farmers tended to reject the technology due to low returns to labor.

There is general agreement on the need for improved livestock nutrition in Africa, but opinions for solutions vary considerably. Breakthroughs will depend on the successful raising of animal productivity from low to intermediate levels rather than endeavoring to provide full nutritional requirements for high productivity over the short range. To this end ILCA's program, which is directed toward low capital investment and minimum labor expenditures seems appropriate. High emphasis in ILCA's programs is placed on evaluating labor availability versus requirements when introducing a proposed change, as pastoralists with no labor available beyond that provided by their families often cannot cultivate forages or prepare hay and silage. In some situations, alternate methods appear

workable. ILCA has been successful, for example, in
central Nigeria in using cows as cultivators of their
own feed crop. The animals graze the pasture and then,
during the early rains, break down the soil's hard
crust with their hooves to prepare a seedbed. During
the grazing and trampling manure is added. Seeded
forage legumes are grown to serve as a fodder bank
during the dry season. This technique makes it possible to grow good quality feed without extra labor or
implements [ILCA 1983b].

Water

Water is an essential nutrient, but little is
known about actual requirements for normal physiological functions or under stress [NRC 1981]. It is known
that supplies to maintain homeostasis is often unavailable in much of Subsaharan Africa. Low water availability and the observed desire of animals for it have
led pastoralists to assume greater water availability
would reduce stress and improve efficiency in animal
performance. On this premise highest priority in livestock development programs has gone to expansion of
watering facilities in several countries. More water
can help, but there may be important tradeoffs. For
instance, low water intake slows the rate of passage of
ingesta through the digestive tract. This ensures
higher levels of digestibility when feed quality is
low, especially in protein content. There is a physiological mechanism to recycle nitrogen which, under high
intakes of water and feed, would be excreted. This
process operates most effectively when the diet is low
in protein, and is more specific when water intake is
low [Pratt and Gwynne 1977].

An animal's normal instinct is to seek water when
feed intake is low and it is under heat stress. Water
helps to maintain comfort and gives a sensation of fill
which is comforting [NRC 1981]. There is also a desire
on the part of animals to maintain a near normal level
of body hydration. The tendency to seek comfort and a
normal state of hydration become major factors in overgrazing around watering points, the general conclusion
of which is that additional watering points may
actually reduce efficiency in the utilization of
rangelands. Borehole wells will ensure water but, here
again, there may be tradeoffs between wells and surface
ponds. Well water may contain compounds, such as sodium
sulphate, magnesium and trace minerals far in excess of
animal needs, and may actually accentuate problems of
malnutrition [Lambourne, unpublished].

Because of their feeding strategy, cattle have a
high need for water [ARC 1965] but, on low feeding,

watering just on alternate days is satisfactory [King 1983]. Water intake per kg of dry matter consumed may be as much as 40 percent less for sheep or goats than cattle [ARC 1965] due to low fecal output and selectivity in feeding. Camels are highly selective feeders, secrete highly concentrated urine, have low fecal water and seldom sweat, therefore are the most water tolerant among domestic animals [King 1983]. Lactating females and young of all livestock species need 30 to 50 percent more water than adult males or non-lactating females [NCR 1981].

Water resource development for livestock has concentrated on expansion in volume. But, evidence clearly shows the need for attention to water quality and location of resources in relation to forage quantity and quality; otherwise, the return to capital investment will be low. It should also be kept in mind that "over use" of water may decrease rather than increase efficiency in rangelands use.

CONCLUSIONS

During the past two decades a great deal of money has been spent by governments of Africa and outside agencies to enhance the well-being of pastoral peoples, but results have been much less than desired. Traditional pastoral systems were viewed as simple in nature when, in actuality, they are quite complex. In numerous situations, single phase technology such as boreholes were introduced with great expense but gave little return in productivity of livestock. Often programs were centered on preservation of the environment with little attention to needs and welfare of herders.

Careful study of pastoralists has revealed their systems are not static but dynamic, and the constraints to higher animal performance are more numerous than determined earlier. It has become clear that malnutrition among livestock and absence of capital or credit and labor shortages are among the primary constraints in Subsahara Africa.

Prospects of increased livestock production in pastoral areas appear promising through recent developments in methodology which will be useful in arriving at better guidelines for obtaining returns from rangelands. Research on improving the nutritive value of agricultural by-products and crop residues shows merit for these resources in supplementary feeding. Results from experiment stations and from limited testing on farms indicate that forage legumes and residues from crop legumes can be employed effectively as supplementary sources of protein feeds for the dry season. New technology to enhance reproductive efficiency and rapid multiplication of animal germplasm could be quite bene-

ficial, but costs for their use remain prohibitively
high for short-range use. In the future, more attention is needed to expanding the supply of milk from
pastoral herds to protein deficient urban areas.
Research on animal traction is also justified for small
holder agricultural systems. It is clear that labor
constraints in pastoral systems should be given major
emphasis in attempts to develop appropriate technology.

REFERENCES

ARC. "The Nutrient Requirement of Farm Livestock." No.
2 (Ruminants) Agric. Res. Counc., London, 1965.
Bayer, W. and E.O. Otchere. "Grazing Time of White
Fulani Cattle Kept by Selected Pastoralists on the
Kaduna Plains." Annual Rpt. ILCA Subhumid Zone
Program, Kaduna, Nigeria, 1982.
Brumby, P.J. "The International Livestock Centre (ILCA)
and Food Production in Africa." Proc. Conf.
Tropical Veterinary Medicine, Orlando, Florida,
1983.
de Leeuw, P.M. and P.A. Konandreas. "The Use of an
Integer and Stochastic Model to Estimate the
Productivity of Four Pastoral Production Systems
in West Africa." Proc. Nat'l. Beef Conf., Kaduna,
Nigeria, July, 1982.
Demment, M.W. and P.J. Van Soest. "Body Size, Diges-
tive Capacity, and Feeding Strategies of Herbi-
vores." Winrock International, Morrilton,
Arkansas, 1983.
Donaldson, T. "A Report on Family Sizes, Domestic
Economics, Livestock Composite and Progeny Histor-
ies of the Afar Production Unit Networks." JEPSS
Discussion Paper No. 1, ILCA, Addis Ababa, Ethi-
opia, 1982.
Eicher, C.K. and D.C. Baker. "Research on Agricultural
Development in Sub-Saharan Africa: A Critical
Survey." MSU Int'l. Dev. Paper No. 1, Michigan
State Univ., East Lansing, 1982.
Gryseels, G. and F.M. Anderson. "Research on Farm and
Livestock Productivity in the Central Ethiopian
Highlands: Initial Results." ILCA Res. Rpt. No.
12., ILCA, Addis Ababa, Ethiopia, 1982.
Haaland, R., "Man's Role in the Changing Habitat of
Mernas under the Old Kingdom of Ghana." ILCA
Working Document No. 2, Addis Ababa, Ethiopia,
1979.
Hopcraft, D. "Productivity Comparsion between Thomson's
Gazelle and Cattle and Their Relation to the Eco-
system in Kenya." Ph.D. Thesis, Cornell Univer-
sity, Ithaca, New York, 1975.
Humphreys, L.R. Tropical Pastures and Fodder Crops.
London: Longman, 1978.

ILCA. "Programme and Budget 1984." ILCA, Addis Ababa, Ethiopia, 1983a.

_____. "Livestock Production in West Africa's Subhumid Zone." ILCA leaflet, Kaduna, Nigeria, 1983b.

_____. "The Programme of Work and Budget for 1982." ILCA, Addis Ababa, Ethiopia, 1982.

_____. "Browse in Africa: The Current State of Knowledge." H.N. LeHonerou, ed. ILCA, Addis Ababa, Ethiopia, 1980.

Jahnke, H.E. Livestock Production Systems and Livestock Development in Tropical Africa. Kiel, Germany: Kieler Wissenschaftsverlag Vauk, 1982.

Kapu, M.M. "Soil, Forage and Animal Mineral Interrelationships on the Kachia Grazing Reserve, Kummin Biri, Kaduna State, Nigeria." Annual Rpt ILCA Subhumid Zone Program, Kaduna, Nigeria, 1982.

King, J.M. "Water and Livestock: Animal water need in Relation to Climate and Forage." ILCA Bull. No. 7, Addis Ababa,Ethiopia, 1983.

McCammon-Feldman, B., P.J. Van Soest, P. Horvath, and R.E. McDowell. "Feeding Strategy of the Goat." Cornell Internatinal Agriculture Mimeograph No. 88. Cornell University, Ithaca, New York, 1981.

McDowell, R.E. "Strategy for Improving Beef and Dairy Cattle in the Tropics." Cornell International Agriculture Mimeograph No. 100. Cornell University, Ithaca, New York, 1983.

_____. "Feed Resources on Small Farms." Proc. Seminar: The Improvement of Farming Systems, Bamako, Mali, February 20 - March 1, 1978.

McDowell, R.E., et al. "Tropical pastures with and without Supplement for Lactating Cows in Puerto Rico." Univ. Puerto Rico Bull. 238, Rio Piedras, 1975.

McDowell, R.E., D.G. Sisler, E.C. Schermerhorn, J.D. Reed and R.P. Bauer. Game or Cattle for Meat Production on Kenya Rangelands? Cornell International Agriculture Mimeograph No. 101. Cornell University, Ithaca, New York, 1983.

Mosienyane, B.P. "Nutritive Evaluation of Botswanan Range Grasses for Beef Production," M.S. thesis, Cornell University, Ithaca, NY, 1978.

Mtimuni, J.P. "Identification of Mineral Deficiences in Soil, Plant and Animal Tissues as Constraints to Cattle Production in Malawi," Center Tropical Agri. Rpt. No. 6, University of Florida, Gainesville, 1982.

Nicholson, M. "Calf Growth, Milk Offtake and Estimated Lactation Yields of Borana Cattle in the Southern Rangelands of Ethiopia." JEPSS Res. Rpt. No.6, ILCA, Addis Ababa, 1983.

NRC. "Effect of Environment on Nutritive Requirements of Domestic Animals." National Academy Press, Washington, DC, 1981.

Nsibandze, E.P. "The Economics of Cattle Ranching in Machakos District, Kenya. M.S. Thesis, Cornell University, Ithaca, New York, 1982.

Oxby, C. "Group Ranches in Africa." FAO, Rome, 1981.

Pratt, D.J. and M.D. Gwynne. Rangeland Management and Ecology in East Africa. London: Hodder and Stoughton, 1977.

Reed, J.D. "The Nutritional Ecology of Game and Cattle on a Kenya Ranch." Ph.D. Thesis, Cornell University, Ithaca, New York, 1983.

Reed, J.D., P.J. Horvath, M.S. Aller and P.J. Van Soest. "The Use of Ytterbim to Isolate Polyphenols (Tannins)." Meet. Abstr. Amer. Soc. Animal Sci. p. 454, 1982.

Talbot, L.M., W.J.A. Payne, H.P. Ledger, L.D. Verdcourt and M.H. Talbot. "The Meat Product Potential of Wild Animals in Africa: A Review of Biological Knowledge." Tech. Commun. Commomw. Bu. Anim. Breed. Genet. 16(1965):1.

Van Soest, P.J. Nutritional Ecology of the Ruminant. Corvallis, Oregon: O & B books, Inc., 1982.

Von Kaufmann, R. "International Livestock Centre for Africa Subhumid Programme Annual Report." Kaduna, Nigeria, 1982.

Water-Bayer, A. "Sedentarisation and the Role of Women in a Pastoral Economy." ILCA Working Document, Addis Ababa, Ethiopia, 1983.

Wilkinson, J.M. "Silages made from Tropical and Temperate Crops." World Animal Review 45 (1983):36-42.

Wilson, R.T. "Livestock Production in Central Mali." ILCA Bull. No. 15. Addis Ababa, Ethiopia, 1982.

World Bank. Accelerated Development in Subsaharan Africa: An Agenda for Action. Washington, DC: The World Bank, 1981.

4
Application of Systems Analysis to Nomadic Livestock Production in the Sudan

Ahmed E. Sidahmed and L. J. Koong

Livestock production is a very complex system which has many interrelated components such as climate, soil, plants and obviously, animals operating with a high degree of interaction within a certain economic and social environment. Traditionally, these components have been studied separately by scientists in animal science subdisciplines such as nutrition, breeding and physiology, and in other disciplines such as range and forage science, health science, and economics. Animal scientists have long recognized that optimization of our resources for livestock production and maximization of production efficiency requires these components to be synthesized and studied in an integrative form. This has usually been attempted in an informal, intuitive and unorganized fashion due to the lack of defined theory and methodology. However, with the emergence of systems theory as a science during the last 20 to 30 years, the use of a systems approach to the study of livestock production has received increased attention.

INTRODUCTION TO SYSTEMS ANALYSIS

The word "system", in its simplest definition, means the grouping of a set of components linked together for some common purpose or function. Systems analysis is the study of how component parts interact and contribute to the whole system [Forrester 1968; Spedding 1975] . It usually begins with a conceptual framework and well defined objectives which identify the area and level of the system to be studied and the goals of the intended activity [Koong et al. 1978]. A flow diagram which describes the system and the interactions among its components qualitatively is frequently employed in this process leading to a mathematical model; that is, a set of mathematical equations which describe the system quantitatively.

During the process of model construction, two types of equations can be used. An empirical equation is one which is fitted to experimental data so as to describe a relationship which has been observed between two or more variables. Such equations carry no implication about the underlying reason for the relationships and cannot be safely applied beyond the range of values for independent variables for which data are available. Theoretical equations are those derived from some theory or hypothesis about the fundamental nature of the system. They take underlying forces and mechanisms into account insofar as feasible, and utilize equation parameters which have definite biological meaning or identity [Riggs 1963]. All models have empirical aspects, but the degree of empiricism can vary greatly. For example, models based on sound theoretical equations provide better projections over a broader range of conditions than do those composed primarily of empirical equations. A systems approach can be applied to modeling animal production at many different hierarchical levels, from the cell, organ or tissue, to the whole animal, herd or region or country. The degree of empiricism usually increases as the modeling activity expands beyond animal and herd levels [Joandet and Cartwright 1975].

In recent years, developed countries have increased their effort in helping less developed countries (LDCs) towards self-sufficiency, especially in the area of agricultural production. However, development projects in LDCs have not all been successful. One of the major causes of lack of success has been the absence of integrated approaches which incorporate the different environmental, social, economic and biological variables that govern the agricultural production system. The pastoral societies of these regions in particular have come under unprecedented political, economic and environmental pressure in recent years [Pendelton and Van Dyne 1980]. A major problem is over-ambitious and inadequately planned development projects implemented by expatriates and local administrators which have emphasized working for the local society rather than with them. For example, Thimm [1979a,b] reviewed several development projects in the arid and semi-arid regions of the Sudan which were plagued by serious problems. Among them were two livestock development projects which were intended to open opportunities for the settlement of nomads in western Sudan. The first project (Babanusa milk production project) planned 44 potential ranch sites of 10,000 ha each as providers of milk for a central milk processing factory. In each ranch 200 nomadic families were supposed to settle, raise animals and cultivate crops. The second project

(Geriehl-Sarha ranch) attempted to enclose an area of 200 km^2 in north-western Kordofan and to run it cooperatively by 50 families of settlers.

The first project was never initiated and no proper settlement was really achieved in the second. According to Thimm [1979a,b] and Khogali [1979] the most important problems were organizational shortcomings and lack of social acceptance. The nomads were suspicious of the government's intentions. In addition, being located in communal grazing areas, opposition arose among those who did not participate in these projects. Moreover, the carrying capacity of the ranches was far below the numbers of animals owned by the families to be settled. Since the nomads were averse to reducing herd numbers, keeping a portion of their animal wealth outside ranch sites was an alternative. This solution was tantamount to a continuation of nomadism. Other problems were basically economic such as marketing of livestock being highly seasonal due to inadequate infrastructure. Also, low nutritional plane during most of the dry season contributed to a low productivity level. These examples demonstrate that introduction of new management practices without consideration of their impacts on the total system and its environment are unlikely to be successful.

Advantages that can be achieved by a systems approach have been listed by various investigators [DeWitt 1969; Baldwin and Smith 1971; Dent and Anderson 1971]. They include: (1) unification and summarization of data and concepts leading to increased understanding of particular systems; (2) identification of critical knowledge gaps; (3) reduction of conceptual difficulties encountered in analyses of systems and interactions too complex for intuitive resolution; (4) formulation and testing of alternative hypotheses and (5) as an aid to the planning and policy decision making processes.

APPLICATION OF SYSTEMS METHODOLOGY AND MODELING IN AFRICA

Systems analysis techniques have been used to identify critical areas of emphasis for improved total production efficiency. For example, the Texas A&M University (TAMU) beef cattle production model has been applied in Subsaharan Africa [International Livestock Centre for Africa (ILCA) 1978; Sullivan et al. 1981; Cartwright et al. 1982; Smith et al. 1982]. The TAMU model application and simulation results in Botswana suggested three innovations for the livestock production system in that country; controlling breeding, weaning, and use of reserve pastures for certain

classes of breeding livestock [ILCA 1978]. Also, these
activities have been useful in determining forage and
cattle data that must be collected to form a basis for
understanding production systems and for formulating
extension recommendations [Cartwright et al. 1982].

However, for maximum efficiency and efficacy,
models should be built with specific objectives well
defined in advance. In the context of this book, if
our objective is to develop a model for livestock production systems in Subsaharan Africa and use the model
for the purpose of identifying critical areas of research emphasis to assist in policy decisions, the
model should be developed with full consideration of
all local production constraints and social, economic
and political environment of the region. Furthermore,
during the modeling processes, all local data and/or
information should be used to the fullest extent possible. For a project of this scope, a research team
consisting of scientists from all disciplines related
to livestock production such as animal science, animal
health, range and forage production, economics and
social science is essential to obtain successful and
meaningful results. In addition, members of this team
should be centrally located in this region with the
opportunity for constant interaction.

The purpose of this chapter is demonstrating how
systems analysis can be applied to the study of problems associated with livestock production in Subsaharan
Africa, with special emphasis on the nutritional constraints of the system. The Sudan will be used as the
example. It is not the intent to present a model which
describes the livestock production systems in that
country. Rather, we will discuss the critical components and their interaction, thereby providing a framework for the construction of a model which could be
useful in the future assessment of alternative production strategies in the Sudan.

LIVESTOCK PRODUCTION IN THE SUDAN

Many observers at the international development
level are looking towards the Sudan as a future major
supplier of food (particularly meat) for Africa and the
Middle East. At a time when hunger is a tragic reality
in many areas of the world, the development of the
Sudan as an agricultural nation is an absolute necessity. Sudan is Africa's largest country (about 2.5
million km^2). Of its 81 million ha of arable land,
little more than 4.5 million ha is actually cultivated
[Sudan Government 1974; United Nations, FAO 1974], and
only half of this (2 million ha) is under irrigation.
Natural rangelands and forests occupy some 23 and 88

million ha respectively. According to the Interdivisional Working Group on Policy for Meat Development [United Nations, FAO 1974], Sudan has the greatest potential for livestock development among the countries of the Near East and Subsaharan Africa. If well-developed, these resources can make an important contribution to the region's meat requirements during the next two decades. Though low in terms of the livestock resources of the Sudan, the contribution of the livestock industry to the national economy, at 10 percent of Gross Domestic Product (the rest of the agricultural sector contributes 25 percent of GDP), is among the highest for developing countries.

The Animal Resource Economic Administration of the Sudan, Ministry of Agriculture [Sudan Government 1979], estimated the livestock population as 18 million cattle, 17.6 million sheep, 12.7 million goats and 2.6 million camels. These numbers represent 21 percent of Africa's livestock population and yet Sudan's human population is only 4 percent of Africa's total. Khogali [1979] considers the official estimates far below the actual number since they are based on tax-list figures. The nomadic tribes to the north (the Baggara, who are cattle owners) and the transhumant Nilotic tribes to the south, who combine farming and animal rearing, make up one-third of the Sudan population and own 90 percent of its livestock [United Nations, FAO 1974].

Sudan has a highly diversified environment ranging from desert in the northern extremes to a semi equatorial climate (1,600 mm annual rain) in the south (Table 4.1). However, the largest portion of the country is classified as savanna (between Latitude 5° N and 14° N) where annual rainfall ranges from 200 to 1,260 mm. Summer rains are predominant except in the Red Sea hills region of northeastern Sudan. The various vegetation types include the scanty vegetation of the desert zones; the sparse shrubs and scanty acacia trees of the semidesert zone; the acacia, short grass, shrubs and herbs of the low rainfall savanna (200-270mm rain) zone; the acacia, tall grass, woodland areas of the tall grass savanna (750-1,250 mm rain) zone; small areas of mountain vegetation; and a narrow strip of tropical rain forest vegetation. The savanna is the main grazing and agricultural area of the Sudan. The annual mean temperature ranges from 32° C in January to 42° C in June. The mean relative humidity ranges from 17 percent in April to 61 percent in August.

Communal grazing systems practiced by nomads and transhumants is the dominant form of animal husbandry in all the zones described [Baashar 1966; Hjort 1976;

Table 4.1. Sudan rangelands

Ecological Zones	Location (Latitude North)	Annual Rain	Carrying Capacity/sq. km	Animal Husbandry
		——mm——	——AU——	
Desert	North of 17	50	–	
Semi-desert	14–17	50–200	3–6	
Savanna:				
Low woodland savanna	10–14	200–750	6–8	
Flood region	8–10	>750		
High woodland savanna	5–10	750–1250	8–11	
Tropical forests	3.5–5	>1250	11–22	

Animal Husbandry bracket (right side): Nomadic, Transhumance, Settled

Source: Compiled from information in Mustafa Baasher [1966], Hjort [1976], and United Nations, FAO [1974].

Le Houerou 1981]. These systems are the result of traditional adjustments to fluctuating forage and water supplies where only extreme mobility can maintain the flocks and herds [French 1966]. Generally, the survival of the indigenous livestock species in such severe environments depends primarily on being sufficiently adaptable to adjust to the local nutritional, management, health and physiological circumstances. While the Nilotic transhumance tribes move short distances (mainly east or west of the river according to flooding), the nomadic Baggara move long distances north to south in the western part of the country. The resources in the low rainfall savanna are more dispersed and the traditional strategy is typical pastoral nomadism [Johnson 1979]. Rainfall characteristics and animal health constraints determine to a major extent the types of livestock in each area. While cattle and sheep exhibit a cyclic movement in the savanna (humid and dry) region and sometimes trespass cultivated areas, camels and goats are predominant in the semi-desert areas.

The rangelands of the Sudan (Table 4.1) provide most of the nutrient resources for livestock production. The quantity and quality of the range species is governed mainly by rainfall, soil distribution and proximity of water courses. In general, the contribution of cultivated forages to livestock production is very limited and neither forage processing nor storage is participated at any significant level. Concentrate feeds and crop by-products available include cotton and other oil crop seed cakes (seseme, ground nuts, etc), sorghum, molasses and wheat bran. However, currently the production is low and exports and human consumption compete with local markets. Even so, some concentrate feeds are used to finish cattle around major cities.

ANALYSIS OF NUTRITIONAL CONTRAINTS TO LIVESTOCK PRODUCTION IN WESTERN SUDAN

Any attempt to analyze the livestock production system in western Sudan must begin with an understanding of the nomadic animal husbandry practices. The grazing areas of the nomads in the western regions broadly include the sandy soils of the semi-desert and the short grass low rainfall savanna, and the clay soils of the northern fringes of the tall grass high rainfall woodland savanna zones (Table 4.1 and Figure 4.1). The grazing pattern is cyclic where early rain season pastures in the semi-desert are utilized by all four species of livestock, with camels moving further northward to the desert. Late in the rainy season cattle, sheep and goats migrate to the short grass savanna zone and stay there until the mostly annual

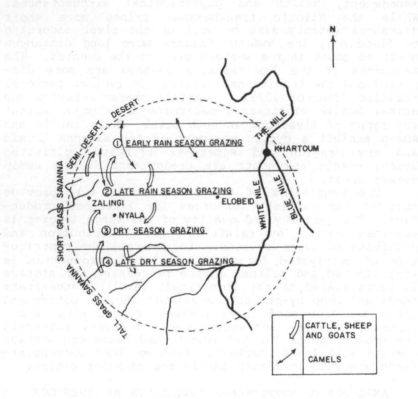

Figure 4.1. Nomadic livestock movements in western Sudan

short grass ranges mature and water resources start to
decline. By the early dry season the muddy fly-infest-
ed rangelands in the tall grass savanna zone of northern Bahr el Ghazal Province become more accessible and
the animals move further south where drinking water in
hafiers (man-made depressions) and wells is still
available. Towards the end of the dry season many
locations in the tall grass savanna regions are under-
stocked because of the scarcity of drinking water,
therefore placing additional strains upon the animals.
During this period in the grazing cycle, large numbers
of young offspring suffer severe nutritional depriva-
tion.

The nomads' strategy aims at securing a rapid conversion of the growing vegetation by their animals,
subject to erratic weather patterns and relatively
unorganized management practices. The result is some
areas being overgrazed while others are scarcely
utilized, as indicated by a quotation from a United
Nations [1974] conference on World Food: "The grazing
lands of the developing countries are the most over-
utilized and at the same time under-utilized resources
in agriculture." The following problems are associated
with this unorganized utilization of range resources in
western Sudan:

(1) Deteriorating Range Quality. Heavy grazing
     pressures during the rainy season have a
     detrimental effect on the young growing
     shoots, particularly of the most desired
     species. For example, in the northwestern
     areas (N. Darfur and N. Kordofan) several
     desirable and highly nutritious species (e.g.
     Aristida plumosa, Belpharis spp. and Monsonia
     spp.) have disappeared from around the water-
     ing points [University of Khartoum 1982].

(2) Soil Erosion. Overgrazing depletes the cover
     and soil erision occurs especially in the
     marginal areas. In some areas of Kordofan
     and Darful (western Sudan) provision of rural
     water supplies has made it possible to expand
     cultivation and increase livestock numbers
     [University of Khartoum 1982]. These factors
     and improved veterinary services have result-
     ed in severe overgrazing, and have caused
     desertization in the northern parts of the
     region.

(3) Fire Destruction. Although fire is one of
     the most effective management tools, its
     irrational use is very destructive. Repeated
     burning replaces the high quality perennials
     by the less nutritious annuals. According to

the United Nations, FAO [1974], between 30 and 80 percent of the rangelands are consumed annually by fire, with the highest losses in the high rainfall zones.

(4) Poor Animal Performance. Animal performance is poor at all levels and for all classes of livestock. Performance data for cattle are listed in Table 4.2. These include low fertility and calving percentage, long kidding and calving intervals, poor health conditions, low growth rates, and high mortality. Most of these problems are caused by inadequacy and seasonality of nutrient supply.

In addition to the above mentioned production problems which are directly associated with nutrient supply, the low productivity of western Sudan's livestock and rangeland systems are caused by the following external constraints:

(1) Social. Traditionally the nomadic owner favors the multiplication of herd numbers, which eventually results in overstocking and rangeland deterioration.

(2) Economic. Isolation, remoteness and primitive marketing systems are some of the handicaps faced in livestock production. In addition, an inefficient marketing system reduces revenue and discourages expanded commercial offtake.

(3) Administrative. The land tenure system recognizes (de facto) tribal ownership of rangelands, preventing legislation which would secure land for commercial investment purposes.

In recent years the "Native and Tribally Based Local Administration System" has been eliminated without being replaced by an effective local administration [University of Khartoum 1982]. As a result, traditional grazing controls have been ignored and internal tribal conflicts over grazing have become more intense. Tribes from less productive grazing areas have migrated to other areas (outside their traditional territories) with richer resources. As a consequence of intertribal conflicts, some of the best Baggara cattle grazing areas in western Sudan have been subjected to severe deterioration.

TOWARDS A SYNTHESIZING STRATEGY

The objective of this section is outlining some

Table 4.2. Performance data of the Baggara cattle

Item	Units	Location	
		Kordofan	Western Darfur
Age at first calving	Yrs.	3½-4	n.a.[a]
Average calving percentage	%	50-70	49
Mortality rate in young stock	%	20	19
Mortality rate in adult stock	%	5	n.a.
Number of breeding bulls	c	4.5	n.a.
Life production	Calves	8-10	n.a.
Lifelong activity of breeding bulls	Yrs.	7-8	n.a.
Lifelong productivity of females	—	up to sterility	n.a.
Estimated herd composition: Females	%	70-75	n.a.
Males	%	25-30	n.a.
Breeding females (of all animals)	%	n.a.	43
Liveweight of breeding females	kg	n.a.	300
Total offtake (sales + consumption)	%	n.a.	16
Slaughter weight	kg	300	n.a.
Dressing percentage	%	50	45

Source: Compiled from information in United Nations [1974] and Wilson et al. [1980].

[a] Not available.

possible management strategies which can improve the current livestock production system. All possible technologies which can be applied to improve the individual components of the system must be considered. These components are biological, social, economic, and organizational, and are highly interactive and interdependent. Implementation should emphasize long-term production efficiency. We feel that introduction of the following management technologies will offer significant improvement in the efficiency of livestock production systems in the Sudan and has relevance to other areas.

Grazing Management and Range Improvement Practices

Grazing management practices which tend to limit seasonal migration and allow redistribution of grazing use among the different ecological zones, optimize grazing pressure, and improve range plant utilization efficiency should receive the highest priority. Stocking rates should be reduced in the early rain grazing areas. These areas are the most arid (250 mm rain) and as such, since the changes for rain-fed crop farming are uncertain, livestock rearing becomes the most suitable development alternative [French 1966]. These are also areas where the young are born, and overstocking is detrimental to animal performance.

The climatic influences are uncontrollable. Soil types vary from poor sandy "Goz" in the north to clay plains. There are also several loamy/clayey depressions [Wilson et al. 1980]. Lack of technical capabilities, uncertainty of rain and vastness of the rangelands may prohibit fertilizer application on a wide scale. However, if limited scale commercial ranches are established, range fertilization as well as herbicide application may be successful. Range reclamation by fencing and rotational grazing is not practical in the communal system and is not financially affordable. Reseeding of viable, desired species may be a possible strategy if strict control measures on grazing could be implemented.

Evaluation of the nutritional changes in range quality from growth to maturity is important. Such information is required for plans aiming at harvesting and processing range vegetation from the under-utilized areas. Processed forages such as hay could be provided to the livestock in their marketing routes or in their concentration points near water sources.

Prescribed Fires

Use of prescribed burns for firebreaks and to revegetate new shoots during the early dry season can be a very effective management practice. Also fires are useful in the process of exterminating ticks and several undesirable insects in the high-rain savanna zones [Magid 1972].

Supplemental Feeding

This practice is important particularly when animals receive below maintenance feed requirements during the long dry months (up to 8 months). Crop residues and concentrates are locally available either in the mixed crop-animal operations by the transhumant or neighboring settled cultivators. However, this strategy requires much improvement in order to minimize competition for export or for human consumption.

Water Development

Water projects have to be developed in combination with effective grazing management in a multiple land-use approach.

Improved Animal Health Conditions

This is especially critical among young animals. Successful seasonal vaccination campaigns and disease control measures by the veterinary authorities have resulted in an explosion in the number of animals in some areas. However, many nomads tend to avoid availing their livestock to authorities for vaccination until it is too late.

Improvement of Herd Structure

Herd structure can be improved by controlled breeding and selection for quality from indigenous breeds. Selection for high producing bulls of the Kennana sub-type of Baggara cattle breeds has been successful for several years in central Sudan as a measure for improving milk production from cows. In addition, an improved marketing system should help in encouraging higher commercial offtake.

SUMMARY AND CONCLUSIONS

The purpose of this chapter is to introduce the application of systems analysis to the study of problems associated with livestock production in Subsaharan Africa using Sudan as an example. Discussion of the critical components of livestock production in western Sudan was undertaken. The elements discussed were biological (animal, plant, soil factors) and external (social, economic, and environmental). In addition,

the problems which have resulted in the failure of some
livestock development projects in western Sudan were
discussed. In conclusion, we propose that any attempts
towards development should be preceded by the construction of a mathematical model which should contain all
the essential elements of the current production systems.   Unavailable information (or data) must be
collected once their order of priority to the systems'
performance is determined. The impact of any new
intervention to the current system components or to the
total system should be examined through simulation in
order to insure the effectiveness of the new strategy.
Additionally, the model could be used to identify the
critical problem areas and suggest alternative strategies for optimal utilization of resources for maximum
livestock productivity.

REFERENCES

Baasher, Mustafa M. "Range and Livestock Problems
    Facing Settlement of Nomads." Proceedings of the
    10th Annual Conference of the Sudan Philosophical
    Society, Khartoum, 1966.
Baldwin, R.L., and N.E. Smith. "Application of a
    Simulation Modeling Technique in Analysis of
    Dynamic Aspects of Animal Energetics." Fed. Proc.
    30 (1971): 1459.
Cartwright, T.C., F.M. Anderson, N.G. Buck, T.C.
    Nelsen, J.C.M . Trail, D. Pratchett, J. Sanders,
    W. Astle, D. Light, T.W. Rennie, T.J. Rose, and C.
    Shorrock. "Systems Modeling in Cattle Production:
    An Application in Botswana." World Animal Review
    41 (1982): 40.
Dent, J.B., and J.R. Anderson. Systems Analysis in
    Agricultural Management. Sidney: John Wiley and
    Sons, Australian Pty., Ltd., 1971.
DeWitt, C.T. "Dyanmic Concepts in Biology." Prediction
    and Measurement of Photosynthetic Productivity.
    Proceedings of IBP/PP Technical Meeting, Trebon
    Symposium, ed. Ivan Stelik, pp 17-23. 1969.
El Hag, G.A., and A..S. Mukhtar, "Varying Levels of
    Concentrates in Rations of Sudan Desert Sheep."
    World Rev. Anim. Prod. 14 (1978):73.
Forrester, J.W. Principles of Systems, 2nd Preliminary
    ed. Cambridge, MA: Wright-Allen Press, Inc.,
    1968.
French, M.H., "Animal Production and Savanna Areas."
    United Nations, Food and Agriculture Organization,
    LA: SF/SD/66/8, No. 4, 1966.
Hjort, A. "Traditional Land Use in Marginal Dry
    Lands." Ecological Bulletins/NFR 24(1976): 24-43.
International Livestock Centre for Africa (ILCA).
    "Mathematical Modeling of Livestock Production

Systems: Application of the Texas A&M University Beef Cattle Production Model to Botswana." Addis Ababa, 1978.

Joandet, G.E. and T.C. Cartwright. "Modeling Beef Production Systems." J. Anim. Sci. 41(1975): 228-2246.

Johnson, D.L. "Management Strategies for Drylands: Available Options and Answered Questions." Proceedings of the Khartoum Workshop on Arid Lands Management, ed. J.A. Mabbutt, pp. 26-5. Tokyo: The United Nations University, 1979.

Khogali, M.M. "Nomads and Their Sedentarization in the Sudan." Proceedings of the Khartoum Workshop on Arid Lands Management, ed. J.A. Mabbutt, pp. 55-59. Tokyo: The United Nations University, 1979.

Koong, L.J., R.L. Baldwin and M.J. Ulyatt, "The Application of Systems Analysis of Mathematical Modeling Techniques to Animal Science Research." Symposium on Use of the Computer in Animal Science Teaching, Research and Extension, Amer. Soc., Anim. Sci., (1978): 9-19.

Le Houerou, H.N. "The Rangelands of the Sahel." J. Range Manage. 33(1980): 41-46.

Magid, el R. A. "Towards an Efficient Range and Pasture Management in Southern Sudan." Ministry of Agriculture, Southern Region, Juba: Educational and Cultural Production Department Printings, 1972.

Meyn K. Beef Production in East Africa. Munich: Info-Institut fur Wirtschaftsforschung, 1970.

Pendleton, D.F., and G.M. Van Dyne. "Livestock and World Grazinglands: Status and Outlook. A working paper developed in the U.S. National Science Foundation Project: Prediction of Grazingland Productivity Under Climatic Variations." Colorado State University, Fort Collins, 1980.

Riggs, D.S. The Mathematical Approach to Physiological Problems. Massachusetts: M.I.T. Press, 1963.

Smith, G.A. Carles, J. Schwartz, J. Blackburn, H. Ruvuna, and T.C. Cartwright. "Analysis and Synthesis of Small Ruminant Production Systems." Proceedings of the Small Ruminant CRSP Workshop, Nairobi, Kenya. pp. 41-54. SR-SCSP Management Entity, U. of California, Davis, 1982.

Spedding, C.R.W. The Biology of Agricultural Systems. London: Academic Press, 1975.

Sudan Government, Ministry of Culture and Information. Sudan Facts and Figures. Khartoum: The Government Printing Press, 1974.

Sudan Government, Ministry of Agriculture. Unpublished report by the Animal Resource Economics Administration, Khartoum, September 1979.

Sullivan, G.M., T.C. Cartwright and D.E. Farris. "Simulation of Production Systems in East Africa by use of Interfaced Forage and Cattle Models." Agricultural Systems 7 (1981): 245-265.

Thimm, H.U. Development Projects in the Sudan: An Analysis of Their Reports with Implications for Research and Training in Arid Land Management. Tokyo: The United Nations University, 1979a.

_____. "Socio-Economic Assessment of Agricultural Development Projects in the Sudan." Proceedings of the Khartoum Workshop on Arid Lands Management, ed. J.A. Mabbutt, pp. 71-75, Tokyo: The United Nations University, 1979b.

United Nations, Food and Agriculture Organization (FAO). "The Livestock and Meat Industry of the Sudan." A study prepared for the FAO Interdivisional Group (IWG) on Policy for Meat Development, Rome, 1974.

United Nations. "Assessment of the World Food Situation--Present and Future." UN World Food Conference, Item 8 of Provisional Agenda E/CONF. 65/3. Rome, 1974.

University of Khartoum, Geography Department with the Assistance from the Program for International Development, Clark University, Worchester, MA. "Environmental Context of Development in the Sudan." Khartoum, 1982.

Wilson, R.T., L. Baily, J. Hales, D. Moles, and A.E. Watkins. "The Cultivation - Cattle Complex in Western Darfur." Agricultural Systems 5(1980): 119-135.

PART II

Aspects of Production and Marketing

An essential part of improving animal productivity is through national and regional animal disease control and eradication programs. William Moulton leads off this section of the book by providing an overview of the major diseases affecting animal production in Subsaharan Africa. These diseases are generally combatted separately and, as one is suppressed, others emerge. Also, when production becomes land intensive, a closer concentration of animals results in some diseases such as brucellosis and tuberculosis increasing in incidence. Consequently, as the livestock industry grows in size and productivity, an effective livestock health program implies continuous funding for new facilities, vaccine production and quality control, vector control, disease surveillance, and required restrictions on animal movement.

Trevor Wilson reinforces the conclusion that the livestock industry is a complex set of interrelated factors through his analysis of the role of goats and sheep, which account for about 17 percent of domestic ruminant biomass in Subsaharan Africa. They are important providers of meat and milk, especially when production of these commodities from camels and cattle is seasonally low. Wilson argues that programs, especially those aimed at improving the welfare of small holders, should emphasize development of low-technology, low-cost packages based on improving existing management practices within traditional systems.

A major concern in livestock development relates to breeds and breeding. Numerous research programs have been conducted in Subsaharan Africa on what is perceived to be the limited genetic potential of indigenous cattle for expanded productivity of beef and milk. John Trail and Keith Gregory assess the value of previous breed productivity research and conclude that the majority of research work has not given very useful

information about comparative performance of cattle
breeds in Africa. In fact, out of a comprehensive
review of more than 500 references, they find that only
5 percent of the reports provide sufficient data to
allow breed comparisons on the basis of a productivity
index. Another major conclusion by Trail and Gregory
is that much more research is needed on trypanotolarant
breeds such as the N'dama and West African Shorthorn,
and their crosses with Zebu. They also argue that
expanded work is needed on breeding to achieve genetic
resistence to other diseases, and on crossbreeding to
optimize the simultaneous use of both additive (breed
differences) and nonadditive (heterosis) sources of
genetic variation. All work must be done on a zonal
basis.

Inadequate marketing systems for livestock can
also be a major constraint to improved production.
Conversely, an efficient market can result in reduced
costs to consumers and increased incomes for producers.
Phylo Evangelou demonstrates an approach to the analysis of market performance by considering local-level
livestock marketing mechanisms in Kenya's Maasailand.
His work indicates that current marketing mechanisms
are not an overriding constraint to the region's livestock development. Market structure and trader conduct
vary across Subsaharan Africa, but at least in this
particular pastoral area the improved efficiency of the
market will depend upon producers becoming increasingly
more market oriented and national pricing policies more
accurately reflecting demand.

The complex theoretical problem of livestock pricing policy is simplified by Greg Sullivan in this
section's last chapter. The short-run and long-run
effects of prices set by governments at low levels in
response to political interests is shown in graphical
terms and by using a comparative case study of Ghana
and Tanzania. He concludes that while some regulation
is necessary to avoid violent price fluctuation, prices
must be sufficiently high to stimulate expanded output
by promoting a greater commercial orientation toward
livestock production, as well as the adoption of
productivity expanding practices.

5
Major Disease Deterrents to Improving the Subsaharan Livestock Industry

William M. Moulton

CURRENT STATUS OF PROGRAMS

With few exceptions disease control programs conducted in Subsaharan Africa are determined by resources available to national governments. The identification of priorities for the application of these resources depends on the need to respond to immediate problems and rarely deals with long-term objectives. This is understandable when the limited extent of these resources is acknowledged. In this context, resources include human, physical and fiscal factors.

Locally generated resources in the field of animal health may be supplemented, usually for short and/or intermittent periods, by international organizations such as the Food and Agriculture Organization of the United Nations (FAO), by regional organizations such as the European Economic Community (EEC), or by bilateral funds, i.e. DANIDA, UK-ODA and USAID. The stability and continuity of external support, particularly regional and bilateral aid, is subject to political considerations at the source.

Examples of disease suppression involving regions of Africa include the multi-agency supported campaigns against rinderpest and contagious bovine pleuro-pneumonia (CBPP), carried out in the 1960s. A thorough evaluation of these programs is critical at this time as there is a stong movement currently gathering support to mount again a regional anti-rinderpest campaign in Africa.

There are several agencies involved in tsetse fly control to reduce trypanosome infestation in vast areas of Africa, in order to make livestock production possible in areas where the blood parasite is at present transmitted by the fly. These efforts have had varying degrees of intensity for several years and require critical evaluation of the social, political and economic involvement necessary to apply known technology effectively.

A unique approach to dealing with African animal diseases has been undertaken by the International Laboratory for Research on Animal Diseases (ILRAD), Kenya, in research on trypanosomiasis and theileriosis, another blood parasite which is highly lethal to cattle. The laboratory is supported by the major international technical assistance agencies under the auspices of the Consultative Group on International Agricultural Research. The objective is to obtain leading world scientists, provide them with modern facilities in a setting where the disease is enzootic so that practical research can be carried out not only on the characteristics of the agent itself but also its behavior in animal populations under varying management conditions.

From the foregoing it is obvious that major emphasis in dealing with livestock diseases is placed on the protection of cattle as compared to other species such as camels, sheep, goats, pigs and poultry. There are several reasons for the great attention paid to cattle, the predominant one reflecting their economic and social significance. Furthermore, veterinary training has been a result of curricula developed for intensive husbandry in Europe and, thus, little attention has been given to diseases of other species until very recently. Similarly, the curriculum does not identify the very different social, economic and technical requirements of animal production and health unique to the region.

INFLUENCE OF DISEASES

Foot-and-Mouth Disease (FMD)

International commerce is the desirable objective in considering the vast marketing potentials available to Africa's animal resources. The major barrier to developing an international market with countries with convertible currencies is foot-and-mouth disease (FMD) even though, compared with many other enzootic diseases of Africa characterized by high mortality and obvious production losses, FMD does not elicit great concern from the average African livestock owner. This is because reduction of milk supply or longer reproductive cycles are not catastrophic to his survival. The major impairment of African agricultural societies resulting from an outbreak of FMD is its possible coincidence with planting or harvesting when draft animals are affected (Table 5.1).

Countries that would be desirable as red meat markets-- Europe, North America and Japan--are meat

Table 5.1. Major disease deterrents to livestock production

Disease	Agent and Species	Area Affected	Vaccine Availability	Immunity Duration	Treatment	Potential Economic Consequences
Rinderpest	Virus, ruminants, occasionally swine	Asia, Africa	Good	Lifelong	None	High mortality, rapid spread
Foot-and-mouth disease	Virus, all cloven-footed species	South America, Africa, Europe, Asia	Fair[a]	3 months to 1 year	None	Major barrier to international commerce
African swine fever	Virus, swine	Africa, Iberia, Brazil, Haiti, Sardinia	None	None	None	High mortality, tick-borne, carrier animals
African horse-sickness	Virus, equidae	Africa	Good	Five years	None	High mortality, vector-borne
Newcastle disease	Virus, avian species	Worldwide	Good	Several weeks	None	Viscerotropic, velogenic, high mortality
Hog cholera, classical swine fever	Virus, swine	Much of the world	Good	2 years +	None	High mortality, rapid spread
Contagious bovine pleuropneumonia	Mycoplasma, cattle	Africa, Asia	Fair	1 year	Not practical	Related to cattle movement
Trypanosomiasis	Protozoa, cattle, horses, pigs, camels, sheep, goats	Africa	None	None	Occasionally practical	High mortality
East Coast fever (Theileria)	Protozoa, cattle	East Africa	None	None	Occasionally practical[b]	High mortality
Sheep & goat pox	Virus, sheep & goat	Africa, Asia	Good	One year	Not practical	Rapid spread and high mortality in young
Peste des Petits ruminants	Virus, small ruminants	Africa	Good	One year	None	Increasing prevalence, some high mortality

Source: Author.

[a]Restraints reflect the need for technical and physical sophistication required for precise virus typing for diagnosis and complexity of vaccine production and testing.

[b]May resort to tick control or treat after infection.

importers. They have however spent vast sums of money over many years to free their livestock populations from FMD, and continue to spend large sums to maintain barriers against its reintroduction.

FMD virus is transmitted not only by contact between live animals but also survives in products, tissues, organs, glands, secretions (including semen), bone marrow, and may also be conveyed by clothing, hay or equipment that has been in contact with animals affected by the disease. For those reasons, countries with efficient livestock production practices and which are free from FMD, place severe restrictions on animal related imports to prevent FMD from gaining access to their livestock populations.

Technology does exist for dealing with the disease. Both Kenya and Botswana have demonstrated that FMD can be managed under African conditions given the political incentive of the national government to provide the legal enforcement necessary to carry out regulation of animal movement. For example, Kenya established an FMD-free zone in collaboration with Swiss animal health authorities whereby it was possible to raise animals and provide for their slaughter and meat shipment in a manner acceptable to the Swiss veterinary authorities. Similarly, the Government of Botswana has developed an export meat trade in collaboration with European Economic Community animal health authorities. Cattle moving from one region to another are quarantined for FMD surveillance. In collaboration with a European commercial organization, a modern vaccine production facility was recently established in Botswana. It is the second of only two such modern diagnostic and vaccine production facilities on the entire continent of Africa, the other one being in Kenya.

Considering the insidious nature of FMD virus, its ability to change its antigenic characteristics through natural passage, and the relatively short protection provided by vaccine (3-12 months), a successful program requires expensive facilities from the standpoint of training individuals, a biologically secure facility for vaccine production and diagnosis as well as the administrative structure necessary to get vaccine into the target animals on a prescribed schedule. It must also be supported by the financial and legal authorities in order to assure utilization of the investment, which can only be assured if associated with the export of beef acceptable to countries that have freely convertible currencies. This assumes that abattoir facilities and inspection procedures are also acceptable to, and approved by, the importing country. It must be pointed out here that potential importing countries will not accept meat from vaccinated animals,

but only from those that are totally susceptible to
FMD, by having been protected from exposure to both
natural infection and vaccine.

Vaccine is used to maintain a barrier of protected animals inhibiting the movement of virus in animal
populations, and is applied only in regions that are
adjacent to what has been legally designated as an FMD-
free zone in which the totally susceptible animals are
maintained. It is obvious that such a program requires
unusual political resolve, mutual understanding, and
support by livestock owners to accomplish the identified objectives.

In considering the deterrents to marketing that
result from the presence of disease in livestock populations, none may have the significance of FMD. The
other major diseases are not affecting international
commerce to the same extent, but are important since
they cause lowered productive efficiency. This may
take the form of loss from mortality or of slower
weight gain even though the animal survives the disease.

Rinderpest

The trans-African campaign carried out in the
1960s known as Joint Program 15 (JP15) was based on
massive vaccination and considerably reduced the
ravages of the disease in domestic livestock. The
policy established at that time was for national
governments to assume responsibility for continued
vaccination of young animals in ensuing years. Since
the infrastructure and resources available to many of
the national governments were not adequate, this policy
was not continued, with the result that gradually
larger numbers of non-vaccinated, susceptible animals
made up the livestock populations which, when exposed
to rinderpest virus, allowed the disease to reappear in
areas where it had previously been suppressed.

The major expense related to the use of rinderpest
vaccine is that associated with assembling animals and
the administration of the product as opposed to the
actual cost of vaccine which is relatively insignificant. Rinderpest vaccine is an attenuated virus
produced in tissue culture and can be made in developing countries if adequate attention is given to proper
quality control. It is inexpensive to produce in large
quantities and can be turned out on a routine basis
given suitable facilities and a moderate level of technical accomplishment. The handling of the vaccine
under field conditions requires constant refrigeration.
The vaccine is one of the best immunizing products
available in that it provides practically a lifelong

immunity and it is free of post-vaccination reactions even in milking and pregnant animals. Also, and in contrast to FMD, the virus is monovalent and the vaccine will protect against rinderpest occurring any place in the world.

It is expected that good herdowners will take it upon themselves to immunize their animals. Long-term rinderpest control is only practical if large areas are incorporated in order to assure that migratory animals do not reintroduce the disease. Along with regional vaccination programs to control rinderpest, it is essential to carry out post-vaccination surveillance to determine antibody levels produced by the administration of the vaccine. It is also necessary to develop diagnostic competence within the region in order to carry out differential diagnosis to identify rinderpest from the many rinderpest-like infections. In protected or partly protected animal populations, rinderpest is not the spectacular highly fatal disease that it is in susceptible populations and consequently the identification of areas of residual virus becomes essential.

In areas where rinderpest has been a problem in cattle production, inauguration and carrying out of a vaccination program is usually not difficult. However it does require a source of vaccine that meets international standards for quality control and tests for adequate virus content in addition to the organization of vaccination crews and assembling of cattle. Also required is a commitment by the respective government to maintain post-vaccination monitoring of antibody levels and annual vaccinations of the animal's progeny in order to assure a level of immunity for a sustained period of time.

There is currently considerable enthusiasm being generated to mount another trans-African rinderpest vaccination program. While donor governments and their respective technical agencies have considerable experience resulting from JP15 and other disease control programs in developing countries, it is essential to get judgments, opinions and evaluations from the actual individuals living within the affected countries who participated in JP15 and the subsequent policies following the initial vaccination phase. Their judgments in the identification of strengths and weaknesses will be invaluable in identifying the direction the currently considered program should take in order to assure more effective continuity of the objectives desired.

Contagious Bovine Pleuro-pneumonia (CBPP)

Among the diseases that frequently receive attention from the standpoint of regional vaccination is CBPP. This disease becomes most obvious if large numbers of animals are congregated and moved over long distances. It is a chronic infection and is spread only by aerosols infected by the organism, a mycoplasm, being directly exhaled from the infected animal in the presence of a susceptible bovine. The course of the infection may be from a few to several months. Some animals function as carriers not showing evidence of the infection and yet are able to transmit the organism. As the disease progresses, the herdsmen become aware that affected animals do not keep up with the rest of the herd, their breathing becomes labored, and they progressively lose weight. Approximately half the animals develop severe lung lesions and have to be slaughtered. Left along the way, they thereby add to the distribution of the disease or they are slaughtered on the spot for human consumption. Vaccines currently available for CBPP produce problems since they cause tissue reactions to which the livestock owners object. The level of protection is relatively low and revaccination is desirable if long exposure is expected. Attempts to control CBPP have depended on the use of serological tests to identify infected animals, their removal from the livestock population and the use of vaccines where exposure is anticipated. Such control is obviously expensive and requires a sustained input of resources if appreciable results are to be obtained.

Other Diseases

The other major bacterial diseases for which control methods are frequently applied include anthrax which affects humans as well as cattle, swine, horses, sheep and goats, and the clostridial infections producing malignant edema, blackleg, enterotoxaemia, botulism and tetanus. These organisms are usually associated with specific locations where they have been periodically identified. The diseases are commmonly known to livestock owners who will make provisions for vaccinating their herds if the disease is known to be a hazard in the region.

Products of animal origin that are used for animal feed supplements, as well as hides, skins, wool and hair are exported from developing countries to other parts of the world. Outbreaks of anthrax have frequently occurred in livestock populations of the developed world as a result of importations of contaminated animal feed supplements. The contamination of any of the above products by anthrax spores may also result in human infection.

The considerable number of tick-borne diseases such as East Coast fever, babesiosis, anaplasmosis and heartwater require continued use of effective acaricides for the dipping of cattle. There is considerable experience within the African continent on the efficacy, problems and economics of carrying out tick control. While immunizing methods against these various blood pathogens are being researched, procedures for using immunizing agents efficiently on large numbers of animals have yet to be developed.

Two-thirds of tropical Africa, about 10 million km^2 are infested by tsetse flies. There are at least 22 species, each with its own specific biological requirements. African trypanosomes are transmitted to, and may infect, all species of domestic animals. The disease may be acute or chronic, leading to progressive anemia and death if the affected animal is untreated. There are three predominant species which affect cattle. While wildlife carry trypansomes, they do not exhibit symptoms of disease as do cattle.

Several techniques have been used and are currently being researched to provide more effective means of reducing the tsetse population as well as directly attacking the blood parasite (the trypanosome), itself. Several countries of Europe are supporting research primarily in Africa. The Food and Agriculture Organization of the United Nations is attempting to coordinate the multiplicity of efforts being directed toward tsetse and trypanosome suppression. A significant component of this effort is based on successful reclamation of land so that areas cleared of tsetse will not become reinfested.

WILDLIFE INFLUENCES

Grazing is frequently shared by domestic stock and wildlife. In addition to the competition for grazing and water, some infectious diseases are common to both groups. They include trypanosomiasis, some of the tick-borne diseases, malignant catarrhal fever, FMD, rinderpest and bluetongue. To control disease transmission, it is obviously advantageous, whenever possible, to separate the wild from the domestic species. With human agricultural pressures demanding more and more land, there is an urgent need for additional provision and protection for wildlife.

Additional research is needed in order to identify the true role of wild species as reservoirs of virus between outbreaks occurring in domestic livestock. For example, it has been shown that the Cape buffalo has the ability to retain FMD virus in its esophageal secretions for several years. Even though the virus isolated from the buffalo was of a virus type that had not been reported in domestic animals over several years, it was fully virulent for cattle when artificially inoculated. Wild species are not a factor in the transmission and dissemination of CBPP.

DISEASE CONTROL BY QUARANTINE

The principles of animal quarantine are frequently referred to as a mechanism for controlling dissemination of disease. In theory, this is a highly rational approach. In practice however, in order to be effective, quarantine requires not only understanding and sympathy of the livestock owners but also the commitment of a stable national government to provide authority and resources so that a quarantine can be maintained effectively and with legal consequences in the event of violations.

As has been shown in the developed as well as developing countries, if there is an economic incentive for violation, quarantines will not be effective without vigorous enforcement and severe penalties. Perhaps the most encouraging example of the use of quarantines in an effective manner is that associated with the establishment of an FMD-free zone and the ability to maintain the disease-free status over a sustained period of time. As has been stated earlier, the justification for acceptance of the restrictions related to quarantine action is the value of red meat that can be exported to a country with a desirable and convertible currency.

SUMMARY

While the more obvious disease related constraints to production have been identified in the chapter, it is inevitable that, when these diseases are suppressed, the endemic diseases will become more apparent. Such diseases include several of the other parasitic diseases. When better livestock production practices are applied with closer concentration of animals, diseases such as brucellosis and tuberculosis become hazards, not only for their effect on livestock production but also because of their public health consequences.

In order to develop an effective livestock health program, including the provision of various animal health services such as diagnostic facilities, vaccine production and quality control, vector control, disease surveillance and the required restrictions on animal movement, continuous funding is necessary. There are frequent problems that often interfere with the efficient meeting of these objectives particularly in developing countries where political instability results in intermittent or altered objectives. Nevertheless, when appropriate diseases are considered for control, with the potential of a viable market, cost effectiveness can be shown.

RECOMMENDATIONS

Once livestock production strategies are identified, it is necessary to program the animal health resources toward accomplishing those objectives in order to ensure as broad a support as possible. Realistic goals and mechanisms for program evaluation are essential for the field oriented disease investigation, diagnosis and control components. Vaccine production should not be undertaken if other sources can be found that are competitively priced, meet international standards of quality control and are readily available.

Training at all levels-- professional, technician, helper--must relate to local circumstances and include exposure to ancillary disciplines such as economics and sociology relating to local animal production practices. Considering the evolutionary stage of present-day African economies and the various means being explored to utilize resources for the maximum benefit of human populations, closer evaluation of the role of livestock is warranted.

SUPPLEMENTAL READING

British Veterinary Association. Handbook on Animal Diseases in the Tropics, ed. Sir Alexander Robertson, British Veterinary Association, 1976.

Cornell University. Game or Cattle for Meat Production on Kenya Rangelands. Report to Zilly Endowment, Inc. March 1983.

Falconer, J. "Disease Situation Report -- Foot and Mouth Disease--Botswana." Bulletin Office International des Epizooties 92 (1980): 655-657.

FAO. Non-Tariff Barriers to International Meat Trade Arising from Health Requirements. Rome, 1973.

_____. Report on the workshop on Emergency Disease Control. Cyprus, December 1982.

FAO-WHO-OIE. Animal Health Yearbook. Rome, 1982.

Hedger, J. "The Isolation and Characterization of Foot and Mouth Disease Virus from Clinically Normal Herds of Cattle in Botswana". Journal of Hygiene 66(1968): 27-36.

Joint Publications Research Service. Worldwide Report: Epidemiology. Numerous issues.

Morgan, D.O., D.M. Moore, and P.D. McKercher. _Vaccination against Foot and Mouth Disease, New Developments with Human and Veterinary Vaccines_. New York: Alan R. Liss, Inc., 1980.

National Research Council. _Animal Health_. World Food and Nutrition Study, National Academy of Sciences, Washington, D.C., 1977.

_____. _Manual of Standardized Methods for Veterinary Microbiology_, ed. George E. Cottral. National Academy of Sciences, Washington, D.C., 1978.

Plowright, Walter. "The Effects of Rinderpest and Rinderpest Control on Wildlife in Africa." Symp. Zool. Lond. (1982) No. 50, 1-28.

UNESCO/UNEP/FAO. _Tropical Grazing Land Ecosystems_. Paris, 1979.

World Bank. "Animal Diseases in Developing Countries Technical and Economic Aspects of their Impact and Control." AGR Technical Note No. 7, Agricultural and Rural Development Department, Washington, D.C" 1983.

6

Goats and Sheep in the Traditional Livestock Production Systems in Semi-Arid Northern Africa: Their Importance, Productivity and Constraints on Production

R. Trevor Wilson

Subsaharan Africa has more than one-third of the goats in the world and almost one-fifth of the sheep. These two species contribute almost 17 percent of the total ruminant biomass (camels, cattle and small ruminants). However, it is rare for them to be accorded the importance they so obviously merit in research or development plans, although they do get rather more management attention by traditional owners than is generally accepted by outsiders.

Table 6.1 shows some pertinent figures for goats and sheep in Africa. These are obviously only gross figures as, for example, the number of goats and sheep combined varies from as low as 0.07 per head of population in Sierra Leone in the forest belt to 6.0 in Djibouti and 5.4 in Mauritania in the dry areas. Meat, milk and skin production from African small ruminants is respectively 16, 14 and 15 percent of total world output from goats and sheep. Goats produce altogether about three times as much milk as sheep.

THE IMPORTANCE OF SMALL RUMINANTS

Ownership patterns are extremely varied and, at least for an outsider, difficult to establish and understand. The ramifications of African kinship systems; the complicated practices of stock friends; the herding-out procedures often involving professional herders of a different ethnic group, all lead to a fluid idea of ownership which often involves many displacements of an animal over its lifetime.

The greatest herd sizes per family are among people in the drier areas. In West Africa and the Sudan this effectively means that flock sizes decrease from north to south. In Ethiopia and Kenya the trend is to smaller flocks at higher altitudes. In the humid coastal zone of West Africa only one or two animals are owned per family. Overlying these general trends goats

Table 6.1. The importance of goats and sheep in Subsaharan Africa

Item	Unit	Africa	Region		
			Semi-Arid and Arid	Humid	Southern Africa
Total area	10^6 km^2	23.1	10.6	9.4	3.1
Agricultural population	10^6	230.2	108.2	107.4	14.5
Number of goats	10^6	128.5	93.5	23.2	11.8
Number of sheep	10^6	127.2	78.0	14.7	34.7
Ratio of goats: sheep		1.01	1.19	1.58	0.34
Density of goats & sheep	head/km^2	11.1	16.2	4.0	15.0
Goats & sheep per person	head per capita	1.11	1.59	0.35	3.21
Goats & sheep as a proportion of ruminant biomass	percent	16.8	16.5	9.3	18.4

Source: Mainly calculated from UN, FAO [1981].

assume more importance than sheep as the management
system moves from nomadic to sedentary, and from purely
pastoral systems of production to mixed crop-livestock
systems. Table 6.2 gives some idea of the varying but
generally important contribution of smallstock to total
holdings. Similar figures could be produced for other
areas, for example in Mali, where 95-100 percent of
farmers own an average of 38 goats while 60 percent own
an average of 18 sheep [Wilson 1982]. In south-west
Nigeria 73 percent of farmers own smallstock, some 90
percent of these owning goats only, with an average
holding of 2.8 head [Velez-Nauer et al. 1982].

With the increasing pressure on land due to higher human population levels, it is probable that numbers
of small ruminants, in part because of their wider
dietary range, are increasing relative to cattle. An
additional advantage accruing to small ruminants is
their early physiological maturity and rapid reproductive rate. Because of their ability to produce a
greater number of twin births, goats are undoubtedly
also increasing in numbers faster than sheep.

Another advantage of small ruminants is their
ability to produce returns rapidly in relation to
cattle and camels. The major offtake for meat from
small ruminants occurs at between 9 and 15 months of
age while for cattle offtake does not begin before 3-4
years and for camels 5-6 years. In addition, on
account of the different breeding seasons (Figure 6.1),
small ruminants are able to provide food for human
consumption at times of the year when cattle and camels
are unable to do this, whether this relates to meat
(Figure 6.2) or to milk (Figure 6.3).

An extremely important part of the role played by
small ruminants is their longer-term ability to with-
stand difficult conditions. This has been amply
demonstrated during the recent series of drought and
dry years over much of Africa. In four West African
countries, while the total annual ammount of meat for
slaughter was maintained after the 1968-1973 drought,
as can be seen from Table 6.3, the contribution of
cattle dropped from 65 percent to 56 percent while that
of goats and sheep increased to meet the former level
of demand. Similar effects have been observed in Kenya
where after the Maasailand drought of the mid-1970s the
percentage sizes of flocks and herds compared with
before were 67 for goats, 61 for sheep and 57 for cattle [Campbell 1978]. In this situation, when climatic
conditions improve there are more small ruminants ready
to begin reproducing within 6-12 months (often even
before a crop can be produced) while cattle may not
reproduce for 12-24 months. In this connection the
place of small ruminants in providing a quick cash

Table 6.2. Mean livestock holdings per household in pastoral and agropastoral societies in Kenya and Chad.

Production System	Kenya		Chad		
	Maasai Pastoral	Karapokot Agropastoral	Zioud Pastoral	Salamat Agropastoral	Gondeye Tcheim Agropastoral
	—head per household—				
Cattle	157.3	11.8	36.4	133.3	2.1
Sheep	44.0	5.4	43.5	2.0	1.3
Goats	83.1	13.6	40.5	46.3	4.7

Source: Kenya, Christie Peacock (pers. comm.); Chad, Dumas [1980].

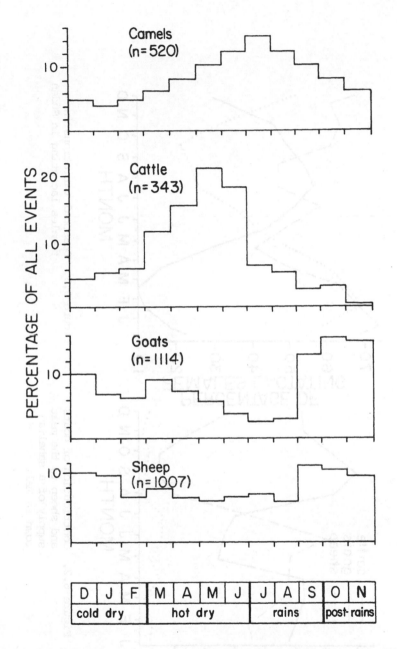

Figure 6.1. Distribution of births for four ruminant species in the Sahel

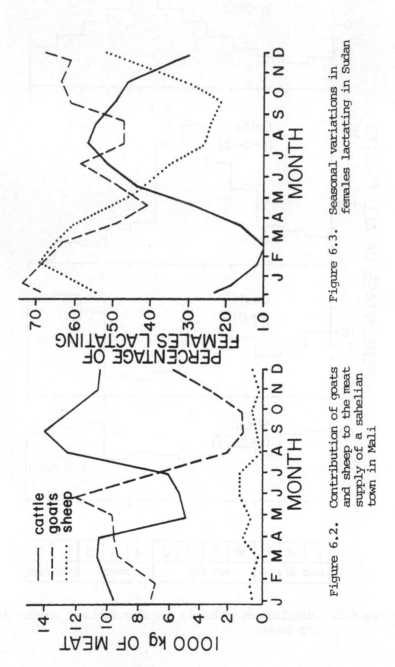

Figure 6.2. Contribution of goats and sheep to the meat supply of a sahelian town in Mali

Figure 6.3. Seasonal variations in females lactating in Sudan

Table 6.3. Contribution of goats and sheep to registered slaughterings in four countries before and after a drought

Country	1970				1976			
	Cattle	Goats + Sheep	Total	Goats + Sheep of total	Cattle	Goats + Sheep	Total	Goats + Sheep of total
	-1000 A.U. equiv.[a]-			-percent-	-1000 A.U. equiv.[a]-			-percent-
Mauritania	21	4	25	16	10	3	13	23
Mali	80	23	103	22	90	38	128	30
Niger	62	79	141	56	33	89	122	73
Chad	56	10	66	15	55	13	68	19
Total	219	116	335	35	188	143	331	43

Source: Various national and FAO statistics.

[a]1000 livestock unit equivalents.

return with which to buy other essential commodities should not be overlooked.

PRODUCTIVITY

Productivity can be expressed in a number of ways. Part of the production concept of goats and sheep has already been shown in Figures 6.2 and 6.3. Goats and sheep are sources for essentially the same products as cattle and camels. They produce some fiber but, apart from some limited scale commercial production in Kenya and in southern Africa, these items are not generally important except to specific groups--for example Moor sheep in Mauritania produce 200-300 grams of hair per head per year which is used for tent making while the Macina wool sheep of the Niger inundation zone provides 600-700 grams of carpet quality wool for local blanket making in Mali [Wilson 1983].

Meat and milk are the main products of all the domestic ruminants. The faster reproductive rate of small ruminants, their more rapid growth and early age at offtake enable them to produce meat on an annual basis at approximately twice the level of their contribution to the herbivore standing biomass. This is clearly demonstrated in Table 6.4 which shows their advantages over cattle and particularly over camels in terms of meat production. It has elsewhere been calculated [Wilson, Diallo, and Wagenaar, 1983] that on an equivalent basis the relative production of meat is 1.0, 1.5, 3.0, and 5.0 for camels, cattle, sheep and goats respectively. However, a fact which is not generally realized is that in traditional production systems camels and cattle are essentially dairy animals. In this production context they are seen in a more favorable light, although still outproduced by sheep and goats in the ratios of 1.0; 0.7; 1.0; and 2.5.

Recent developments in calculating productivity have been aimed at quantifying the differences within and between species either in the same or in different situations. To this end a series of Production Indices has been constructed, these being:

Index I,   total liveweight of young produced per breeding female per year;

Index II,   liveweight of young produced per unit weight of breeding female per year;

Index III, liveweight of young produced per unit metabolic weight of breeding female per year.

It is understood that these Indices, which are expressed in terms of weight but can, in fact, be taken to be simple ratios, are imperfect in that they take account of only one production item (liveweight which can be converted easily to a meat potential). Other

Table 6.4. Liveweight biomass and meat production of domestic ruminants in eight African semi-arid countries[a]

Item	Units	Cattle	Camels	Goats+Sheep
Total liveweight biomass	MT.	14,000,000	1,900,000	2,800,000
Offtake rate	percent	14.5	6.2	30.0
Total carcass weight available at 50% dressing percentage	MT.	990,000	60,000	410,000
Species as a proportion of liveweight biomass	percent	75	10	15
Contribution of species to total meat availability	percent	68	4	28

Source: Author.

[a]Chad, Gambia, Mali, Mauritania, Niger, Senegal, Sudan, Upper Volta.

items which might need to be considered would be milk, hides and skins, draft power, fiber, manure and such probably unquantifiable variables as social values. Nevertheless, they do allow some of the factors influencing productivity to be identified. Table 6.5 shows some of the effects of environmental variables on sheep productivity in two different countries while Table 6.6 compares the Indices of four ruminant species in one country. Various significance levels are attached to the different variables considered.

It is not possible to do more than generalize about individual situations without a thorough initial study. Perhaps one of the most surprising facts to emerge from these analyses is that for sheep a twin birth results, in most cases, in lower productivity than a single birth. This stems from a very high death rate in twins and a reduced genetic potential for growth in the remaining individual. The statement is not true for every situation, however, and in Mali twins are advantageous. All analyses so far have shown twins to be advantageous in goats. The flock factor--

Table 6.5. Production Indices I and II[a] for sheep in two countries with effects of different environmental variables

Variable	Kenya		Sudan	
	Index I	Index II	Index I	Index II
	-kg-	-g-	-kg-	-g-
Raw mean	14.6	512	22.2	598
Least square mean	12.7	402	20.1	547
Sex				
female	12.6	381	15.1	399
male	12.7	422	25.0	695
Parity				
1	8.9	293	17.6	564
2	12.2	399	18.3	471
3	15.9	536	33.7	939
4+	12.2	399	NA[b]	NA
Birth type				
single	14.0	490	22.6	616
twin	11.3	313	19.4	477
Season of birth				
short dry	14.6	461	19.4	500
long rains	13.8	403	23.6	634
long dry	8.6	242	17.2	506
short rains	13.6	500	NA	NA
Flock				
best	17.2	563	44.8	1347
worst	9.3	270	8.4	214

Source: Author

[a]See text for definition.

[b]No data available.

which is, in fact, a reflection of individual management ability--has so far always shown significant
differences with the best flocks achieving up to five
times the output of the worst ones. Table 6.6 shows
the superior productivity of sheep and goats over the
two larger species although, whether goats or sheep
attain greater output, depends on the country and,
where different managements such as transhumant and
sedentary or rainfed and irrigated agriculture are
considered, on the system.

CONSTRAINTS TO INCREASED PRODUCTIVITY

Productivity of small ruminants in existing
traditional systems is probably much better than has
generally been accepted. The potential for improving
it further using only limited and appropriate technological improvements is considerable. Unfortunately,
political and institutional constraints will be more
difficult to overcome.

Pre-weaning mortality is probably the major problem
acting on the present level of output. As many as 35
percent of kids and 30 percent of lambs die before
weaning. It is the death rate rather than reproduction
interval or weight increase which is the proximal
factor in the great variation in Production Indices

Table 6.6. Production Indices[a] for the four domestic
ruminant species in Mali

Species	Index I	Index II	Index III
	-kg-	-g-	-kg-
Goats	18.7	565	1.47
Sheep	29.5	888	2.31
Cattle	41.2	173	0.76
Camels[b]	43.6	125	0.61

Source: Author.

[a]See text.

[b]Estimated.

Table 6.7. Pre-weaning mortality rates due to
environmental factors in small
ruminants on a Maasai group
ranch in Kenya

Environmental variable	Goats	Sheep
	----percent----	
Least square mean	22.3	19.6
Birth type		
single	18.8	7.7
twin	25.8	31.4
Season of birth		
short dry	23.8	17.2
long rains	14.6	13.9
long dry	31.4	22.4
short rains	19.3	22.6
Flock		
best	12.4	12.2
worst	40.0	25.8

Source: Author.

between owners as the death of a single born young
means an Index of zero and the death of one of a pair
of twins greatly lowers the Index. Table 6.7 shows
some of the variations in death rate in a study
undertaken in Kenya. The causes of a total of 45
percent of all the deaths in goats and 41 percent in
sheep were reported as unknown. Of all other causes of
death only 12.5 percent in goats and only 3.5 percent
in sheep were attributed to malnutrition or starvation.
Some 56 percent of other deaths in goats and 29 percent
in sheep were attributed to disease.

Unlike cattle, where the principal epidemic
diseases such as rinderpest and contagious pleuro-pneumonia have now been overcome (or at least the means of
overcoming them are well known and proved) disease is
undoubtedly the main constraint to improved output in
small ruminants. Small ruminant pest, pasteurella and
other pulmonary diseases are major killers of goats and
sheep while foot rot and parasites are of lesser importance in the dry areas. Where, in small samples,

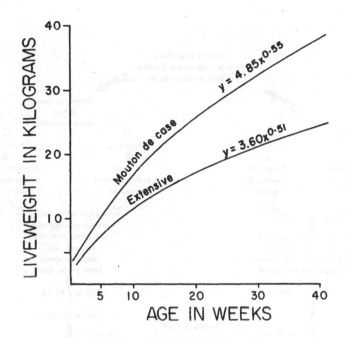

Figure 6.4. Comparative growth curves for "Mouton de Case" and sheep reared under extensive conditions

disease has been overcome, deaths following a package of veterinary interventions have been reduced from 40 to 13 percent in goats with Index I being increased correspondingly by 40 percent [Mack 1982]. Such interventions, however, are costly and need close and constant supervision.

Nutrition, which is now the major problem in cattle productivity, is less of a constraint to small ruminant productivity. This is supported by the highly seasonal nature of births in cattle associated with conception in the rainy season, as well as by the great variations in weight amounting to as much as 20 percent loss in the dry season. In small ruminants seasonality is much less marked with some births occurring throughout the year (Figure 6.1) and percentage weight changes being much less. The wider dietary preferences and greater selectivity of small ruminants probably have a strong influence on this factor. However, nutrition could be improved and, where supplementary feeding is practiced in traditional societies, weight gains, as can be seen in Figure 6.4, can be very impressive when

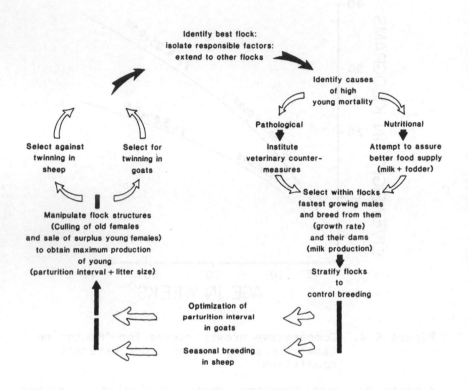

Figure 6.5. Potential improvement pathways for traditionally managed small ruminant flocks on Maasai group ranches

Note: Open arrows indicate alternative or secondary pathways.

compared with contemporary animals allowed to feed only on the open range.

These gains, coupled with the nature of the reproductive cycle which is correlated with seasonal factors only to a slight degree, show that local African breeds are not necessarily of low genetic potential. In fact, it is not likely that African breeds in themselves are a constraint at the present time. Breed limitations may become important, although this is by no means certain, when current disease, nutritional and above all management constraints have been overcome.

Using management as a starting point it is possible to construct a set of improvement pathways for traditional African small ruminant production. Figure 6.5 shows such a scheme designed for a pastoral system in Kenya. In its essence it builds upon the best features of the existing system. It should be perfectly feasible to operate as it is based on an analysis of results obtained over a three year period of study, all of the components of the improvement pathways thus already being attained to some extent by some of the flocks involved. Similar schemes could be designed for other countries and at later stages, and particularly where herders' organizations such as group ranches or co-operatives are operating, may include the establishment of nucleus breeding flocks and the reduction of possible inbreeding by male exchange schemes.

SUMMARY

Goats and sheep total some 17 percent of the domestic ruminant biomass of Subsaharan Africa. In the northern drier areas there are 170 million goats and sheep equivalent to 1.2 animals for each human at 16 animals per km^2. They are important providers of meat and milk, particularly to small holders and during periods of the year when production of these commodities from camels and cattle is low. Resistance to and recovery from drought conditions is an additional important factor in dry areas. In development plans and when livestock projects are planned they seldom receive the emphasis that their role and productivity should ensure for them. Increased productivity with concomitant greater contributions to human welfare could be achieved by low-technology, low-cost packages based on improving existing management practices and existing biological potential within the traditional systems already found in Africa.

REFERENCES

Campbell, David J. "Coping with Drought in Kenya Maasailand: Pastoralists and Farmers of the Loitokitok Area." University of Nairobi, Institute for Development Studies, Working Paper No. 337, 1978.

Dumas, Raymond. "Contribution a l'etude des petits ruminants du Tchad." Rev. Elev. Med. Vet. Pays Trop. 33(1980): 215-33.

Mack, Simon. "Disease as a constraint to productivity." Small Ruminant Productivity in Africa, ed. Ruth M. Gatenby and J.C.M. Trail, pp. 81-84. Addis Ababa: International Livestock Centre for Africa, 1982.

United Nations, Food and Agriculture Organization. Production Yearbook, Volume 35. Rome, 1981.

Velez-Naur, Miguel, et al. "Productivity of the West African Dwarf Goat at Village Level in South-West Nigeria." Proceedings of the Third International Conference on Goat Production and Disease, p. 356. Scottsdale, Arizona: Dairy Goat Journal Publishing Co., 1982.

Wilson, R. Trevor. "Husbandry, nutrition and productivity of goats and sheep in tropical Africa." Small Ruminant Breed Productivity in Africa, ed. Ruth M. Gatenby and J.C.M. Trail, pp. 61-76. Addis Ababa: International Livestock Centre for Africa, 1982.

_____. "Livestock Production in Central Mali: the Macina Wool Sheep of the Niger Inundation Zone." Trop. Anim. Hlth. Prod. 15(1983): 17-31.

Wilson, R. Trevor, A. Diallo and Klaas T. Wagenaar. "L' elevage mixte et les bilans demographiques des animaux domestiques dans les zones aride et semi-aride de l'Afrique nord tropicale." Paper presented at the Seminar on the Demography, Social Structure and Well-Being of Pastoral and Agro-pastoral Communities in the Sahel, held at Bamako, Mali, 24-27 January, 1983.

7
Animal Breeding in Subsaharan Africa: Toward an Integrated Program for Improving Productivity

J.C.M. Trail and K.E. Gregory

Subsaharan Africa generally has a great need and, provided specific production constraints can be alleviated, has a great potential to increase both milk and meat production [Trail and Gregory 1981a]. In planning increased animal productivity in subsaharan environments, a principal requirement is for information allowing the usefulness of major animal types to be confidently predicted for different ecological zones, production systems, management levels, disease situations, nutritional resources, etc. Expense is a major problem in this day and age in setting up from scratch operations to collect this information.

INTRODUCTION

The objectives of this paper are to indicate the main ecological zones of Africa as they have been described by several authors in the past and to mention the importance of cattle in each, the production systems involved and the general types of cattle that can be maintained. The limited genetic potential of indigenous cattle has often been quoted as a major constraint to beef and milk production in Africa. Numerous research programs have been conducted in many countries of the continent. Thus, an attempt is made to determine the value of previous breed productivity research in assessing current decision making. The challenge in most areas of Subsaharan Africa is to synchronize germ plasm resources with the environmental resources most favored by economic considerations. The possible approaches are explored in this chapter.

CATTLE

Cattle in Africa are used for several purposes and in most cases cannot be separated into classes of beef, dairy and draft animals. In many situations the same animals are milked, used for draft, and finally slaugh-

107

tered for consumption. Well defined beef or dairy industries are rare, although around many of the larger population centers there are improved dairies and milk collection schemes where milk is collected from small-holder producers. The levels of husbandry and management and the attitude of many owners to their cattle are such that productivity is often extremely low, and little or no increase in production is possible through the introduction of potentially more productive breeds until the traditional methods of husbandry are changed.

The indigenous breeds predominate and their description and classification are well covered by Mason's The Classification of West African Livestock [1951], Mason and Maule's, The Indigenous Livestock of Eastern and Southern Africa [1960] and Epstein's The Origin of the Domestic Animals of Africa [1977]. They fall into three very broad groups - the humped Zebu, the humpless or taurine, and the small cervico - thoracic humped Sanga - with further possible subdivisions. Existing indigenous cattle populations are generally well adapted to survive and reproduce in their environment, possessing qualities such as mothering and walking abilities, water economy, heat tolerance, disease tolerance, and ability to exist on low quality feeds. Usually, however, they are late maturing, have poor growth rates, low milk yields and produce small carcasses.

Exotic cattle introduced to Africa also fall into three broad groups - Zebu cattle from Asia and America such as the Sahiwal and Brahman, Zebu/European hybrids such as the Santa Gertrudis, Droughtmaster and Bonsmara, and European beef, dual-purpose, and dairy cattle of numerous different breeds. With increasing proportions of exotic Bos taurus breeds, the levels of nutrition and management generally need to be raised to achieve satisfactory performance.

ECOLOGICAL ZONES

For our purposes, Africa south of the Sahara is subdivided into five ecological zones following Meyn [1978]. The five zones are:

    very arid,
    arid to semi-arid,
    semi-arid to humid, without tsetse,
    temperate highland,
    humid, testse infested

These categories reflect elements of climate, elevation and the occurrence of disease, and each is relatively uniform in terms of livestock production problems.

In the very arid areas (less than 400 mm rainfall) the natural vegetation permits only low stocking rates (over 15 ha/head), and the livestock production system generally is based on nomadism. About 8 million head of primarily Zebu and Sanga cattle utilize the very arid areas seasonally, in the subsahara, in the Horn of Africa and in the southwestern corner of the continent. Water shortages and lack of pasture usually force live-stock producers to move to areas with better grazing during the dry season. Beef produced in this area balances between liveweight gains during the short flush period and losses during the dry season; produc-tion depends most on the ability of cattle to survive, reproduce and minimize weight losses during the dry season. There is strong natural selection for survival with little opportunity of artificial selection.

The arid to semi-arid areas receive from 400 to 600 mm annual rainfall, which is still too dry for crop production. Beef production problems are similar to those in the very arid areas; however, the growing sea-son is relatively longer and the dry season shorter, thus turning the balance in favor of production. The cattle population in these areas is estimated at 33 million head.

The semi-arid to humid areas (over 600 mm annual rainfall) which are free from tsetse are suitable for agriculture. Cattle either compete with or support other farm enterprises. Natural pastures offer satis-factory levels of nutrition for part of the year; the low nutritive value of grasses during the dry season poses problems but pasture improvement is feasible. There is little difficulty in keeping improved Zebu or Sanga cattle and their crosses with European breeds, but purebred European cattle are often unsuitable because of high temperatures and humidity. The cattle population in this zone is about 66 million head.

The temperate highlands (with more than 600 mm annual rainfall) are suitable for all types of cattle in terms of climate and basic pasture quality. The main cause of poor nutrition in these areas is over-stocking. With about 33 million head, this zone has the largest concentration of cattle per sq km on the African continent.

The tsetse fly which transmits animal trypano-somiasis is a major constraint in expanded cattle pro-duction in the humid African lowlands. Only the hump-less cattle indigenous to West Africa (N'dama, West African Shorthorn) and their crosses with the Zebu show a degree of trypanotolerance, but they are few in numbers (about 8 million). In most tsetse-infested areas, cattle production is possible only after com-plete eradication of the tsetse flies or under treat-ment with prophylactic or curative trypanocides, which may be uneconomical if the tsetse challenge is high.

PREVIOUS BREED PRODUCTIVITY RESEARCH

Considerable efforts have been made to increase output through breeding programs in many countries of the continent. These efforts included the introduction of cattle from America, Asia and Europe for maintenance as purebreds and for use in upgrading and crossbreeding schemes, and also the operation of selection programs within indigenous cattle populations.

Logical decisions on selection between breeds requires both comparisons made in the same environment, and information on a sufficient number of performance traits to allow an acceptable index of overall productivity to be constructed. An assessment of past research work which satisfies these criteria can save on future inputs. In a bibliography [Trail 1981] covering performance aspect of indigenous, exotic and crossbred cattle in Africa south of the Sahara, approximately 500 references are listed, of which 50 are review or descriptive papers. The remaining 500 contain objective original data on some aspect(s) of reproductive performance, growth, viability or milk production. This bibliography was built up over a three-year period as a by-product of the studies on production systems being carried out in various zones of Africa by the International Livestock Centre for Africa. It covers the 30-year period, 1949 to 1978. Overall analysis of this bibliography shows that only about 20 percent of the references contain information on the three or more performance characters sufficient to allow characterization of breed types through a productivity index. For example, one simple productivity index used is 'weight of calf plus liveweight equivalent or milk produced per unit weight of cow maintained per year', this index being extended to cover more traits if information is available. Additionally, only 20 percent of the references contain comparative information on two or more breed types. When these two necessary attributes are put together, only 5 percent of the reports provide sufficient data to allow breed comparisons on the basis of a productivity index. This illustrates that the majority of research work does not give really useful information about the comparative performance of cattle breeds in Africa.

ENVIRONMENTAL CONSIDERATIONS

Climatic, nutritive and disease/parasite environments characteristic of much of Subsaharan Africa generally favor the use of cattle with varying percentages contributed by indigenous breeds. Except for a relatively small number of indigenous Bos taurus breeds in parts of central and West Africa, indigenous breeds are

Bos indicus. The indigenous breeds have a higher level of overall adaptability than introduced Bos taurus breeds to the environmental stresses. Response capability for both milk and meat production characteristics is generally low in most breeds of indigenous cattle even when environmental constraints are alleviated. Many breeds of introduced Bos taurus cattle have the additive genetic merit to respond for both milk and meat production characters when environmental stresses are minimal. However, in Subsaharan Africa, economic and technological factors generally do not permit sufficient modification of natural environments to realize as high a percentage of the genetic potential as can be exploited in temperate zones.

The optimum degree of modification of natural environments is determined by the value of the increased output in relation to cost of improving the environment. The relatively great demand for and high value of fresh dairy products and the greater efficiency of nutrient conversion to milk relative to beef, favors a higher degree of environmental modification to increase milk production than to increase beef production. The most feasible approach to increasing both milk and beef production is to improve the natural environment to the level favored by economic factors and to use cattle that possess the most nearly optimum additive genetic (breed) composition for the improved environment. This most optimum breed composition will be reached by using a contribution from both Bos taurus and Bos indicus breeds through either organized crossbreeding systems or through composite populations or breeds [Gregory et al. 1982].

BREED CHARACTERIZATION

Breed characterization in the tropics is directed towards achieving effective selection among breeds either for use as straightbreds, or for use in crossbreeding systems to utilize heterosis and complementarity, or for use as contributors to new breeds to simultaneously utilize heterosis and arrive at and maintain the most optimal breed composition. This requires breed evaluation for both additive direct genetic effects and additive maternal genetic effects for the biological traits of major economic value in the ecological zone of interest. There are important differences in climatic, nutritive and disease-parasite environments among the major ecological zones. These differences are the result of variations in such factors as amount and distribution of precipitation, elevation, soil fertility, solar radiation, temperature and relative humidity [Trail and Gregory 1981a]. Thus, simultaneous characterization of environments in which breeds are evaluated is necessary. This will extend the value of breed characterization efforts to the broadest range of production situations relating to both environmental conditions and production goals and objectives.

Major differences among breeds of cattle (Bos taurus and Bos indicus) have been demonstrated in both temperate and tropical zones for most characters that contribute to production efficiency [Cartwright et al. 1964; Touchberry 1970; Koger et al. 1975; Koch et al. 1976, 1979; Koch, Dikeman, and Crouse 1982; Koch, Dikeman, and Cundiff 1981, 1982; Koch and Dikeman 1977; Laster et al. 1976, 1979; Smith, Laster, and Gregory 1976; Smith et al. 1976; Gregory et al. 1978, 1979a, 1979b; Cundiff et al. 1981; Trail and Gregory 1981a, 1981b; Gregory and Trail 1981].

Breeds of Bos indicus cattle differ in response capability for both milk and meat production in tropical environments. For example the Sahiwal breed is considered to be unequaled in transmitted effects for milk production among Bos indicus breeds of cattle [Trail and Gregory 1981a]. Breeds of Bos taurus cattle differ greatly in performance characters associated with milk and meat production in improved environments in temperate zones [Touchberry 1970; Koch et al. 1976, 1979; Koch, Dikeman, and Cundiff 1981, 1982; Koch, Dikeman, and Crouse 1982; Koch and Dikeman 1977; Laster et al. 1976, 1979; Smith, Laster, and Gregory 1976; Smith et al. 1976; Gregory et al. 1978, 1979a, 1979b; Cundiff et al. 1981]. Similarly, there may be important differences among breeds of Bos taurus cattle in general adaptability to tropical environments and to some of the factors associated with these environments.

The exploitation of livestock possessing genetic resistance to disease is being given increasing consideration in livestock development programs particularly where conventional disease control measures are too costly, too complex to implement or, as is also common, drugs and vaccines are not available. This is particularly the case in developing countries. Such an approach is applicable to animal African trypansomiasis, a disease that certain indigenous Bos taurus breeds of cattle are able to survive in tsetse fly endemic areas without the aid of treatment but to which other breeds rapidly succumb.

While trypanotolerant breeds are a well recognized component of livestock production in certain areas of Africa, they represent only about 5 percent of the total cattle population in the 36 countries where tsetse occur [ILCA 1979]. Failure to exploit these breeds can possibly be attributed to the belief that because of their small size they were not productive and that their trypanotolerance was a characteristic which had been acquired to the local trypanosome population. However, it has now been confirmed that trypanotolerance is an innate characteristic and may, therefore, be genetically exploited (reviewed by Murray, Morrison, and Whitelaw [1982]). Furthermore, in a recent survey of the status of trypanotolerant livestock in 18 countries in West and Central Africa [ILCA 1979], indices of productivity were examined using all the basic production data that could be found for each region, each management system and for different levels of tsetse challenge. The results indicated that in areas of no or low tsetse challenge the productivity of trypanotolerant cattle relative to other indigenous breeds was much higher than previously assumed. Comparative data between breeds were not available in many areas because the level of trypanosomiasis risk was such that breeds other than trypanotolerant ones could not survive. As a result of these findings, there is currently considerable interest in the use of trypanotolerant breeds in tsetse-infested areas of Africa.

CROSSBREEDING

The challenge in use of crossbreeding systems is to optimize the simultaneous use of both additive (breed differences) and nonadditive (heterosis) sources of genetic variation. From the standpoint of additive genetic composition, the optimum contribution by Bos indicus breeds should be no more than required to achieve general adaptability to the production environment. A greater contribution would be expected to result in reduced production because the production potential of Bos indicus breeds, generally, is lower than Bos taurus breeds. From the standpoint of utilizing nonadditive genetic variation (heterosis) the optimum contribution by Bos indicus breeds is one-half, because of the higher level of heterosis from crosses of Bos indicus breeds with Bos taurus breeds than from among crosses of Bos taurus breeds [Cartwright et al. 1964; Koger et al. 1975]. Because of the importance of synchronizing general adaptability and performance characteristics of genetic resources with other production resources we believe that compromises should generally favor optimizing breed composition. The potential production increase through optimizing additive genetic (breed) composition is likely greater than the differences in heterosis between crosses of Bos indicus breeds with Bos taurus breeds relative to crosses among Bos taurus breeds. The objective is to identify and use the genetic resources that have optimum response capability for milk and meat production in the modified climatic, nutritive and disease-parasite environment that is favored by economic factors [Gregory et al. 1982].

Heterosis

The cumulative effects of heterosis on characters that contribute to weight of calf weaned per cow exposed to breeding have been shown to be 23 percent for crosses among breeds of Bos taurus cattle [Gregory et al. 1965; Wiltbank et al. 1967; Cundiff et al. 1974a, 1974b] and 50 percent or more for crosses between Bos taurus and Bos indicus breeds of cattle in subtropical environments [Cartwright et al. 1964; Koger et al. 1975]. Trail et al. [1982] reported heterosis effects for individual traits in Subsaharan Africa that were closer to the lower level for crosses of a Bos indicus breed (Boran) and a Bos taurus breed (Red Poll). Results show that 60 percent or more of the observed cumulative heterosis is the result of heterosis effects on maternal characters. In a comprehensive two-breed rotational crossbreeding experiment that included the specialized Bos taurus dairy breeds of Holstein and Guernsey and spanned four generations, Touchberry [1970] estimated the effects of heterosis to be 21.7 percent for an index of performance that included measures of growth, milk production and viability. Heterosis effects reported by McDowell et al. [1976] on characters relating to reproductive efficiency in a specialized dairy program were not consistent among different characters for crosses of a Bos indicus breed with Bos taurus breeds.

Rotational Crossbreeding Systems

Experimental evaluation of two- and three-breed rotational crossbreeding systems has indicated that high levels of heterosis are sustained in successive generations and that the relationship between loss of heterosis and loss of heterozygosity approaches linearity in rotational crossbreeding systems [Gregory and Cundiff 1980].

Heterozygosity reaches equilibrium after seven generations for both two-breed (67 percent relative to F_1) and three-breed (86 percent relative to F_1) rotational cross-breeding systems; heterozygosity after equilibrium is reached, as a percentage of F_1 for n breeds in a rotation, is equal to $(2^n-2)/(2^n-1)$ [Dickerson 1969, 1973]. In a two-breed rotational crossbreeding system, after equilibrium is reached, 67 percent (2/3) of the inheritance is contributed by the breed of the sire and 33 percent (1/3) of the inheritance is contributed by the breed of the maternal grandsire. In a three-breed rotational crossbreeding system, after equilibrium is reached, 57 percent (4/7) of the inheritance is contributed by the breed of the sire, 29 percent (2/7) of the inheritance is contributed by the breed of the maternal grandsire and 14 percent (1/7) of the inheritance is contributed by the breed of the maternal great-grandsire.

Because of wide fluctuation in additive genetic composition between generations in rotational crossbreeding systems, the breeds used should be reasonably comparable in characteristics such as birth weight in order to reduce the level of dystocia [Laster et al. 1973]. They should also be compatible in performance characteristics such as size and lactation potential to facilitate common management of all breed of sire groups [Cundiff 1977]. Wide fluctuation in additive genetic composition between generations in rotational crossbreeding systems greatly restricts the synchronizing of genetic resources to environmental resources. For example, if the optimum percentage contribution by a Bos indicus breed should be either 25 percent or 50 percent in order to achieve greatest harmony between genetic resources and other production resources, the optimum is approached infrequently if sires of pure breeds of Bos indicus and Bos taurus cattle are used in rotation [Gregory et al. 1982].

Composite Breed Formation

A high percentage of cattle in Subsaharan Africa are in herds too small to use well organized crossbreeding systems on a self-contained basis. The wide fluctuation in breed composition between generations, characteristic of rotational crossbreeding systems, is an even greater constraint to optimizing additive gentic composition in Africa than in temperate zones, because of the need for a contribution by breeds of Bos indicus cattle in order to achieve adaptability to the stressful environment of the tropics [Gregory and Trail 1981]. There is therefore need for experimental evaluation of the potential of composite breed formation as an alternative to continuous crossbreeding systems.

Retention of initial heterozygosity following crossing and subsequent random mating within the crosses (inter se) is proportional to $1 - \Sigma_i^n P_i^2$, where P_i is the fraction of each of n breeds in the pedigree of a composite breed [Wright 1922; Dickerson 1969, 1973]. This loss of heterozygosity occurs between F_1 and F_2. Retention of heterozygosity favors the inclusion of a large number of breeds in composite populations. The increased heterosis retention resulting from including additional contributing breeds must be balanced against possible loss of average additive genetic merit from the additonal breeds. If retention of heterosis is proportional to retention of heterozygosity in composite populations, the percentage of F_1 heterozygosity retained in composite populations based on an approximately equal contribution by either three or four breeds is equal to, or exceeds, the percentage of F_1 heterozygosity retained in a two-breed rotational crossbreeding system after equilibrium is reached. However, there is need to determine experimentally if retention of heterosis in composite populations is proportional to retention of heterozygosity, as was indicated for rotational crossbreeding systems by Gregory and Cundiff [1980].

A major advantage of composite breeds over rotational crossbreeding systems is the opportunity to achieve and maintain the most optimum contribution by each breed available for use in the system. Another potential advantage of composite breeds relative to rotational crossbreeding systems is that response to selection may be greater in composite populations than in parental breeds contributing to them. This is due to increased genetic variation expected as a result of differences in gene frequencies among the parent breeds, and greater selection intensity possible because of a higher reproduction rate as a result of heterosis [Dickerson 1973; Cundiff 1977].

An important consideration in the development of composite breeds is to maintain population size large enough so that the initial advantage of increased heterozygosity is not dissipated by early re-inbreeding of composite populations. Linearity of association of retention of heterosis with retention of heterozygosity in rotational crossbreeding systems [Gregory and Cundiff 1980] indicates that heterosis is due primarily to dominance effects of genes and suggests that heterosis may be largely accounted for by recovery of accumulated inbreeding depression that has occurred in breeds since their formation [Dickerson 1969, 1973; Cundiff 1977; Gregory and Cundiff 1980].

CONCLUSIONS

Integrated dairy-beef systems for producing both dairy products and beef are expected to continue to predominate in Subsaharan Africa because efficiency of food production for smallholders generally favors cattle that have the capability to produce milk in excess of requirements of offspring. In the very arid zone with 6 percent of the cattle, it is apparent that, in practice, little can be achieved through the introduction of new genotypes or attempted selection within indigenous populations. In the temperate highland zone, with 22 percent of the cattle, it appears that the importation and use of other indigenous, crossbred and exotic cattle types, based on their evaluation elsewhere, is completely feasible as long as due consideration is given to such factors as the production system, level of management and feeding practices, etc.

In the humid, tsetse infested zone with 6 percent of the cattle, the exploitation of trypanotolerant breeds of cattle offers one of the most important approaches to the control of the continental problem of animal African trypanosomiasis. Trypanotolerance can be reduced under certain adverse conditions such as high levels of tsetse challenge or it can be supplemented by previous exposure. Therefore, to realize the full potential of trypanotolerant breeds, it is important that the main environmental factors which affect trypanotolerance be identified and the extent of their influence quantified. Full comprehension in trypanotolerant animals of the factors which control parasite growth and allow the development of an effective immune response might provide marker(s) for selective breeding of trypano-resistant livestock or might allow methods to be devised for enhancing resistance to trypanosomiasis in more susceptible breeds.

In the arid to semi-arid and semi-arid to humid zones with the remaining 66 percent of the cattle, the climatic, nutritive and disease-parasite environment generally favors the use of cattle with varying percentages contributed by Bos indicus breeds because of their adaptability. Breeds of introduced Bos taurus cattle have the additive genetic merit to respond for both milk and meat production characters only when environmental stresses are mimimal. However, in these zones, economic and technological considerations generally do not permit modification of the natural environment to permit as high a percentage of their genetic potential to be exploited as in temperate situations. The most feasible approach to synchronizing cattle genetic resources with other production resources is thus to achieve the level of improvement in the natural environment that is favored by economic factors and to use, through crossbreeding systems or through composite breeds, cattle that possess the most nearly optimum additive genetic composition contributed by both Bos taurus and Bos indicus breeds.

A comprehensive program of characterization for major economic traits in candidate breeds of both Bos indicus and Bos taurus cattle in these ecological zones is necessary to provide the basis for effective selection among breeds for use in rotational crossbreeding systems and/or as contributors to composite breeds.

In this day and age, the setting up from scratch of operations to collect needed information on breeding improvement on the ground is expensive. Thus all

efforts that are contributing towards the building up
and tying together of information necessary to achieve
increased animal productivity in Subsaharan Africa need
to be maintained and supported. During the past decade, in addition to continued national and bilateral
supported breeding research programs and expanded performance recording schemes, there have been some other
significant inputs. FAO has launched a number of
operations in the areas of evaluation, conservation and
utilization of animal genetic resources. The Consultative Group on International Agricultural Research has
established its International Laboratory for Research
on Animal Diseases (ILRAD) and its International Livestock Centre for Africa (ILCA). An important approach
of ILCA is to complement and link together national
research operations in specific fields. A current
example is a network of trypanotolerant livestock
situations being built up in West and Central Africa,
where ILCA is coordinating a study over five years
involving 10 nationally operated situations where work
is in progress and where more definitive data can be
collected with relatively little additional input.

REFERENCES

Cartwright, T.C., G.F. Ellis, Jr., W.E. Kruse and E.K.
    Crouch. "Hybrid vigor in Brahman-Hereford
    crosses." Texas Agr. Exp. Sta. Tech. Monogr. 1,
    1964.
Cundiff, L.V. "Foundations for Animal Breeding re-
    search." J. Anim. Sci. 44 (1977):311.
Cundiff, L.V., K. E. Gregory and R.M. Koch. "Effects of
    Heterosis on Reproduction in Hereford, Angus and
    Shorthorn Cattle." J. Anim. Sci. 38 (1974):711.
Cundiff, L.V., K.E. Gregory, Frank J. Schwulst and R.
    M. Koch " Effects of Heterosis on maternal perfor-
    mance and milk production in Hereford, Angus and
    Shorthorn cattle." J. Anim. Sci. 38 (1974):728.
Cundiff, L.V., R.M. Koch, K.E. Gregory and G.M. Smith.
    1981. "Characterization of biological types of
    cattle - Cycle II: IV. Postweaning growth and feed
    efficiency of Steers." J. Anim. Sci. 53 (1981):
    332.
Dickerson, G.E. "Experimental Approaches in Utilizing
    Breed resources." Anim. Breed. Abstr. 37 (1969):
    191.
Dickerson, G.E. "Inbreeding and Heterosis in Animals."
    Proc. of the Animal Breeding and Genetics Symp.
    in Honor of Dr. Jay L. Lush, pp. 54-77. ASAS,
    Champaign, IL, 1973.
Epstein, H. The Origin of the Domestic Animals of
    Africa. Volumes 1 and 2. New York: Africana,
    1971.

Gregory, K.E. and L.V. Cundiff. "Crossbreeding in Beef Cattle: Evaluation of Systems." J. Anim. Sci. 51 (1980):1224.

Gregory, K.E., L.V. Cundiff, G.M. Smith, D.B. Laster and H.A. Fitzhugh, Jr. "Characterization of Biological Types of Cattle - Cycle II: I. Birth and Weaning Traits." J. Anim. Sci. 47 (1978):1022.

Gregory, K.E., D.B. Laster, L.V. Cundiff, G.M. Smith and R.M. Koch. "Characterization of Biological Types of Cattle - Cycle III: II. Growth Rate and Puberty in Females." J. Anim. Sci. 49 (1979a):461.

Gregory, K.E., G.M. Smith, L.V. Cundiff,R.M. Koch and D.B. Laster. "Characterization of Biological Types of Cattle - Cycle III. I. Birth and Weaning Traits." J. Anim. Sci. 48 (1979b):271.

Gregory, K.E., J.C.M Trail. "Rotation Crossbreeding with Sahiwal and Ayrshire Cattle in the Tropics." J. Dairy Sci. 64 (1981):1978.

Gregory, K.E., J.C.M. Trail, R.M. Koch and L.V. Cundiff. "Heterosis, Crossbreeding and Composite Breed utilization in the Tropics." Proceedings 2nd World Congress on Genetics Applied to Livestock Production. VI: (1982):279.

Gregory, K.E., L.A. Swiger, R.M. Koch, L.J. Sumption, W.W. Rowden and J.E. Ingalls. "Hetrosis in Pre-weaning Traits of Beef Cattle." J. Anim. Sci. 24 (1965):21.

ILCA. Trypanotolerant livestock in West and Central Africa. Monograph 2, Addis Ababa, 1979.

Koch, R.M. and M.E. Dikeman. "Characterization of Biological Types of Cattle. V. Carcass Wholesale Cut Composition." J. Anim. Sci. 45 (1977):30.

Koch, R.M. M.E. Dikeman, D.M. Allen, M. May, J.D. Crouse and D.R. Campion. "Characterization of Biological Types of Cattle. III. Carcass Composition, Quality and Palatability." J. Anim. Sci. (1976):48.

Koch, R.M., M.E. Dikeman and J.D. Crouse. "Characterization of Biological Types of Cattle (Cycle III): III. Carcass Composition, Quality and Palatability." J. Anim. Sci. 54 (1982):35.

Koch, R.M., M.E. Dikeman and L.V. Cundiff. "Characterization of Biological Types of Cattle (Cycle II): V. Carcass Wholesale Cut Composition." J. Anim. Sci. 53 (1981):992.

Koch, R.M., M.E. Dikeman and L.V. Cundiff. "Characterization of Biological Types of Cattle (Cycle III). V. Carcass Wholesale Cut Composition." J. Anim. Sci. 54 (1982):1160.

Koch, R.M., M.E. Dikeman, R.J. Lipsey, D.M. Allen and J.D. Crouse. "Characterization of Biological Types of Cattle (Cycle II). III. Carcass Composition,

Quality and Palatability." _J. Anim. Sci._ 49 (1979):448.

Koger, Marvin, F.M. Peacock, W.G. Kirk and J.R. Crockett. "Heterosis Effects on Weaning Performance of Brahman-Shorthorn Calves." _J. Anim. Sci._ 40 (1975):826.

Laster, D.B., H.A. Glimp, L.V. Cundiff and K.E. Gregory. "Factors Affecting Dystocia and Effects of Dystocia on Subsequent Reproduction in Beef Cattle. "J. Anim. Sci. 36 (1973):695.

Laster, D.B., G.M. Smith, L.V. Cundiff and K.E. Gregory. "Characterization of Biological Types of Cattle (Cycle II). II. Postweaning Growth and Puberty of Heifers." _J. Anim. Sci._ 48 (1979):500.

Laster D.B., G.M. Smith and K.E. Gregory. "Characterization of Biological types of Cattle. IV. Postweaning Growth and Puberty of Heifers." _J. Anim. Sci._ 43 (1976):63.

Mason, I.L. _The Classification of West African Livestock._ Technical Communication No.7. Edinburgh, Commonwealth Bureau of Animal Breeding and Genetics, 1951.

Mason, I.L. and J.P. Maule. _The Indigenous Livestock of Eastern and Southern Africa._ Technical communication No. 14. Farnham Royal, CAB, 1960.

McDowell, R.E., B.F. Hallon, J.K. Camoeus and L.D. Van Vleck. "Reproductive Efficiency of Jersey, Red Sindhis and Crossbreds." _J. Dairy Sci._ 59 (1976):127.

Meyn, K. "Constraints of Beef Production in Africa - Breeding Aspects." ILCA Livestock Development Projects Course, Nairobi, 1978.

Murray, Max, W.I. Morrison and D.D. Whitelaw. "Host Susceptibility to African Trypanosomiasis: Trypanotolerance." _Adv. Parasit._ 21 (1982):1.

Smith, G.M., D.B. Laster, L.V. Cundiff and K.E. Gregory. "Characterization of Biological Types of Cattle. II. Postweaning Growth and Feed Efficiency of Steers." _J. Anim. Sci._ 43 (1976):37.

Smith, G.M., D.B. Laster and K.E. Gregory. "Characterization of Biological Types of Cattle. I. Dystocia and Preweaning Growth." _J. Anim. Sci._ 43. (1976):27

Touchberry, Robert W. "A Comparison of the General Merit of Purebred and Crossbred Dairy Cattle Resulting From Twenty Years (four generations) of Crossbreeding." _Proc. 19th National Poultry Breeders' Round Table,_ Poultry Breeders of America, Kansas City, MO, 1970.

Trail, J.C.M. "Merits and Demerits of Importing Exotic Cattle Compared with the Improvement of Local Breeds. Cattle in Africa South of the Sahara." _Intensive Animal Production in Developing Coun-_

<u>tries</u>. Occasional Publication No. 4. Harrowgate, British Society of Animal Production, 1981.

Trail, J.C.M. and K.E. Gregory. "Sahiwal Cattle: An Evaluation of Their Potential Contribution to Milk and Beef Production in Africa." ILCA Monograph 3. International Livestock Center for Africa, Addis Ababa, 1981a.

Trail, J.C.M. and K.E. Gregory. "Characterization of the Boran and Sahiwal Breeds of Cattle for Economic Characters." <u>J. Anim. Sci</u>. 52 (1981b):1286.

Trail, J.C.M., K.E. Gregory, H.J.S. Marples and J. Kakonge. "Heterosis, Additive Maternal and Additive Direct Effects of the Red Poll and Boran Breds of Cattle." <u>J. Anim. Sci</u>. 54 (1982):517.

Wiltbank, J.N., K.E. Gregory, J.A. Rothlisberger, J.E. Ingalls and C.W. Kasson. "Fertility of Beef Cows Bred to Produce Straightbred and Crossbred Calves." <u>J. Anim. Sci</u>. 26 (1967):1005.

Wright, S., "Effects of Inbreeding and Crossbreeding on Guinea Pigs, III." USDA Bull. 1121, 1922.

8
Cattle Marketing Efficiency in Kenya's Maasailand

Phylo Evangelou

Kenya's meat supply is not keeping pace with rising levels of demand and shortages are likely to occur within the decade if present trends continue [Kenya, Republic of 1980]. The net meat surplus (production minus consumption) in the higher-potential regions of the country, the principal areas of national supply, is declining as human populations mount and land pressures necessitate the cropping of pasturelands. Consequently, the production and offtake potentials of the nation's rangelands will need to be more effectively exploited if the burgeoning demand for beef, mutton and goat meat is to be satisfied.

Kenya's Maasailand is one of the pastoral areas upon which the country and especially the urban market is increasingly dependent for the supply of cattle and small stock. This chapter focuses upon the region's local-level marketing mechanisms, with the objectives of (i) examining the extent to which marketing inefficiencies hinder livestock development in Maasailand and (ii) identifying constraints to the emergence of a more efficient marketing system. Its broader purpose is to demonstrate a generally applicable approach for assessing the role of markets in Subsaharan Africa's livestock development.

APPROACH

The performance of a marketing system has two aspects, commonly specified as technical efficiency and price efficiency [Purcell 1979]. Technical efficiency is attained when goods and services are provided at minimum average cost, that is, when the least-cost combination of marketing activities are employed. As the name suggests, technological improvements are usually the source of increased technical efficiency. Price efficiency refers to the capacity of the marketing system to adjust to changing supply and demand conditions.

The market is considered relatively price efficient if
there is a smooth flow of information along marketing
channels and participants are able to readily modify
their allocation of resources in response to price
signals.

This examination of market performance, or efficiency, in Maasailand centers on the practices of sampled
livestock traders. The analysis is approached in terms
of the structure-conduct-performance (S-C-P) model
originally developed in the study of industrial organization [Scherer 1970], in which market performance is
attributed to the conduct of buyers and sellers in such
matters as pricing and degree of cooperation or collu-
sion. Conduct in turn is related to the market's
structure, that is, such characteristics as the number,
size, and spatial distribution of buyers and sellers,
and the relative presence or absence of barriers to
entry into the trade. Recognition of the underlying
influence upon both market structure and participant
conduct of various basic conditions affecting supply
and demand, from the availability of substitute products to laws, regulations, and dominant socioeconomic
values, completes the S-C-P theoretical construct. A
detailed analysis of all of these factors for Maasailand's livestock markets is beyond the scope of this
chapter. Nevertheless the S-C-P model provides a useful framework for considering market efficiency. The
qualitative assessment of performance is subsequently
reconsidered in terms of sampled traders' actual costs
and returns.

Samples of traders in two ecologically disparate
areas of Maasailand have been chosen for the analysis.
One of the areas is at the extreme western end of Narok
District, a region of relatively high productive potential, while the other sample is drawn from the more
arid eastern side of Kajiado District (Figure 8.1).
The marketing activities within these two areas and
trade links to the urban (Nairobi) market are examined,
using primary data acquired between mid-1980 and mid-
1982, while the writer was working with the Kenya
Country Programme of the International Livestock Centre
for Africa (ILCA). The chapter's format matches the
theoretical approach. Discussion begins with descriptions of market structure and trader conduct. Costs
and returns of sampled traders are then considered, and
lastly, conclusions regarding market performance are
drawn and possibilities for increased efficiency
proposed.

MARKET STRUCTURE

The Maasai pastoral economy is only secondarily
market oriented. Livestock production is principally

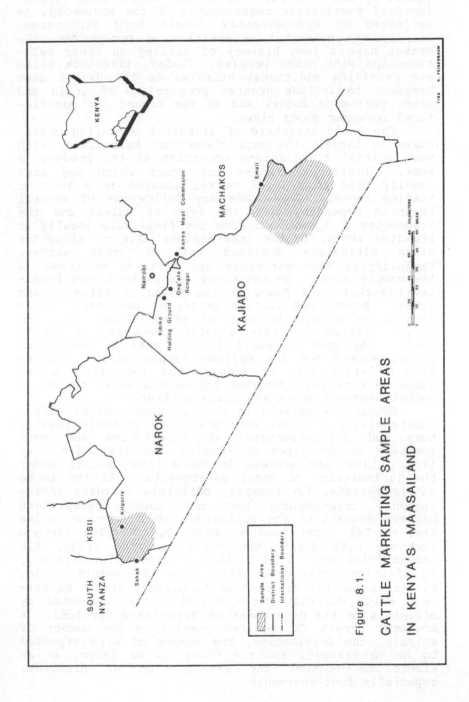

Figure 8.1.

CATTLE MARKETING SAMPLE AREAS
IN KENYA'S MAASAILAND

guided by the immediate (milk) and longer-term (reproductive) subsistence requirements of the household, as
reflected in predominantly female herd structures.
Despite the nonmarket orientation of production, the
Maasai have a long history of trading in their relationships with other peoples. Today, livestock sales
are receiving additional stimulus as the Maasai diet
broadens to include greater proportions of grain and
other purchased foods, and as the demand for manufac-
tured consumer goods rises.

The basic structure of livestock marketing mechanisms is largely the same throughout Maasailand, with
most initial transactions occurring at the producer's
home. Livestock traders buy stock which are then
usually sold at a local market located in a town or
trading center. The trade may follow one of several
channels depending upon the type of animal and the
transactor's objectives, from the final sale locally of
breeding stock, to the intermediate sale of slaughter
stock ultimately destined for an urban market.
Exemplifying this structure are patterns of trade in
the sample areas. In the Narok sample area, two local-
level markets are found at the towns of Kilgoris and
Sakek. Buyers and sellers gather in customary clear-
ings twice a week in each of these towns, with the
cattle offered for sale usually numbering 30 to 70
head. The town of Emali is the site of the first
formal market for the Kajiado sample area. Drawing
from a larger area of supply than the Kilgoris and
Sakek markets and scheduled but once a week, the Emali
market averages about 280 head on offer.

Though the markets in the two sample areas differ
substantially in size, they are alike in their regulatory and infrastructural characteristics and more
innately in the types of trading channels which they
link. Livestock markets in Maasailand operate under
the jurisdiction of local governments. At the three
sample markets, for example, officials validate interregional transactions (but not local transactions
between Maasai) by the collection of a cess, or sales
tax, of Ksh 7 per head or about $.74. (The average
exchange rate during the period of the study, 9.5
Kenyan Shillings = 1 US Dollar, is assumed throughout.)
In addition, livestock taken from the sample areas,
either to another district or to markets within Maasailand which directly serve Nairobi, are inspected by
officials of the Department of Veterinary Services. A
movement permit is required specifying the number of
animals, the destination, the number of days expected
to be in transit, and the route to be taken, in an
effort to control the transmission of diseases,
especially foot-and-mouth.

There is little direct governmental involvement in Maasailand's local-level livestock markets beyond revenue collection and disease control measures. The Livestock Marketing Division (LMD) is relatively inactive in this part of the country (in contrast to its quarantining activities in northern Kenya) and several of the holding grounds established in Maasailand by LMD have since come under the authority of local governments. Infrastructural development of the sample markets ranges from minimal to nonexistent. The Kilgoris and Sakek marketplaces are no more than open fields. The Emali market, slightly better equipped, has a corral where cattle are kept temporarily following sale and a crush used in the examination of stock by veterinary officials.

More intrinsic to the market structure are the underlying trade patterns that prevail. Sakek and Emali straddle district (tribal) boundaries, a common feature of Maasailand's markets which underscores the importance of localized, interregional trading networks. Neighboring and more densely populated non-Maasai districts are sources of breeding stock for the Maasai, as well as outlets for young Maasai-bred bulls and steers desired as draft animals. Feeder stock are traded in both directions, but with the major flow outward from Maasailand.

While a small portion of the cattle marketed for slaughter are sold to local-level butchers, this trade is mainly in sheep and goats, with cattle principally destined for the Nairobi market. During the late 1970s Nairobi depended upon cattle supplied from Maasailand for an estimated two-thirds of its beef supply [UNDP/FAO 1980]. The destinations of 2,584 head of cattle sold at Emali over a one-year period beginning September, 1981 (one-fourth of the estimated number of total transactions) are given in Table 8.1. Sixty-two percent of the cattle were sold for slaughter, with the majority channeled to the urban markets of Nairobi and Mombasa. Thirty-eight percent were sold to producers in Maasailand and neighboring Machakos District for breeding or draft purposes. Though comparable data were not collected for the Narok sample area, similar local and interregional trade patterns involving South Nyanza and Kisii Districts exist.

Cattle marketing in the sample areas, then, can be visualized as a funneling of stock, usually by traders, from households to local markets at which local-destination transactions principally involving nonslaughter stock are overshadowed by the sale of slaughter stock destined for Kenya's urban markets. It is a flexible structure, involving differentiated products (by sex and age of stock), numerous buyers and sellers, and

Table 8.1. Destinations of cattle traded at Emali, 1981/1982

Purpose	Destination	Number[a] ——Head——	Proportion of Total ——Percent——
Slaughter	Ong'ata Rongai	510	20
	Other Nairobi Markets	214	8
	Kenya Meat Commission – Athi River	242	9
	Mombasa Markets (on the coast)	220	9
	Emali	25	1
	Other	380	15
	Subtotal	1,591	62
Breeding or draft	Machakos District	612	24
	Kajiado District	381	14
	Subtotal	993	38
	Total	2,584	100

Source: Adapted from Bekure, Evangelou, and Chabari [1982], Table 2.

[a] One-fourth of estimated total transactions, September, 1981 to September, 1982.

little direct regulation beyond veterinary inspection
and the collection of a sales tax. The major regulatory influence is, in fact, indirect, namely, the
impact of administered meat prices. The rationing and
resource allocating functions of prices are distorted
when imposed controls hold prices below equilibrium
levels. The effect of price controls on market performance in Maasailand is considered following an examination of the conduct, and costs and returns of livestock
traders.

TRADER CONDUCT

Livestock traders in Maasailand are generally
young men in their twenties for whom "producer-trader"
would be a more accurate name, since they invariably
own nontraded herds and flocks. Adding to this indis-
tinction between trader and producer, transactions may
be sporadic and take place solely within a local
market's catchment area, comprise a full-time and
interregional business, or constitute any temporal-
spatial combination between these extremes. Local
governments require livestock traders to be licensed,
charging an annual fee averaging Ksh 100, but enforcement is lax. Number of transactions is also not a
meaningful criterion by which to distinguish the Maasai
trader from the Maasai producer. For example, one
household head in the Narok sample area is known to
have bought and sold over 100 head of cattle during
1981 for a cumulative gross margin (selling price minus
buying price) of over Ksh 10,400, and yet did not
consider himself a trader.
The lack of uniformity among traders' operations
is partly attributable to the ease of entry into and
exit from the trade. There are no barriers to becoming
a livestock trader, as long as one has the time and
money (though it is unlikely that a successful career
could be established by a non-Maasai). For some
traders, once a local trading network is set up there
is little incentive to alter it, given the advantages
of familiarity and routine. Others, less restricted by
homeherd responsibilities and with greater financial
resources, chart wider-ranging trading paths. Thus,
one finds overlapping trading spheres, with traders
drawing from pools of producers and frequenting markets
in a variety of patterns.
Established trading spheres do not signify or
imply sole rights to a purchasing area. On the contrary, traders emphatically acknowledge that a trader
may buy livestock anywhere and from whomever has animals for sale, producer or fellow trader. At the Emali
market, of the 2,981 head of cattle for which it was
recorded whether the seller was a trader or producer

(about one-fifth of the 13,500 head estimated offered for sale during the survey year), 96 percent were brought to market by traders, and 41 percent of these had not been purchased directly from producers, but rather from other traders [Bekure, Evangelou, and Chabari 1982]. For the Kilgoris and Sakek markets, a similar high percentage of market participants are traders, and pre-market transactions between traders are equally prevalent.

When asked why they trade livestock, most traders describe consumption demands, with food purchases most commonly mentioned. Second among reasons for trading is the opportunity, using the profits, to increase the size of the home herd. As noted by Doherty [1979], trading can be a means of quickly acquiring a personal herd for a young man. But the attraction of trading is more than economic, as most traders view it as also simply a way of making life more enjoyable than it would be otherwise, "staying at home looking after cattle". Contributing to this positive aura, livestock trading is one of the few forms of regular employment which does not compromise the Maasai life-style. One's cultural identity is fully retained, and for young Maasai men even enhanced.

Traders usually operate individually in the purchase of stock. Once livestock have been bought, however, there is a high degree of coordination in the actual movement of animals to markets. For example, in the Kajiado sample area, traders will often group purchased cattle at a customary site and have them trekked by herdsboys to Emali, with arrival timed for the day of the market. This coordination lowers operating costs for the traders as well as reduces the risks associated with trekking. Three-fourths of a sample of traders (n=41) interviewed at Emali employed herdsboys, though a few indicated that they did so only occasionally. Similar arrangements characterize the movement of stock within the Narok sample area to a lesser extent, while the trekking of stock from the area to the Nairobi market is invariably coordinated by groups of traders.

Price discovery, whether at a producer's home, at a market, or in transit, is by one-to-one negotiation. There are no auctions at markets, but rather numerous individual transactions taking place simultaneously on a willing buyer - willing seller basis. An animal may be sold more than once on the same market day, adding complexity to the trade. Given the individualized nature of the transactions, not to mention the absence of liveweight sales, traders' profits are earned by a keen eye and bargaining acumen.

Credit is used extensively in the trade, especially between traders but also between trader and producer. Frequently transactions involve payments which are either partially or wholly deferred until a later date. The fact that credit is so readily extended among traders reflects the personal friendships which permeate the trade, and suggests default is not common (although extensions beyond the agreed-upon period for repayment may well be). Widespread use of credit allows on average more transactions to occur than would otherwise take place.

Market articulation, then, depends upon Maasai traders, whose trading activities are competitively independent and yet incorporate coordinating cost and risk reducing arrangements and the informal but frequent use of credit. There are no barriers to entry into the trade, as is signified by the blurred distinction between producer and trader, and scope of operation is limited only by one's efforts, home-herd responsibilities, financial capability, and trading skills.

COSTS AND RETURNS

The Kajiado Sample Area

An equivocal picture of market performance is inferred from this overview of market structure and trader conduct. Conditions both favorable and unfavorable to market efficiency exist, as summarized in Table 8.2. Additional insight regarding performance levels can be gained by the analysis of prices which characterize the trade. For the Kajiado sample area, mean costs and returns for slaughter cattle purchased from producers and marketed at Emali and Ong'ata Rongai were calculated by Bekure, Evangelou, and Chabari, for a sample of 152 weighed Small East African Zebu (Table 8.3). This breed comprised over 99 percent of the 3,051 sales recorded at Emali for which breed was clearly identified. The sampled animals were purchased from producers at a mean price equivalent to $106.50 per head, and sold at Emali for $147.00 which resulted in a return to the trader after marketing expenses (net income) of $33, or 30 percent of the original purchase price. The mean price at Ong'ata Rongai, $202, allowed a return after expenses again of about 30 percent on the Emali purchase price. The percentage added value between purchase from producer and final sale for slaughter averaged nearly 90 percent. Prices and costs per kg in Table 8.3 are based on a 260 kg animal throughout. Since some weight loss probably occurs during movement from point of initial purchase to point of final sale, the indicated prices per kg are only approximations.

132

Table 8.2. Inferred cattle marketing performance in Maasailand, early 1980s

Criterion	Existing Condition	Impact on Performance
Market Structure		
Numbers of buyers and sellers	Many	Favorable
Entry and exit	No barriers	Favorable
Flow of information	Mixed	Mixed
Regulatory controls		
Direct	Jurisdictional, disease control	Favorable
Indirect	Meat price controls	Unfavorable
Trader Conduct		
Price discovery	One-to-one negotiation	Unfavorable
Provision of credit[a]	Extensive	Favorable
Coordination of stock movement	Frequent	Favorable

Source: Unpublished research.

[a]Informal credit arrangements among traders and between traders and primary producers.

Average returns of 30 percent to the trader's capital, management and personal labor over a mean period between purchase and sale of a week or less seem unexpectedly high, given the competitive nature of the trade. The trader may be able to exert a degree of bargaining power but probably not overly so, since the producer has alternative marketing possibilities, from bargaining with competing traders to himself taking, or having a member of his household take, the animal to market. The producer's trading options, however, may be limited and his bargaining position weakened if he has urgent cash demands. In some instances the producer may attempt to time the sale in order to take advantage of seasonal fluctuations in demand, but in general expected price is not the major determinant of the decision to sell an animal.

A partial explanation of the large trading margins found at Emali can perhaps be found in the frequency of nonsales. Bekure, Evangelou, and Chabari estimate that one out of five head of cattle supplied to the Emali market is not sold. Table 8.4, which shows the responses of a sample of Emali traders when asked about the frequency of their nonsales, supports this estimation. Post-nonsale options may result in wider margins than would otherwise occur. The trader can return home with the stock (or arrange for someone else to care for them) and then offer them for sale again the following week, or simply carry them on to Ong'ata Rohngai, bypassing the Emali transaction.

Table 8.3. Mean costs and returns to cattle traders at Emali and Ong'ata Rongai, 1981/1982

Item	Cost or Return	
	per head	per kg liveweight
	------U.S. Dollars------	
Purchase price paid producer	106.50	.41
Marketing costs to Emali		
Direct		
Trekking fee	2.11	
Watering fee	.21	
Loss - Trading	1.05	
Death (1/60)	2.45	
Indirect		
Food and Lodging	1.26	
Personal transport	.42	
Total	7.50	.03
Sale price at Emali	147.00	.57
Trader's net income at Emali	33.00	.13
Purchase price paid at Emali	147.00	.57
Marketing costs to Ong'ata Rongai		
Direct		
Cess	.74	
Trekking fee	2.11	
Watering fee	.21	
Loss - Trading	1.47	
Death (1/60)	3.37	
Indirect		
Food and lodging	2.11	
Personal transport	1.26	
Miscellaneous	1.26	
Total	12.53	.05
Sale price at Ong'ata Rongai	202.00	.78
Trader's net income at Ong'ata Rongai	42.47	.16

Source: Adapted from Bekure, Evangelou, and Chabari [1982], Table 7.

134

The Narok Sample Area

Data on the transactions of three traders operating in the Narok sample area permit additional insight into traders' costs and returns. The sampled traders, designated A, B, and C, follow distinct trading patterns. Trader A engages almost exclusively in the trade to Nairobi, while B and especially C also rely upon sales at the Sakek market, within the Narok sample area. Table 8.5, which summarizes their household and home-herd circumstances, shows that Trader A comes from a larger and more wealthy household than do Traders B and C and attained a higher level of formal education, factors which help to explain Trader A's more extensive involvement in the Nairobi trade. As in the Kajiado sample area, their transactions principally involve adult steers, with adult females the next major age-sex category traded. The mean buying and selling prices, gross margins, and percentage gross margins for the three traders are found in the first three columns of Table 8.6. As expected, greater returns are acquired at the Kibiko Holding Ground than from sales within the Narok sample area.

Table 8.4. Survey results on nonsales of cattle at Emali, 1981

Numbers of Traders (n=40)	Stated Frequency of Nonsales
14	Once a month
11	Twice a month
6	Not frequently
3	One-third of the time
2	Once or twice a month
2	Less than half of the time
1	More than half of the time
1	Frequently in times of drought

Source: Unpublished research.

Table 8.5. Background information on Traders A, B, and C, Narok sample area, 1981/1982

Item	Trader		
	A	B	C
Age	27	23	26
Formal education	4 Years Secondary	None	Primary
Dependents			
Adults	4	5	2
Children	8	4	5
Home herd			
Cattle			
Female			
Adult	156	60	48
Immatures			
Calves			
Male	⎱260	⎱40	
Adults			⎱16
Immatures			⎱15
Calves			
Steers	114	15	16
Total	530[a]	115	151
Sheep	30[a]	5	8
Goats	40[a]	6	15

Source: Unpublished research.

[a]Owned jointly by Trader A and his father.

Table 8.6. Prices, gross margins, and percentage gross margins for Traders, A, B, and C, and one small-scale Kajiado sample area trader, 1981/1982

Place of Sale, Item	Narok Sample Area Traders			Kajiado Sample Area Trader
	A	B	C	
	----------U.S. Dollars----------			
Kibiko Holding Ground				
Number of traded cattle included in survey	32	22	56	-
(1) Mean buying price	140.50	127.10	126.10	-
(2) Mean selling price	165.90	153.70	142.60	-
(3) Gross margin (2 - 1)	25.40	26.60	16.50	-
Standard deviation of gross margin	10.90	19.60	13.30	-
Sakek				
Number of traded cattle included in survey	-	8	59	-
(1) Mean buying price	-	67.70	70.80	-
(2) Mean selling price	-	77.00	77.00	-
(3) Gross margin (2 - 1)	-	9.30	6.20	-
Standard deviation of gross margin	-	3.20	5.00	-
Emali				
Number of traded cattle included in survey	-	-	-	56
(1) Mean buying price	-	-	-	83.30
(2) Mean selling price	-	-	-	93.10
(3) Gross margin (2 - 1)	-	-	-	9.80
Standard deviation of gross margin	-	-	-	6.90
	----------Percentage----------			
Kibiko Holding Ground				
Percentage gross margin[a]	18.1	21.0	13.1	-
Standard deviation of percentage gross margin	8.5	16.7	14.7	-
Sakek				
Percentage gross margin[a]	-	13.7	8.6	-
Standard deviation of percentage gross margin	-	10.9	7.1	-
Emali				
Percentage gross margin[a]	-	-	-	11.8
Standard deviation of percentage gross margin	-	-	-	8.6

Source: Unpublished research.

[a]Percentage gross margin = $\frac{(3)}{(1)}$ X 100.

Comparing Tables 8.3 and 8.6, the mean sale price per head at Ong'ata Rongai is notably higher than that at the Kibiko Holding Ground. This discrepancy may be largely due to differences in the sizes of cattle slaughtered. An average carcass weight for cattle slaughtered in the Narok sample area was found to be 85 kg (n=69), whereas in the Kajiado sample area it was 114 kg (n=145). Differences between traders' returns in the two sample areas are even more unexpected. Whereas traders' margins after expenses approach 30 percent between Emali and Ong'ata Rongai, returns before expenses between the Narok sample area and the Kibiko Holding Ground for Traders A, B, and C were 18.1, 21.0, and 13.1 percent, respectively. While recognizing the limited confidence, statistically, which can be attached to findings based upon such a small sample, the transactions of the three traders are considered representative.

The difference between the two sample areas in average returns to traders is especially surprising given the relative distances livestock are moved. The trek from Emali to Ong'ata Rongai and other Nairobi markets takes about three days whereas from the Narok sample area to the Kibiko Holding Ground requires eight to nine days. In the latter case, cattle bought over a period of several weeks are accumulated until their number justify the trek, and a trader must have the financial capability to postpone the sale of purchased cattle for up to several weeks. Hence, the financial requirements and physical risks for Narok traders engaged in the Nairobi trade are greater than for their Kajiado counterparts.

Upon arrival at the Kibiko Holding Ground, the Narok traders' post-nonsale options are limited, which may partly explain the lower gross margins. They can pay to water and graze their stock at the holding ground for a nominal fee (Table 8.7), and hope for a more favorable seller's market in the immediate future. But, the uncertainty of future prices and the cost in trading time foregone can make this an unattractive alternative, as suggested by the fact that sales at less than purchase price do occur--2 out of 22 recorded sales for Trader B and 4 out of 56 for Trader C.

Clearly, profits can be sizable for traders in both sample areas but especially those in Kajiado. However, the frequency of nonsales and possible post-nonsale options need to be included in a realistic evaluation of traders' margins and market performance. Net incomes as represented in Table 8.3 may be on the high side, with the distribution of trader margins skewed by occasional, exceptionally large profits. Foremost, it should be kept in mind that traders' costs and returns, as is true for all aspects of their opera-

138

Table 8.7. Grazing fees at the Kibiko Holding
Ground, 1981

Period	Fee	
	Herded by Owner	Herded by LMD
	----U.S. Dollars per head----	
Day	.04	.05
Week	.32	.39
Month	1.26	1.58

Source: Unpublished research.

tions, vary widely. The last column of Table 8.6 shows
the mean buying and selling prices, gross margin, and
percentage gross margin for a trader in the Kajiado
sample area who sells two or three cattle per week at
Emali. Clearly his margins do not approach those
indicated in Table 8.3.

MARKETING EFFICIENCY AND PRODUCTION

Shortness of space precludes examination of a
number of additional important components of Maasailand's livestock marketing systems, from the influence
of nonmarket transactions, to the sale of steers by
commercially oriented private ranchers, to the interregional trade in sheep and goats and the operations of
local-level butchers. Still, the observations presented are sufficient for drawing general conclusions
regarding the relative efficiency of the cattle trade
in Maasailand. Returning to the two aspects of market
performance defined in the beginning, the question
remains whether technical and price inefficiencies are
constraining livestock development in Maasailand. The
evidence presented suggests not. In terms of the S-C-P
model, a generally efficient level of performance is
inferred from the market's structure and traders'
conduct. Though mean returns per sale for traders,
especially those operating in the Kajiado sample area,
appear excessive, trading risks and post-nonsale
options help to explain the large trading margins.
Improvements in market performance can be achieved, but only if there are concurrent changes in the
circumstances surrounding the supply and demand for

livestock, in particular national pricing policies and
Maasai producers' objectives. For example, the use of
vehicles for transporting cattle might be proposed as a
means of improving market flow. At the time of the
study, there was in fact one individual who was
trucking cattle, albeit irregularly, from the Narok
sample area to the Nairobi markets. Renting a lorry,
he would transport 14 to 16 head at a time from Emarti
(one day's trek from the Narok sample area), at a cost
of about $180. He would sell standard and commercial
graded cattle to the Kenya Meat Commission which pays
on a dressed weight basis, while higher-graded animals
were sold to private wholesalers and butchers willing
to pay more for the better beef. At times this individual even used vehicles to transport feeder cattle to
private Kajiado ranches located near Nairobi.

This single transport operation, however, can be
expected to develop into a widespread marketing alternative only if prevailing production conditions change.
As long as meat price controls prevent cattle/transport
price ratios from rising to levels whereby the regular
use of vehicles would improve service and reduce
costs--improve the technical efficiency of the market--
the commitment of vehicles to the full-time transport
of cattle is unlikely to occur. At the producer end,
with commercial objectives secondary at best and sale
levels low and erratic, the regular supply of stock
necessary for a major capital investment in transport
vehicles remains uncertain. Economic comparisons of
trekking and trucking have been documented for other
areas of Subsaharan Africa. As reported by Simpson and
Farris [1982], for example, Josserand and Sullivan's
[1979] study of traders' margins in the 765-800 km
movement of cattle from Upper Volta to Togo revealed a
return of 12 percent on total investment for trekking
and 8 percent for trucking. In Maasailand, the
emergence of trucking as the economically favored means
of transport will depend upon changes in national pricing policies and producers' production objectives.

As another example of the impact upon market
efficiency of production conditions one might cite the
general lack of information on market prices which
characterizes the trade. It could be argued that
individual negotiation of prices is inefficient,
encouraging speculation and not yielding a level of
market stability which could be attained by public
auction, which presumably would reduce the costs of
price discovery and streamline the flow of information.
Yet the assumption that the additional information
provided by publicized prices would result in more
efficient resource allocation by producers--increase
the price efficiency of the market--is unfounded, given
producers' prevailing nonmarket objectives. Market

intelligence would be improved by the public auction of animals, but whether there would be an increase in production for the market would depend upon producers' responsiveness to this additional information. Unused weighbridges installed by the LMD at various markets in Maasailand exemplify the futility of providing the conditions for improved market performance if market participants do not perceive the benefits that could be thereby derived.

SUMMARY

Local-level trading mechanisms in Maasailand are not an overriding constraint to the region's livestock development. Apparent sources of inefficiency, from nonpublic price negotiations at poorly equipped markets to the long-distance trekking of stock, do not create undue costs or constrict product flow, given present production conditions. Market efficiency will improve, but only in combination with (i) producers becoming increasingly market oriented and thereby providing a reliable supply, and (ii) national pricing policies accurately reflecting demand, the two conditions needed to induce positive change. In the balance between production and marketing advances, the relatively efficient performance of the market in Maasailand implies that the developmental lead needs to be taken on the production side.

In other parts of Kenya, as well as elsewhere in Subsaharan Africa, livestock are marketed under conditions significantly different from those described here. In Kenya's northern rangelands, for example, disease control measures more stringently restrict stock movements and traders generally operate on a much larger scale. Importantly, the demonstrated approach for examining marketing constraints has general applicability. Regardless of the particular circumstances, market performance can be assessed by considering market structure, trader conduct, and the costs and returns involved in the trade. In the case of Maasailand, improved market performance will depend upon more favorable national pricing policies and increasingly market oriented production. Increased marketing efficiency, in turn, can then be expected to provide additional incentive for expanded production.

REFERENCES

Bekure, S., P. Evangelou, and F. Chabari. "Livestock Marketing in Eastern Kajiado, Kenya." International Livestock Centre for Africa, Kenya, Working Document 26 (Draft), Nairobi, 1982.

Doherty, D. "Factors Inhibiting Economic Development on Rotian Olmakongo Group Ranch." University of Nairobi, Institute for Development Studies, Working Paper No. 356, 1979.

Josserand, H., and G. Sullivan. Livestock and Meat Marketing in West Africa. Vol. 2. Ann Arbor: University of Michigan, Center for Research on Economic Development, 1979.

Kenya, Republic of. National Livestock Development Policy. Nairobi: The Government Printer, 1980.

Purcell, W. Agricultural Marketing: Systems, Coordination, Cash and Futures Prices. Reston, Virginia: Reston Publishing Company, 1979.

Scherer, F. Industrial Market Structure and Economic Performance. Chicago: Rand McNally College Publishing Company, 1970.

Simpson, J., and D. Farris. The World's Beef Business. Ames: The Iowa State University Press, 1982.

UNDP/FAO. "Meat Wholesale Marketing in Nairobi." Ministry of Agriculture, Marketing Development Project, Phase II (Ken 78-006), Nairobi, 1980.

9
Impact of Government Policies on the Performance of the Livestock-Meat Subsector

Gregory M. Sullivan

The livestock-meat subsectors in many subsaharan countries are operating at low levels of technology, productivity and commercialization. Reasons for this phenomenon involve political, social, physical and economic factors. This chapter focuses on the economic factors as they relate to government pricing policies.

THE PROBLEM

Pricing policies are a mechanism by which governments in less developed countries (LDC's) try to achieve specific economic and social objectives. Sectors of the economy with the greatest potential are strategically targeted for increased economic growth and, in some cases, policies are designed to stabilize incomes to attain greater equity. But, as Krishna [1967] found, pricing policies in the developed countries which successfully stabilize prices and incomes often have the opposite effect in LDC's. For example, pricing policies for livestock-meat can have distributional benefits which accrue to the urban and growing industrialized sectors, but work to the disadvantage of the agricultural and livestock subsectors. These distributional effects are solidified because of the political environment in which the pricing policies are made.

In general, pricing policies in LDC's have caused large annual losses to livestock producers due to a misallocation of resources [Bale and Lutz 1981; Peterson 1979]. They cause production to be undervalued which compounds an inherent problem of maintaining cattle longer than required on a common property resource [Sullivan 1980]. Furthermore, producers who could invest to commercialize livestock production have less incentive to do so.

143

Another factor is that, in most LDC's, a high proportion of disposable income is largely spent on food [Pinstrup-Andersen and Caicedo 1978; Mellor 1976]. Rising food prices are a harsh reality for a large portion of consumers and especially for low income segments of the population. However, a government policy of price subsidization targeted to low income consumers is often an inefficient mechanism for alleviating malnutriton [Reutlinger and Selowsky 1976]. Food price policies can also have negative distributional benefits that favor middle and upper-income urban groups at the expense of lower-income rural poor [Brown 1978].

These policies can hinder the development of the agricultural sector resulting in agricultural output being significantly smaller than it would be in the absence of price distortions [Peterson 1979]. It is well documented that distortions in economic incentives have been some of the greatest causes for malfunctions in market systems [Schultz 1978; Newberry and Stiglitz 1981]. Governments' efforts to restructure the market result in high economic, social and political costs. Institutions such as marketing boards have been a common instrument for restructuring the market system which have generally resulted in the misallocation of resources [Abbot 1962].

Market failure is not commodity specific, but relates to the political economy in which government policies are designed and carried out. The overriding premise is that policies which stabilize prices and incomes eventually lead to negative impacts on producers, consumers and taxpayers, and ultimately lead to a spiralling upward of higher marketing costs, and even greater government involvement in the market system.

CONSTRAINTS

Several constraints impede livestock development in Subsaharan Africa. The political environment in many of these countries is such that price distortions in agriculture persist because the political power base lies with the urban population and the industrial sector. This phenomenon is very similar to changes in the political structure in many developed countries where shifts from a rural to urban poltical power base have occurred.

Government attempts to establish effective pricing policies have not been successful because they do not account for the environment in which producers make their marketing decisions. For example, livestock are held for various reasons by producers, such as a convenient repository of wealth [Doran, Low, and Kemp

1979]. It is important to note that livestock have a perceived value because of their utilization of by-products in the village, i.e., a traditional value. This value is greater in many respects than the commercial market value, one that in most cases is an administered price by the government [Sullivan et al. 1978]. Government price policies then become an ineffective mechanism for stimulating commercial supplies. This problem is further complicated because communal grazing of rangelands allows perceived costs of production to seem minimal to individual producers.

Another evident constraint is a bias in favor of the public sector for fostering development of the livestock-meat subsector. The private sector, especially middlemen, are viewed with suspicion in terms of their contribution to the functioning of the market. Consequently, governments' desire to control part or all of the marketing system is pervasive in many of the subsaharan countries. The establishment of institutions, either totally or quasi-public, to control segments of the livestock-meat marketing system can be detrimental to improved productivity. These government agencies, which take ownership of agricultural products, isolate producers and consumers from the market clearing force.

The performance levels for these institutions in many cases do not enhance the efficiency of the market system. A government institution, especially one such as a livestock marketing company, does not have an incentive structure that can keep operating costs to a minimum. Liebenstein [1978] has referred to this phenomenon as "x-efficiency". High operating costs, because of inefficiencies, result in lower prices being paid to producers by the institution purchasing the commodity.

SHORT-RUN AND LONG-RUN EFFECTS OF PRICE CONTROLS

A graphical display of the impact of government price controls illustrates the short-run and long-run effects on the development of the livestock-meat subsector (Figure 9.1). The industry retail supply curve is S_O and the industry retail demand curve is D_O for a constant-cost competitive industry (Panel a). The market clears at an equilibrium price of P_O with no government price controls, where P_O equals the average total cost (ATC) for the industry. If government sets a price ceiling of P_C below equilibrium price, producers supply n_Oq_1 while consumers demand Q_1 (n_O is number of producers in market). A shortage is created of $Q_1-n_Oq_1$ in the short-run.

In the long-run shortages increase for several reasons. First, firms unable to produce because $P_C <$

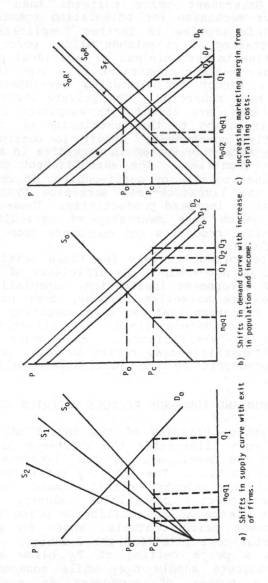

Figure 9.1. Short-run and long-run effects of price ceiling set below market equilibrium price

ATC exit from the industry. The supply curve shifts to the left (S_1 and S_2) over time. If no change occurs in industry demand, deficits increase. A second scenario (Panel b) is one in which the supply curve remains unchanged and the industry demand shifts (D_1 and D_2) because of increase in population and/or income. Deficits also increase over time. A third scenario which is common to many marketing systems in LDC's is high rates of inflation causing the marketing margins to increase (Panel c) with no changes in the structural parameters of the primary demand (D_R) and supply curves (S_f). The derived demand at the farm level (D_f) and the derived supply at retail (S_{oR}) fall and the margin increases. Because the retail price is held constant, the same level of absolute quantity is demanded, Q_1, causing price at the farm level to fall even lower. The quantity supplied by producers declines and deficits increase ($Q_1 - n_oq_2$). The middlemen pass increases in marketing margins to producers in the form of lower prices paid when retail prices are controlled by government. The full effect of an increase in marketing costs is felt by producers who, over time, receive lower prices and consequently supply less. In all three scenarios described, the net effect is greater meat shortages in the government controlled markets.

Another consideration is that costs for middlemen increase because of threats of fine or imprisonment for not adhering to government control when deficits become very large. These additional marketing costs would be passed on to consumers in the form of higher prices or to producers in lower prices received. As shortages of meat continue, and a "black market" price occurs, a general market failure in the livestock-meat subsector develops. A dual system with a public and a private market for retailing meat then emerges with consumers uncertain about available supplies and expected prices. The government's ability to maintain stability in the markets with subsidies consequently fails because foreign exchange is usually limited for importing the meat necessary to satisfy increasing deficits.

TWO CASE STUDIES: THE LIVESTOCK-MEAT INDUSTRY

The red meat subsector is important in Subsaharan Africa because of its supply of protein to household consumption as well as widespread demand for the product. Ghana and Tanzania are two countries for which recent assessments of their red meat subsectors have been conducted [Sullivan 1979; Texas A&M University 1976]. Information collected for each country permits a synthesis of similarities in government policies as regards the livestock-red meat subsector because both countries imposed price controls on red meat during a common time period.

Production

Production of livestock in each of the two countries is comprised of many producers who maintain small herds grazing on communal pasture. Ghana's national herd was smaller than Tanzania's at the time of the studies. Ghana was also a net importer, in contrast to Tanzania which was a net exporter (Table 9.1). Sheep and goats in both countries have a higher market offtake than cattle.

Productivity and commercialization are at low levels largely because efforts to institute policies to correct the effect of communal grazing on low productivity have been minimal. Government pricing policies have also exacerbated the overgrazing problem in the two countries by keeping output price below free market levels [Sullivan, Farris, and Stokes 1981]. Government attempts to increase commercial supplies of red meat by livestock production projects have had an insignificant impact. The offtake rates in both countries for cattle, sheep and goats were lower in 1979 than in the base period of 1969-71 (Table 9.1).

Distribution of Live Animals

The distribution system for live animals varies in each country because of geographic and infrastructural constraints. The primary modes of transport used are trekking, truck or rail. In each country, livestock in transit have high rates of mortality and shrinkage. Sanderson and Stokes [1976] estimated 18 percent shrink and 7 percent mortality rates in Tanzania. In both countries, the governments use marketing companies to purchase and transport a portion of the livestock marketed. The high costs of these operations are passed on to producers in the form of lower prices paid.

Slaughter and Processing Industries

These two stages in the marketing system in each country varied during the period studied because of the end uses of the by-products. In the rural areas in each country, abattoirs were maintained by urban councils, but facilities were generally rudimentary and quality of health inspection varied from location to location. Slaughter plants in both Ghana and Tanzania were for hot carcass trade to urban consumers. In Tanzania, the largest slaughter and processing plant was in the capital of Dar es Salaam. Tanganyika Packer Ltd.(TPL) supplied meat to the city as well as canned beef for

Table 9.1. Livestock and meat statistics for Ghana and Tanzania, 1969-71 and 1979

Country	Year	Product	Inventory --hd--	Production --mt--	Offtake[a] --%--
Ghana	1969-71	Cattle	902,000	19,000	19
		Sheep	1,324,000	4,000	28
		Goats	1,408,000	4,000	32
	1979	Cattle	930,000	13,000	12
		Sheep	1,198,657	5,000	25
		Goats	927,439	6,000	30
Tanzania	1969-71	Cattle	11,753,000	111,000	10
		Sheep	2,823,000	10,000	23
		Goats	4,441,000	18,000	27
	1979	Cattle	12,388,000	136,000	9
		Sheep	3,716,000	12,000	26
		Goats	5,574,000	19,000	27

Source: United Nations, Food and Agriculture Organization. Production Yearbook and Trade Yearbook. Rome, 1969, 1970, 1971 and 1979.

a Commercial offtake slaughtered.

local and export markets. Government institutions in
each country controlled the flow of inputs, processing,
and sale of by-products from the processing plants. In
both countries the canning facilities were grossly
under utilized.

Wholesale and Retail Institutions

Private individuals and government agencies participated in varying degrees in the wholesaling and
retailing activities in each country. In Ghana, the
Meat Marketing Board (MMB) was the sole agency responsible for importation of live animals and beef carcasses. The distribution of these commodities was regulat-
ed as well as the prices charged for them. Marketing
of domestic livestock was carried out by the private
sector. Government operated retail shops and was the
sole retailer of imported red meat.

In Tanzania, the Government created the Livestock
Development Authority (LIDA) which had many responsibilities. Two companies operating within LIDA were the
Tanzania Livestock Marketing Company (TLMC) and Tanganyika Packers, Ltd. (TPL). The District Development
Councils (DDC), a separate government entity from LIDA,
had the responsibility for fixing prices, supervising
the operation of retail shops, and selling retail
licenses in urban areas.

A similarity in both countries was the expansive
role that the public sector maintained over the operation of the marketing systems. The political beliefs
were that government intervention was necessary to
smooth out perceived dislocation in supplies and
instability in market prices. The private sector
activities were pushed to the point of having to
operate as an "illegal market" if supplies were to be
made available.

In the cases of Ghana and Tanzania, initiatives
by the private sector were dampened by government
intervention. Priorities were placed on maximizing
consumer welfare with disregard to producers and
middlemen in the marketing system. In both Ghana and
Tanzania, the Government institutions continuously
expanded their authority over greater segments of the
industry by assimilating day-to-day marketing functions. A common goal of the government in both countries was controlling the livestock-red meat industry.

IMPACT OF PRICING POLICIES FOR RED MEAT

Government's role in the livestock-red meat subsector of each country varied in degree. Each country
undervalued its domestic production to supply inexpensive animal protein to urban consumers by instituting

price ceilings on meat as graphically shown in the
panels of Figure 9.1. Retail meat prices were controlled and based on a formula pricing mechanism. The
relationship between border and domestic prices at
different levels of the marketing system are illustrated in Table 9.2   to show the distortion between
market prices and each government's controlled prices.

An overvalued exchange rate in Ghana distorted the
market price for imported beef. The border price for
bone-in beef was N¢1.26 per kg C.I.F. The border price,
using a shadow exchange rate price should have been
N¢5.49 per kg or even higher. The Government's retail
price was N¢3.35 per kg while the actual market price
for bone-in beef from domestic supply was approximately
double at N¢7.00 per kg or even higher at the black
market rate. Beef was undervalued compared to other
scarce goods available. The shortage of red meat in
the government shops was evidence of the distortion of
price as a rationing mechanism. When the government
shops had meat, it was usually sold out early in the
morning.

In Ghana, scarce foreign exchange was used to
purchase imported beef to satisfy urban consumer
demand. With price disincentives for increased domestic production, imports were required to satisfy meat
shortages. Ghana's imports of beef and veal increased
from an average of 106 mt in 1967, to 1,414 mt in 1973,
up to 7,800 mt in 1977 (Table 9.3). The overvalued
exchange rate made the domestic price for imported meat
about one-half of the actual black market price for
beef. The deficits increased at a rapid rate as illustrated in Figure 9.1.

In Tanzania the live animal price for cattle paid
by the government marketing board was a flat price
based on a weight range with no differential for quality, season, or location of purchase [Sullivan and
Farris 1982]. Government also fixed the price at retail for beef. The government retail prices were not
competitive being about 65 percent of the actual retail
market price (Table 9.2). The canned beef F.O.B. price
at Dar es Salaam was only slightly higher than the
market retail price. This disequilibrium caused distortion in the distribution of meat. Large subsidies
were paid to the packing plant because the plant was
unable to purchase enough quality cattle from domestic
sources to keep the plant at full capacity because of
competition from the private sector. Exports of canned
beef were erratic during the period of 1973-1979 (Table
9.3) partly due to world demand for canned meat, but
also because Tanzania's production costs were not competitive with other major exporting countries of canned
meat.

Table 9.2. Meat price statistics for Ghana and Tanzania, 1979

Country	Meat product	Border price at govt. exchange rate	Border price at shadow exchange rate	Govt. retail price	Market retail price
		Currency/kg			
Ghana	Beef	N¢1.26	N¢5.49	N¢3.35	N¢7.00
	Lamb and mutton	--	--	N¢4.32	N¢20.00
	Goat	--	--	N¢4.32	N¢20.00
Tanzania	Beef	--	--	Tsh 5.90	Tsh 9.32
	Lamb and mutton	--	--	--	Tsh 9.00
	Goat	--	--	--	Tsh 9.00
	Canned beef	Tsh 10.10	--	--	--

Source: Sullivan, G.M. "Livestock and Meat Marketing in West Africa: Ghana," 1979 and Texas A & M University. Tanzania Livestock-Meat Subsector, 1979.

Note: Exchange rates are: Ghana - (new cedi) N¢1.15 = U.S. $1.00
Tanzania - (shilling) Tsh 8.28 = U.S. $1.00
Prices for Ghana are for carcasses C.I.F. major port of entry. Prices are for bone-in meat.

Table 9.3. Imports of livestock and meat by the meat marketing board in Ghana and exports of canned beef by Tanganyika Packers Ltd. in Tanzania

| Year | Ghana | | | Tanzania |
	Imports of cattle	Imports of sheep and goat	Imports of meat	Canned meat exports
	- - - - - - hd - - - - - - - -		- - mt - -	- - - - mt - - - -
1973	34,125	24,017	1,414	1,520
1974	48,582	26,113	2,995	2,448
1975	9,218	18,949	1,475	782
1976	--	25,662	4,300	1,191
1977	--	16,443	7,800	1,132
1978	--	25,000	7,300	401
1979	--	25,000	7,300	380

Source: G. M. Sullivan. "Livestock and Meat Marketing in West Africa: Ghana" 1979 and FAO Trade Yearbook, 1970-1980.

Note: Small number of cattle imported in 1976 thru 1979 for breeding purposes.

Table 9.4. Marketing performances of the livestock-red meat subsectors in Ghana
and Tanzania

Ghana	Tanzania
1. Government set artificially low producer and retail prices.	1. Distortions in regional producer and retail prices.
2. Black market sales of meat.	2. High marketing costs for government purchases of livestock.
3. Poor allocative efficiency resulting from operation of Meat Marketing Board.	3. High amounts of shrink and mortality in transportation of livestock.
4. Distribution of benefits accrued to high income consumers.	4. High cost of production of canned meat for export-not competitive in world market.
5. Market failure with rationing of meat.	5. Government prices not differentiated for quality of cattle, spatial location and season of year.
6. Scarce foreign exchange used to purchase imported meat.	6. Livestock exported illegally.
7. Large subsidies paid to idle processing plants because of ban on livestock imports from Sahel.	7. Meat shortages in urban areas.
8. Imported beef used as raw input for canning plant.	8. Large subsidies paid to processing and canning facilities.
9. High marketing cost in importing, processing, distributing and retailing red meat.	

Source: Author.

PERFORMANCE OF THE LIVESTOCK-MEAT SUBSECTORS

The overall performance in terms of productivity, allocative efficiency, equity and progressiveness has been poor for the livestock-meat subsector in each country (Table 9.4). The impact of government price policies has created distortions in supply and demand conditions for red meat. Producers have received lower prices than if the market was in equilibrium.

Ghana

A general market failure existed in the livestock-red meat subsector in Ghana by 1979. Government encroachment into the private sector, in terms of prices and quotas for livestock supplies to butchers, led to recurring meat shortages. Government purchases of imports increased significantly during the study period. The benefits from the imports, which were undervalued because of the distorted exchange rate, accrued primarily to middle and high income consumers. The Meat Marketing Board in Ghana eliminated private butchers from the sale of imported red meat and controlled distribution of the imports by selling mainly in the larger urban areas.

The Government paid large subsidies to its marketing company for importing and processing meat. This policy continued even when the slaughter and canning facility was not operating because of unavailable supplies of live animals from domestic herds and neighboring Sahelian countries due to import restrictions. The Government imported boneless and carcass beef from South America and Western Europe to supply the canning plant. This strategy by the Meat Marketing Board was not economically feasible because of the inaccessible location of the plant in Bolgatanga near the northern border with Upper Volta. These deliveries were finally curtailed.

Tanzania

The Tanzanian Government, facing market failure in the supply of meat, restructured its meat price policies in 1979. Restrictions on the private trade had forced up local prices, resulting in greater black-market activities and shortages of meat. The supply of lower quality cattle for the TPL plant, as well as high marketing and operating costs, resulted in canned beef products not being competitive at current world market prices. The industry experienced increasing inventories of canned beef in a falling period of the world's cattle price cycle causing increased operating losses for the government. Large subsidies were paid to continue operating these government projects.

A high incidence of illegal smuggling of cattle from Tanzania into neighboring countries was reported to be occurring. The illegal trade was because producers could export their cattle and then convert currency into Tanzanian currency at a higher "unofficial" exchange rate. Tanzanian domestic livestock prices were not competitive; therefore, quality livestock were diverted away from the Government's established market procurement channel and into other distribution outlets.

GOVERNMENT POLICIES AND THE POLITICAL ECONOMY

The political economies of Ghana and Tanzania were such that the livestock-meat industry was largely controlled by government. The premise that middlemen who handle livestock-meat exploited the consumer with higher prices was a common reason that each government gave for creating government institutions to eliminate these middlemen.

Each government felt that further control at the retail level was required to insure that government prices were adhered to and, hopefully (but wrongly), shortages prevented. The final result was general market failure because of severe meat shortages accompanied, in the case of Tanzania, by an increase in the illegal export of livestock. In the case of Ghana, equity in distribution of meat diminished because the only buyers available were middle and high-income urban consumers or government workers with prior information and access to purchases of scarce meat from MMB's offices. Limited foreign exchange was used to purchase these products for a small segment of society.

The labor productivity in the slaughter and processing industries was low in each country. A low level of progressiveness was present in the industry, and there was a large amount of unused plant capacity. In the case of Tanzania, exports of canned beef to EEC markets were restricted because of poor sanitation conditions. The Government had not made timely investments of capital to upgrade the canning line in the plant.

Price policies in both countries were used as a mechanism for stablizing consumer prices. The role of the MMB in Ghana and the DDC in Tanzania was to set ceilings on urban and rural prices for meat. By 1979, a general market failure had occurred in both countries in terms of meat supplies. Tanzania restructured its price policies in 1978-79 to shift from government control of meat markets to a more laissez-faire policy. In Ghana, a change in the government in 1979 has made it difficult to assess what is currently occurring. Instability in the political process continues to have an impact on economic conditions in Ghana.

A similarity in both countries was the pervasive influence of marketing boards for livestock and meat. These marketing boards controlled purchases and distributions of livestock and meat. From their inception, these marketing boards expanded their control of activities downstream in the marketing channel. The results in each case were a greater misallocation of resources, higher operating costs, larger government subsidies to keep these operations afloat, and increased meat shortages in the marketplace.

These results indicate that governments' intervention in the marketplace through their use of pricing policies and marketing boards created disincentives and resulted in poor economic performance. Competition and innovativeness among producers and merchants in the private sector were eliminated and varying degrees of market failure occurred in the livestock-meat subsectors in these two countries.

RECOMMENDATIONS ON THE ROLE OF GOVERNMENT POLICIES

Government price policies can have two effects as seen from these case studies: (1) create incentives or disincentives for producers to increase output and off-take from their herds and (2) promote the allocation of resources and products efficiently or inefficiently in the marketing system. The effectiveness of an administered price to elicit greater commercial supplies from the traditional as well as commercial production sector herd is based on the level to which price is allowed to rise. With a high traditional value for livestock and livestock by-products for family requirements, market prices would have to become more attractive to elicit greater commercial supplies.

Government prices can cause distortions in allocative efficiencies in the market system. Commercial market channels could be closed to producers because of inappropriate pricing policies. Supplies of livestock are directed away from least cost marketing channels, so consumers have to pay higher prices to obtain the product. These distortions hinder the establishment of least cost marketing operations and the timely flow of the product.

The potential for administered prices to improve the orderly development of the livestock industry in Subsaharan Africa is perceived as low. Administered prices are usually based on political and social concerns rather than on efficiency principles. It is for this reason that closer study needs to be given to a country's domestic price policies to determine the effects on producers, marketing agents and consumers. The impact of a policy on the operational and pricing efficiency of the livestock-meat marketing system needs to be assessed to determine all ramifications on the industry.

A government's plan to create marketing institutions to assist in development of the livestock-meat industry needs to be closely evaluated. Reasons for the existence of these institutions could be to monitor and oversee the operation of the industry without the need for price controls or excessive market regulations. The private sector needs to take the lead in industry development with government institutions insuring that adequate levels of competition are present at all stages in the marketing system. Government marketing boards, for example, can act as facilitative agents if bottlenecks develop. Performance criteria for these government institutions need to be based on economic efficiencies, and not established for political or social reasons.

158

REFERENCES

Abbot, J.C. "The Role of Marketing in the Development of Backward Agricultural Economies." Journal of Farm Economics 44 (1962):349-62.

Bales, M.D. and E. Lutz. "Price Distortions in Agriculture and Their Effects: An International Comparison." Amer. Jour. Agr. Econ. 63(1981): 8-21.

Brown, Gilbert T. "Agricultural Pricing Policies in Developing Countries." Distortions in Agricultural Incentive ed. Theodore W. Schultz, pp. 84-113. Bloomington: Indiana University Press, 1978.

Doran, M.H., A.R.C. Low and R.L. Kemp. "Cattle as a Store of Wealth in Swaziland: Implications for Livestock Development and Overgrazing in Eastern and Southern Africa." Amer. Jour. Agr. Econ. 61(1979):41-47.

Krishna, Raj. "Agricultural Price Policy and Economic Development." Agricultural Development and Economic Growth eds. H.M. Southworth and B.F. Johnston, pp. 497-540. Ithaca: Cornell University Press, 1967.

Leibenstein, Harvey. General X-Efficiency Theory and Economic Development. London: Oxford University Press, 1978.

Mellor, John W. The New Economics of Growth--A Strategy for India and the Developing World. Ithaca: Cornell University Press, 1976.

Newberry, David M.G. and Joseph E. Stiglitz. The Theory of Commodity Price Stabilization: A Study in the Economics of Risk. Oxford: Clarendon Press, 1981.

Peterson, Willis L. "International Farm Prices and the Social Cost of Cheap Food Policies." Amer. Jour. Agr. Econ. 61(1979):12-21.

Pinstrup-Anderson, Per and Elizabeth Caicedo. "The Potential Impact of Changes in Income Distribution on Food Demand and Human Nutrition." Amer. Jour. Agr. Econ. 60(1978):402-415.

Reutlinger, Shlomo, and Marcelo Selowsky. "Malnutrition and Poverty, Magnitude and Policy Options." Baltimore: Johns Hopkins University Press, World Bank Staff Occasional Paper No. 23, 1976.

Sanderson, K.T. and K.W. Stokes. "Pricing Proposals Submitted to the Ministry of Agriculture for the Annual Price Review." Ministry of Agriculture, Dar es Salaam, Tanzania, July 1976.

Schultz, Theodore W. "On Economics and Politics of Agriculture." Distortions of Agricultural Incentives. ed. Theodore W. Schultz, pp. 3-23. Bloomington: Indiana University Press, 1978.

Sullivan, Gregory. "Livestock and Meat Marketing in West Africa: Ghana." Livestock and Meat Marketing in West Africa Vol. II. University of Michigan. Center for Research on Economic Development, 1979.

Sullivan, G.M., D.E. Farris, M.J. Yetley, and W.J. Njukia. "A Socio-Economic Analysis of Technology Adoption in an African Livestock Industry." Qtr. Jour. of Int. Agric. 17(1978):150-160.

Sullivan, G.M. and D.E. Farris, and K.W. Stokes. "A Theoretical Model of Overgrazing in Traditional Livestock Economies of Africa: A Paradox of Perceived Values." Paper presented at the AAEA annual meetings at Clemson University, July, 1981.

Sullivan, G.M. and D.E. Farris. "The Role of Government Policies in Agricultural Development: The Case of the Cattle Industry in Tanzania." Qtr. Jour. of Int. Agric. 21(1982):52-61.

Texas A&M University. Tanzania Livestock - Meat Subsector. International Programs Office. College Station, TX. Vol. 4. 1976.

United Nations, Food and Agriculture Organization. FAO Production Yearbook. Rome, 1970-1980.

_____. FAO Trade Yearbook. Rome. 1970-1980.

PART III

Development Projects and Experiences

Increasing livestock production in Subsaharan Africa, while necessarily requiring a technological foundation, is first and foremost a people problem. The way in which livestock, land and other resources are used is a direct function of individuals' and groups' perceptions about the way in which these factors can and/or should be utilized to achieve their needs and goals. Population growth and a determination by governments to develop (as defined by steady growth in per capita GNP) has led to radically increased intervention in the livestock industry by national governments. The result has been myriad projects designed to bring about a host of changes ranging from increased output by improved nutrition to the altering of herd compositions.

The use of projects to foster economic development has many pros and cons since it is a "top-down" approach regardless of how well tuned the designer might be to client wants and desires. But, as Joan Atherton points out in the first chapter of this section, a development strategy inherently must embody projects. She traces the evaluation of a single bilateral donor policy and strategy, in response to many projects which had less than satisfactory results, and presents components of two recent project papers, for Mali and Niger, to illuminate implementation of the strategy.

Land use is perhaps the fundamental factor which must be considered in improving subsaharan livestock productivity and, consequently, in many areas, especially arid and semi-arid zones, communal land tenure is a major concern in design of projects and policy. Dirck Stryker provides a theoretical framework for understanding "the tragedy of the commons" as it applies to overgrazing. He then looks at land use planning in the pastoral zone of West Africa and concludes that land use schemes have failed due to the

162

inherent need, given zonal conditions, for spatial
mobility and flexibility in herding activities.

Harold Schneider reinforces the need to couch land
use programs within a historical and people perspective. In his chapter, which focuses on pastoralists in
East Africa, he questions the identification of "development" of the livestock industry with the shifting
from a traditional management mode to a commercial
integration with national and international economies.
He concludes that constraints on development must be
defined in terms of economic rationale, and that more
research is needed to discover how and why producers
have made, and are continuing to make, the transition
to increased production for the market.

Peter Little points out that analysis of sociological dimensions of African pastoralism is a formidable
task, but neatly confronts the problem by providing a
description of the most significant socio-economic
indicators that should be examined at the planning or
feasibility state of a livestock development project.
As an example, he notes the importance of identifying
the appropriate livestock management units, as they can
vary greatly even within communities and between seasons. Failure to correctly identify socio-economic
variables along with technical ones has led, and will
continue to lead, to project failure.

Livestock projects, while dependent upon people as
the action mechanisms are, by their nature, built
around major technical areas such as improvement of
animal breeds or feedstuffs. One of the principal
project areas is range management. Alex Dickie and
James O'Rourke provide an applied approach to this
important topic by discussing the way in which viable
rangeland utilization strategies can be developed. In
their opinion, there is a great deal of information
available and constraints to livestock development in
Subsaharan Africa are surmountable providing emphasis
in given to including user groups as full participants
in the process of planning, implementing and evaluating
projects. Thurston Teele, in a very comprehensive
analysis, points out that the real or perceived failure
of livestock projects can be attributed to two main
factors; lack of agreement about what should be done
(project design), and implementation (project management). He carefully outlines the numerous components
for successful project management, concluding that lack
of attention to any of them will prevent a project from
being successful.

10
The Evolution of a Donor Assistance Strategy for Livestock Programs in Subsharan Africa

Joan S. Atherton

INTRODUCTION

Over the past decade, donor livestock policies and programs in the less developed countries (LDCs) of Subsaharan Africa have moved toward a balance between meeting national economic goals and enhancing the individual well-being of the livestock producer. Such a balance represents a distinct change from an earlier period in Africa's development history, in which large-scale schemes for increasing production frequently overlooked the alternative and sometimes conflicting objectives of the livestock owners or keepers. This chapter will discuss the recent trends in policy and strategy in the United States Agency for International Development (A.I.D.) and examine two current project designs to illustrate implementation.

BACKGROUND

In 1979, A.I.D. began to formally reexamine its approach to livestock development[1] in Subsaharan Africa. A workshop, jointly sponsored by A.I.D.'s Bureaus for Program and Policy Coordination and for Africa, was held in Harper's Ferry, West Virginia in May of that year. This workshop focused on those projects that were geographically located in arid or semi-arid zones, were demographically targeted at pastoralist producers who ranged from fully nomadic to fully sedentary, and were characterized by highly capitalized interventions intended to dramatically alter the indigenous production systems. As Horowitz wrote in a discussion paper for the workshop, "(l)ivestock projects have been supposed, simultaneously, to increase productivity, reverse the ecological deterioration of the

This chapter represents the views of the author and does not represent an official position of the U.S. Agency for International Development.

163

range, shift production from a dairy to a beef orientation, improve producer income and quality of life,
maintain a regular supply of cheap meat for the internal market, and increase the supply of high quality meat
to the export market" [1979, p.l].

In other words, the A.I.D. livestock interventions
in Subsaharan Africa that the workshop considered had
probably been overambitious, overoptimistic and perhaps
worst of all, unclear about the development objectives
to be attained. The Harper's Ferry Workshop also
reached some tentative conclusions on recommended
actions for subsequent interventions. These included:

-- improving the reliability of quantitative data
   by exploring alternative methodologies and
   looking at different variables such as
   seasonality;
-- following indigenous systems of management in
   the design of project interventions (This
   almost always implies diminution in size as
   well as changes in composition of the management unit vis-a-vis developed country
   practices.);
-- understanding seasonal exploitation of
   resources as both a survival and conservation
   mechanism;
-- acknowledging that evidence on range degradation is contradictory at best, and therefore
   the need to conserve rangelands by conventional
   means (i.e., lowering stock numbers) is still
   an open question;
-- emphasizing that consolidation of existing
   production activities rather than significant
   expansion will lead to more successful donor
   interventions; and
-- acknowledging that the continued need for reliable information dominates the livestock
   subsector program, so monitoring and evaluation
   should receive attention in every project
   design.

If the intent of the 1979 workshop was to critically evaluate the preceding 20 or so years of A.I.D.
(and other donor) livestock development interventions,
then it could be construed as a success. The findings
and recommendations of the workshop were, however, a
distinct repudiation of the past, and as such were
difficult for many in the donor community to accept.
The workshop had lavished much time and attention on
criticizing biological technicians and technology-
driven solutions without going the further step of
offering a solution to the methodological problem of
integrating the biological and socioeconomic aspects of

livestock development. We were left with recommendations that were searching for a mode of implementation.
Two years after the Harper's Ferry Workshop , a
second Workshop on Development and African Pastoral
Livestock Production was held in Mariottsville, Mary-
land, sponsored by the same two A.I.D. bureaus. The
meeting was considerably more optimistic in tenor,
though very few new projects that reflected the thinking of the 1979 meeting could be cited. There was more
of a sense, however, that the knowledge accrued in the
1970s had begun to be codified into some coherent,
actionable recommendations for donor policy, and that
critics and proponents of various different approaches
had modified their views and found ways to work
together.
The results of the Marriottsville workshop formed
the basis for the Africa Bureau's 1982 Livestock
Development Assistance Strategy (Appendix 12).
Essentially, the second workshop reinforced the
findings of the first, but instead of considering donor
interventions as part of the problem, it concentrated
on the development constraints in the livestock
subsector and on the positive actions that donors could
reasonably take to address them. The constraints were
recognized to be quite complex, and the formulation of
the livestock development issues in the Draft Report
from the workshop [Institute for Development
Anthropology 1981], reflects this complexity:

    1.1 Low productivity, given its potential.
    1.2 Constraints to livestock development.
        (a) Inadequate host government policies and
             programs.
        (b) Inadequate information base for
             technical packages, particularly in
             respect to range management.
        (c) Paucity of animal feed.
        (d) Inadequate animal health services.
        (e) Inadequate market access.
        (f) Limited skills for new technologies.
        (g) Inadequate understanding of the
             rationale of existing systems.
        (h) No clearly defined economically feasible
             technical package which is presently
             applicable on a general basis.
        (i) Poor institutional capacity.
    1.3 Choice of appropriate investments in animal-
        based versus mixed production enterprises.

The problem formulation proved a turning point in
A.I.D.'s thinking and action on Subsaharan African
livestock development. Instead of simply criticizing
prior interventions, the definition of the problems

allowed a consensus to be built, and different disciplines[2] could begin to work together constructively
to suggest means of addressing the constraints.

THE POLICY FRAMEWORK

The two workshops on African livestock development
raised consciousness and identified the problems. The
strategy solution had to be consistent with A.I.D.
policy[3] on food and agricultural development.

All of A.I.D's policy statements are founded on
the Foreign Assistance Act (FAA) of 1961, as amended.
The FAA does not speak directly to the development of
the livestock subsector, but more generally to the
objectives of alleviating hunger, promoting broadly
based, self-sustaining economic growth and improving
the well-being of the rural poor majority. The A.I.D.
agricultural policy is embodied in two statements, a
1978 Agricultural Development Policy Paper and a subsequent 1982 Food and Agricultural Development Policy
and Paper. The 1978 paper emphasized a broadly participatory strategy aimed at the dual objectives of
production and employment. The paper then identified
the functional areas of programming through which the
policy was to be implemented:

1. Asset Distribution and Access -- including
land and water rights, and the strengthening of local
participatory institutions;
2. Planning and Policy Analysis -- pointing out,
among others, the importance of decentralized planning,
the appropriate role of the public sector in agricultural institutions, price policy and government's role
in land tenure reform as areas of concern;
3. Development and Diffusion of New Technology:
Research, Education, and Extension -- indicating that
institution-building is a long, slow but ultimately
worthwhile endeavor;
4. Rural Infrastructure -- including land and
water development; and
5. Marketing and Storage, Input Supply, Rural
Industry and Credit.

All of these functional areas are relevant to
interventions in the livestock subsector. It is interesting to note that in 1978, a year before the first
workshop on Subsaharan African livestock development
programs was held, the Agricultural Policy Paper
stated:

Unfortunately, allocation of scarce capital and
management expertise in the most disadvantaged,
fragile, or damaged areas may not be justified by

short-term economic and financial returns.
Furthermore, the gains from controlled grazing,
better terracing, or afforestation may become
obvious to the individual farmer or herder only
after an extended period, although the costs of
adopting modified systems to permit such gains may
be evident immediately. In addition, the generation of local revenue from, and allocation of
government revenue to, these areas is likely to be
very low. Therefore, institutional development
approaches should concentrate on group management
and bottom-up community self-help activities which
are less management and resource intensive.
[U.S.A.I.D. 1978, p.44]

The 1982 paper augmented the 1978 policy and
focused attention on the means by which the objectives
of production and employment were to be reached in
donor assistance programs. The four major elements of
the 1982 Food and Agricultural Development Paper are:

-- improve country policies to remove constraints
   to food and agricultural production, marketing
   and consumption;
-- develop human resources and institutional
   capabilities, especially to generate, adapt
   and apply improved science and technology for
   food and agricultural development, and conduct
   research on developing country food and
   agriculture problems;
-- expand the role of developing country private
   sectors in agricultural and rural development,
   and the complementary role of the U.S. private
   sector in assisting this expansion; and
-- employ all available assistance instruments in
   an integrated manner, including provision of
   PL 480 food aid in a way that contributes to
   the other three strategy elements as well as
   meeting food security and nutritonal needs.
   [U.S.A.I.D. 1982, p.2]

The 1982 paper goes on to say that:

Each of these elements is intended to increase the
effectiveness to U.S. economic assistance resources. While the mix of these elements will
vary according to the differing needs and problems
among countries, the achievement of lasting food
security requires greater attention to each of
them in most developing countries.

Given this policy context, and the very real constraint of limited resources allocated to the Subsa-

haran African regions in which livestock development interventions predominate, A.I.D.'s Bureau for Africa set out to build on the 1979 and 1981 workshops' technical information base and prepare a livestock development assistance strategy. The policy framework indicated that the assistance should be targeted to poorer producers, could incorporate both production and employment (and therefore income) objectives, should encourage participation, and could look to national policies as well as technical interventions and institution-building to achieve its objectives.

THE STRATEGY

The 1982 Africa Bureau Livestock Development Assistance Strategy (Appendix 12) strikes a balance between interventions designed to have a long-term impact and those with more immediate results. This is in keeping with the conceptual balance between the long-term development objectives of enhancing the well-being of the rural poor producer and the national economic growth requirements. These objectives are treated as interdependent, complementary aspects of livestock development in the subsector assistance strategy.

The strategy has two main areas of concentration, the first of these being the integration of livestock subsector development policy and planning into overall agriculture sector objectives. This is necessary because, as already noted, livestock development tends to be located in geographically less well-endowed areas that are likely to receive proportionately fewer national resources, whether human or financial, than more promising agricultural regions. Further, countries with burgeoning populations and limited resources tend to concentrate on improving the productivity of crop agriculture as a means of providing a cheap and plentiful food supply. Even in areas of mixed crop/ livestock farm operations, food and commercial agricultural crops tend to attract the bulk of the development attention. Recognizing that large land areas and substantial numbers of the rural poor in Subsaharan Africa fall in the categories most likely to be over-looked by hard-pressed LDC governments, the Africa Bureau's livestock development strategy identifies assistance in formulating appropriate national policies, plans and projects that include livestock in the agriculture sector as one of its two major prongs.

The second prong of the assistance strategy is the development and improvement of institutions and delivery systems. This prong acknowledges the paucity and low quality of data on livestock production systems and the need for better research-producer communication. The strategy to address these weaknesses is to enhance the ability of host country institutions and systems to carry out livestock development efforts. Primary among the institution-strengthening targets is research on production and disease constraints. Delivery systems include health interventions, production improvements and systems management advice as well as more conventional items such as inputs and credit. These may take the form of private sector initiatives as well as public sector options. The strategy also encourages A.I.D. Africa missions and regional programs to take advantage of special development opportunities such as the adaptation of "high" technology (e.g., genetic engineering or vaccine development).

Throughout, a participatory approach to livestock development is emphasized because the lessons drawn from prior experience clearly indicate that intervention success depends on meeting producers' perceived needs perhaps more strongly than in any other subsector of agriculture. Another theme of the strategy is the need for selectivity in interventions, choosing for broad support, activities such as veterinary care that have a proven demand and that allows for relatively equitable access to the benefits; and for more limited or "pilot" support to those interventions that show promise but need further research, such as group management of land areas or water points. Finally, the strategy takes note of the very long time horizon needed for interventions in the livestock subsector to achieve their development objectives.

CASE EXAMPLES OF THE STRATEGY IMPLEMENTATION

Two recent A.I.D. project paper designs in Sahalian countries reflect the implementation of the Africa Bureau's Livestock Development Assistance Strategy. Each project follows on an earlier livestock development effort in Mali and Niger, respectively. This continuity in itself signifies a recognition that livestock interventions take longer to fully realize their objectives than most other types of assistance programs, as noted in the strategy statement. In the first instance, the Mali Livestock Sector II Project Paper states specifically that the project is part of a twenty-year development program in assistance to the livestock subsector. The Niger Integrated Livestock Production Project continues the work of a five-year research project, and takes advantage of the earlier data collection and analyses in the plans for the action interventions in the current project. Another common characteristic of the two projects is the selectivity of interventions. Both concentrate on activities that have had a past record of success, measured in terms of increases in output and acceptance by producers. Both designs include a number of experimental elements with concomitant arrangements for monitoring and evaluation, thus combining production and research aspects in the respective projects.

Briefly, the Mali Livestock Sector II Project incorporates the following components:

-- Animal health including continued support for the Central Veterinary Laboratory and greater assurance of vaccine viability in transportation to and storage in regional distribution centers;

-- On-farm cattle feeding, which is a continuation of a pilot activity to encourage some small farm operators to participate in a fattening-out scheme, with changes in the focus of Malian institutional support from the prior project activity, and close monitoring of the economic viability of the scheme;

-- Agricultural research on smallholder livestock systems, such as provision of technical assistance, training and commodities to institute a farming systems research component within the livestock research unit; and

-- Management support which, with initial assistance in the financial management of the project itself, is intended to be the precursor of a longer-term effort to assist in policy and program planning and analysis in the livestock subsector.

The Niger Integrated Livestock Production Project has similar components to the Mali project, but each component is considerably more complex and has a research aspect built into it. It is difficult to do justice to each component in a short description, but the main activities are:

-- Herder organization development for continued establishment of herder associations for common resource management, joint receipt of credit, etc.;

-- Animal production and health, to include development of a cadre of veterinary auxiliaries drawn from herder groups to dispense animal health and nutritional care and advice; continued research on animal health and improvement of vaccine viability in the distribution system;

-- Natural resources management, including testing and extension of range management interventions developed under the earlier project, including

water points development, grazing trials and an
early warning system for environmental stress;
-- Marketing research, with analysis and dissemination of marketing information;
-- Human resource development for improvement of
basic literacy and numeracy, health and nutrition among the herder population; and
-- Project management for institutional strengthening to improve the Government of Niger's
ability to establish and use a management
information system and other tools for program
planning and evaluation in the livestock
subsector.

The Mali and Niger livestock project designs
accord with the principal elements of the Africa
Bureau's Livestock Development Assistance Strategy.
Both contain components to assist the host countries in
policy and program analysis in the subsector. Each has
a research aspect and incorporates improvement in
delivery systems. The Niger project, particularly,
attempts to strengthen local institutions through the
development of herder groups and the use of veterinary
auxiliaries, thereby relying heavily on a participatory
approach to livestock development.

These two projects are representative of the
evolving trend in livestock development assistance that
has characterized A.I.D. policy and programs over the
past five years. The process has been a cumulative
one, and will presumably continue as the new generation
of projects is evaluated, and as their research
components begin to bear the fruit of new knowledge.

FUTURE DIRECTIONS IN LIVESTOCK ASSISTANCE

This chapter has largely ignored some of the
evolving livestock strategy facets that are more difficult to explicate at present. Chief among these is the
attention being turned to mixed farm operations. This
is being done in several ways, through projects or
components that take the farming systems approach to
research or through the introduction of specific live-
stock-oriented interventions such as animal traction,
poultry production and veterinary care for small stock.

Although animal production scientists agree that
animal nutrition is perhaps the most pressing bio-
technical constraint to livestock production, less is
known about the socioeconomic constraints to improving
animal nutrition. Research concerning on-farm labor
allocation must be carried out to evaluate interventions such as forage crop production, fattening-out
operations and even collection of crop residue for use
as fodder. We have yet to turn substantial attention

to dairy activities and the commercial viability of small-farm production enterprises in this realm. Small ruminants have received little attention, and may be particularly important as a means of income for women farmers in Subsaharan Africa, as women and children are often the principal keepers of small stock in mixedfarm operations. There likely are a large number of other interventions to which donors such as A.I.D. might turn attention in the future. One of the lessons that appears to have been learned is that we must proceed cautiously, generating knowledge about present-day production patterns before we offer "improvements" on the system. Thus, the above areas will probably only slowly enter the repertoire of livestock interventions, building incrementally on the successes of pilot activities and knowledge gained through applied research.

SUMMARY

The evolution of a bilateral donor policy and strategy for development assistance to the livestock subsector of agriculture is traced in this chapter. Motivated by the perception that the interventions of the 1950s and 1960s had not proven cost-effective or sustainable in most instances, A.I.D. embarked on a period of intensive review and criticism through its 1979 workshop on pastoral development in Subsaharan Africa. This eventually led to the convening of a second workshop in 1981 whose purpose was to consolidate past learning and recommend new directions for future livestock development activities. Following policy guidance issued throughout this period on the general topics of food and agricultural development, a strategy statement for livestock development assistance was drafted, reviewed extensively and ultimately approved in December, 1982. Components of two recent A.I.D. project papers from Mali and Niger were then discussed in this chapter to illustrate strategy implementation. Future implementation directions, which the strategy touches upon, but into which A.I.D. has yet to move on a large scale, were then listed with the expectation that these areas might be more fully explored at a later date as the experience from pilot versions of interventions begins to accumulate.

NOTES

[1]For purposes of this chapter, programs that are livestock-only interventions will be used as the basis for discussion. Livestock-led mixed farming interventions in Subsaharan Africa remain relatively rare, with the possible exception of a small group of activities introducing animal traction. These are for the most

part too new to evaluate in terms of impact but may hold some promise for future directions in livestock development activities.

[2]The term "disciplines" connotes here a broad range of skills from conventional academic fields, such as range science, forage agronomy, agricultural economics, animal science and sociology, to donor project design, management and policy and program expertise.

[3]In A.I.D.'s terminology, policy serves to explicate the objectives of the U.S. foreign assistance program while strategy outlines the manner in which the program will be carried out to reach those objectives.

REFERENCES

Galaty, John G., Dan Aronson, Philip Carl Salzman, and Amy Choinard, eds. The Future of Pastoral Peoples. Ottawa, Canada: International Development Research Centre, 1981.

Horowitz, Michael M. The Sociology of Pastoralism and African Livestock Projects. Washington, D.C.: A.I.D. Program Evaluation Discussion Paper No. 6, May, 1979.

Institute for Development Anthropology. Draft Report: The Workshop on Development and African Pastoral Production. Binghamton, NY: Report prepared for the U.S. Agency for International Development, 1981.

_____. The Workshop on Pastoralism and African Livestock Development. Washington, DC: A.I.D. Program Evaluation Report No. 4, May, 1980.

Little, Peter. "Issues Paper: For the Workshop on African Pastoral/Livestock Development." Binghamton, NY: Institute for Development Anthropology, Nov. 1981.

Raun, N.S., and K.L. Turk. "International Animal Agriculture: History and Perspective." Journal of Animal Science, Vol. 57, Supplement 2 (1983): 156-170.

Tufts University. Niger Integrated Livestock Production Project Paper. Niamey, Niger: Paper prepared for the U.S. Agency for International Development/ Niamey, May 1983.

U.S. Agency for International Development. Africa Bureau Food Sector Assistance Strategy Paper. Washington, DC: Africa Bureau, Office of Development Resources, Agriculture and Rural Development Division, Oct. 1981.

_____. Africa Bureau Livestock Development Assistance Strategy Paper. Washington, DC: Africa Bureau, Office of Technical Resources, Agriculture and Rural Development Division, Dec., 1982.

_____. Agricultural Development Policy Paper. Washington, DC: Program and Policy Coordination Bureau, June, 1978.

_____. Food and Agricultural Development. Washington, DC: Program and Policy Coordination Bureau, May, 1982.

Winrock International Livestock Research and Training Center. A Winrock International Draft Position Paper on Livestock Program Priorities and Strategy. Morrilton, Arkansas: Paper prepared for the U.S. Agency for International Development, Dec., 1981.

11
Land Use Development in the Pastoral Zone of West Africa

J. Dirck Stryker

This chapter examines land use development in the pastoral zone of West Africa in terms of the relevance and impact of overgrazing. The principal conclusion is that land use schemes have failed in the pastoral zone because of the need for a high degree of spatial mobility and flexibility in herding activities. Despite this, land productivity is high because of the skill and hard efforts of the herders, suggesting that the problems of overgrazing may be less serious than is often supposed.

The first section provides a theoretical framework for looking at "the tragedy of the commons" as it applies to overgrazing. This is followed by an assessment of measures that have been used to control the problem. Next, some results of recent research in Mali and Niger are described. The final section presents the principal conclusions and some of their implications.

THE TRAGEDY OF THE COMMONS

The problem of overgrazing is said to arise principally because of a difference in the incentives facing individual livestock owners and the costs and benefits to the pastoral society as a whole resulting from grazing the rangeland [Hardin 1968]. The individual owner sees the pasture essentially as a free good. If that person does not exploit it, someone else will. Intense competition may thus ensue for the use of this scarce resource. Overuse can eventually result in social costs imposed by deterioration and even destruction of the rangeland [Simpson and Sullivan 1983].

In more technical terms, concentration of excessive numbers of animals in one area for too long results in disappearance of the more palatable and valuable forage species, and their replacement by less nutritious vegetation. Perennial grasses give way to annuals, and these in turn to weeds. Extreme overuse may eventually result in erosion and the exposure of bare hardpan. None of this is taken into account by the individual herder, who sees his or her contribution to the overall problem as negligible.

It is important to distinguish between two economic concepts involved in this discussion. The first is the notion of negative externalities, i.e., social costs associated with the deterioration of the rangeland that are greater than the private costs incurred in using that land that are taken into account by the individual herder. The second concept is that of economic rent. This is defined as the value of the benefits resulting from use of the rangeland after all returns to capital labor, and other costly inputs, whether private or social, have been deducted. If rangeland belongs to individual landowners instead of being exploited in common, this rent accrues to the owners, either explicitly if the land is used by others, or implicitly if used by the owners. If there is a market for land, the economic rent is reflected in the sales value of the land so that herders can move into or out of the livestock sector without gaining or losing; any loss in rent is compensated by the sales price received, and vice versa. The allocation of resources is thus unaffected. Most important, all social costs are counted as private costs in determining the net return to land, and users of the land are so charged by the owner. If the owner uses the land, that person's perception of these costs, including those associated with overgrazing, influences how resources are allocated.

In the absence of a sales market for land, allocation of resources is still efficient as long as there is a lease market and private property rights are guaranteed so that social and private costs are identical. However, once the owners of land see their property rights threatened, so that they cannot be assured of retaining the land or of continuing to receive rent from its users, social costs might diverge from private costs and over exploitation of the land might ensue.

The extreme case of this occurs when land is held in common and there are no individual owners. This situation is represented in Figure 11.1, which depicts marginal costs and benefits of a pastoral economy. The marginal benefits that accrue to herders' labor, skills, capital, and other inputs are shown by the curve MB. These marginal benefits decline as the herders devote more of their inputs to exploiting the same land more intensively, or to using less productive pasture at the extensive margin. Private marginal costs are given by the PMC curve and social marginal

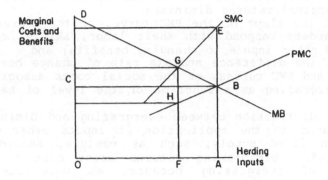

Figure 11.1. Marginal costs and benefits of a pastoral economy

costs by the SMC curve. The difference between the two
is the social cost associated with overgrazing.

If there is no problem of overgrazing, social marginal costs are the same as private marginal costs, and
the equilibrium level of inputs is OA. Total benefits,
equal to the area under the MB curve, can be divided
into the return to herder inputs OABC and the residual
economic rent to land CBD. But to the extent that
social marginal costs diverge from private marginal
costs because of overgrazing, social marginal costs
exceed marginal benefits by EB when herders continue to
equate private marginal costs and benefits. In order
for social marginal costs to be covered by marginal
benefits, the level of inputs must be reduced to OF.
This can be accomplished by a number of means, such as
imposing a grazing tax equal to the difference between
social and private marginal costs, which will cause
herders to reduce the number of animals on the rangeland. In effect, the grazing tax GH substitutes for
the rent that would be received by private owners to
compensate them for the costs of overgrazing if the
land were not communally owned.

The economically optimal solution at an input
level of OF does not call for the complete elimination
of overgrazing. Because marginal benefits increase and
private marginal costs decrease as inputs are reduced,
stocking rates should be maintained at levels that
result in pasture of lower quality than would exist if
these rates were lower. The amount by which inputs
should be reduced below the level OA that exists in the
absence of a tax, or other control measures, depends
on:

1. The slope of the MB curve, or the rate at which marginal returns diminish;
2. the slope of the PMC curve, or the degree to which herders respond with their labor, skills, capital, and other inputs to changing benefits; and
3. the difference and its rate of change between the SMC and PMC curves, or the social costs associated with overgrazing as a function of the level of herder inputs.

The distinction between overgrazing and diminishing returns to the application of inputs other than those in fixed supply, such as pasture, should be emphasized. Diminishing returns exist even in the absence of overgrazing because, as more pasture resources are used, increasing amounts of labor and capital have to be applied at the margin to obtain equal increments of production. This constrains the number of animals that the rangeland will support economically before the technical limits of production are reached. It does not in any way depend upon a difference between the private and social costs associated with grazing.

MEASURES TO CONTROL OVERGRAZING

Most livestock raising in the pastoral zones of Africa is done under conditions in which the land is used in common by numerous herders. Where there is neither a fee nor a grazing quota, this purportedly leads to a situation in which each herder maximizes the number of animals grazed on the rangeland [Hopcraft 1980]. This results in degeneration of the vegetation and degradation of the soil.

To prevent this, range management techniques may be introduced that allow desirable forage species to be grazed but not destroyed. While this can be accomplished fairly easily under private ranch conditions, it is much more difficult where the areas to be managed are vast, the range resources are publicly owned, and the animals are owned individually by semiautonomous producers. Nevertheless, most projects in the pastoral zone consider as a precondition that some method be devised for introducing collective control over range resources. This is held to be especially important when the project is designed to improve animal health and nutrition and thus potentially will increase livestock numbers and the incidence of overgrazing. Although improved marketing opportunities might result in greater offtake, and thus ameliorating the overgrazing situation somewhat, this effect is thought to be lessened to the extent that animals are held as a store of wealth rather than simply for their value in commercial production [Hopcraft 1980; Simpson and Sullivan 1983].

One attempted solution has been to establish commercial and government ranches. This was tried in numerous countries during the period 1960-1975, with financial support from the World Bank and several bilateral donors. The results have been disappointing because of high capital and recurrent costs, and low returns [Eicher and Baker 1982]. In addition, the limiting boundaries of ranches put them at a disadvantage vis-a-vis pastoral herders, who are free to search out the best pasture for their animals in the face of highly variable rainfall.

The failure of numerous commercial and government ranch projects over the past few decades has led governments to shift in the direction of group ranches and grazing blocks. These have usually involved some infrastructural development, especially improved water sources, and exclusive rights to the use of land where rotational grazing or other range management techniques have been employed. The implementation of these projects, however, has been fraught with difficulties, partly because there has been a failure to undertake the required predevelopment analyses, including surveys of water use, socioeconomic studies of traditional grazing systems, and legislation necessary to adjudicate group landholding rights [Eicher and Baker 1982].

A second problem has been that the concept of establishing rights to the use of specific areas of land is inconsistent both with traditional systems that provide for the sharing of pasture resources by different social groups, and with government policies proclaiming the openness of the rangeland to all. In years of drought, any system based on fixed grazing rights tends to break down rapidly as a result of overwhelming demand for available pasture. As Hopcraft notes,

> The general pattern within the group ranches has been that, when the need arises, adjudicated boundaries are virtually ignored and traditional notions of reciprocal obligations to those from other areas reassert themselves. The fundamental logic of a group restricting its livestock numbers to the carrying capacity of its own ranch and, in turn, benefiting from the grazing that it has been able to preserve is thus subverted. [1981, pp. 232-233]

A third difficulty has been that of organizing and managing individual herders that are not already bound to each other by strong social ties. The authority of elected committees is too often negligible when it comes to the critical questions of control over livestock numbers, ownership, and grazing. If the size of the group is restricted to the point that social ties among the members are already strong, the area of land over which the group has control is too small to exploit optimally given spatial variations in rainfall.

RECENT RESEARCH IN MALI AND NIGER

Research has been conducted over the past few years that has an important bearing on the issue of overgazing and how it is to be controlled. Some of this work, undertaken by the AID financed Niger Range and Livestock (NRL) project, is summarized in Aronson [1982] and in USAID/Niamey [1983]. Other important research has come out of the Production Primaire au Sahel (PPS) project in Mali and is reported in Penning deVries and Djiteye [1982].

Of major interest is the finding of the PPS project, shown in Table 11.1, that the production of animal protein per hectare in the livestock sector of Mali is about equal to that found in similar zones (500 mm of rainfall of less) of Australia and the United States. Only in areas where rainfall is more plentiful does milk and meat production per hectare in Australia and the United States exceed that in Mali. Table 11.1 also shows that whereas livestock production in the semi-arid areas of Australia and the United States is very intensive in the use of fossil fuel energy, particularly for transport, Mali instead uses a great deal of labor. This is substantiated by preliminary NRL findings suggesting that herders in Niger spend a great deal of time searching out and moving their animals to the best pasture during the rainy season, whereas during the dry season much labor is devoted to watering. If overgrazing is indeed a major problem in Africa under communal grazing systems, its effect appears to be offset by the skills and labor of the herders.

In addition to labor, herders' command over capital also appears to be an important constraint on development. NRL research strongly suggests, for example, that the pastoral economy is highly monetized and that the relative need for capital is very great [USAID/Niamey 1983, Annex 7]. Much of this is working capital required to tide the herding family over the end of the dry season, when milk supplies have run low and cash is needed to purchase millet. If herders have to sell animals at this time, they incur substantial losses because of the animals' poor condition. Yet many are forced to do this because their herds are too small to support them. It is estimated, for example, that a minimum of 30 cattle are necessary to supply the consumption needs of a WoDaaBe family of seven persons. Of the WoDaaBe family herds recently surveyed in Niger,

Table 11.1. Production of protein with extensive raising of livestock

Region	Rainfall	Specie	Animal Protein		Ratio of Fossil Fuel Energy to Labor
			Per Unit Area	Per Unit of Labor	
	-mm/per year-		kg/ha/yr	kg/man hr	1 million J/man hr
USA					
Utah	<200	lambs	0.3	0.9	105
New Mexico	200–500	diverse	0.5	1.4	142
Texas	500–900	cattle	4.5	4.3	172
Australia					
Pastoral zone	200–500	sheep	0.4	1.9	628
Wheat/Sheep zone	500–1000	sheep	5.5	1.0	218
Mali					
Transhumance	300–Delta	cattle	3.2	.07	0
Sahel	<300	diverse	0.4	.01	0
Savannah	300–800	diverse	0.3 –0.6	0.01–0.04	0

Source: F.W.T. Penning deVries and M.A. Djiteye [1982], p. 467.

only about 7 percent were of this size [Wilson and Wagenaar 1982].

The elasticities of supply of the herder's major inputs labor and capital, therefore appear to be relatively low and the opportunity cost associated with their use to be, at the margin, high. In terms of Figure 11.1, the PMC curve rises quite steeply. This contradicts the common assumption that pasture is the scarce resource and other traditional inputs are available at minimal cost. This has implications, to be spelled out later, for the optimum level of overgrazing.

Equally important are some of the findings related to the extent of overgrazing. The NRL range research team has found, for example, little or no evidence indicating that Sahelian rangelands were ever closer to presumed climax vegetation than they are now. As Aronson observes:

> The reality is that animals and vegetation are in such dynamic interaction that heavy grazing may yield real long-term benefit: browsing animals are one major form of control of bush encroachment onto grass pastures, for example, while heavy grazing may crop certain species and therby enable other species, of potentially higher value, a benefical 'place in the sun'. [1982, p. 11]

Some of the reasons for this absence of signs of overgrazing can be better understood from the research in Mali, the results of which have application over a much wider area. Of major importance are the findings that (1) the quality, as well as the quantity, of pasture is a major constraint on animal nutrition and (2) the quality of pasture diminishes as the biomass, or quantity, increases [Penning deVries and Djiteye 1982]. The annual grasses in the northern part of the Sahelian zone, in fact, are highly nutritious in relation to the perennials to the south. The movement of animals during the rainy season is, therefore, an important means of exploiting the potential productivity of the rangeland. Furthermore, overgrazing of these annuals does not appear to be a serious problem since "... the number of seedlings at the beginning of the growing season is usually not a limiting factor for biomass production" [Penning deVries and Djiteye 1982, p. 486]. Elimination of seeds is important chiefly around permanent water points, though it may occasionally be a more widespread problem in the north. The PPS research also verifies the theoretical conclusion presented earlier that some degree of overgrazing is desirable.

A most important aspect of the PPS research is the conclusion that sedentarization of the nomadic population and management of dry season grazing areas around permanent watering points would probably result in a decline, not an increase, in production since pasture in the dry season areas is of much lower quality than that available in the north during the transhumance. The research results, in fact, are relatively pessimistic concerning finding an economically viable means of increasing animal production by manipulating range resources. Lowering of stocking rates could contribute to a more stable environment, but at the cost of diminished productivity. "Replacing cattle by other species might increase productivity temporarily. However, goats, sheep, cattle and camels combined exploit the Sahelian environment already very efficiently" [Penning deVries and Djiteye 1982, p. 150].

CONCLUSIONS

This review of the theoretical justification for grazing control, the general experience of projects that have attempted it, and recent research on it in Mali and Niger lead to a number of conclusions. The first is that some degradation of the rangeland is to be expected if the land is being optimally exploited from an economic point of view. This implies that the social marginal costs of grazing will be greater than their private marginal costs, so that, even under optimal conditions, grazing will destroy some of the pasture that would otherwise be available to all pastoralists. Second, most projects that have tried to introduce grazing control in the pastoral zone have failed, principally because they have not been able to deal effectively with the need by herders to be highly mobile and flexible in their movements. Third, the labor and capital of herders appear to be important and relatively inelastic constraints on herd size. This implies that the optimum level of herding inputs in Figure 11.1 is probably not too far to the left of the level that would occur in the absence of any controls, though this also depends on the slopes of the SMC and MB curves in that diagram, about which less is currently know than the PMC curve. Fourth, overgrazing, to the extent that it exists in the Sahel, does not appear to have seriously decreased production per hectare compared with other areas of the world where overgrazing is presumably less of a problem. Fifth, except around watering points and marketing trails, there is little direct evidence of extensive rangeland degradation due to overgrazing in Mali and Niger; in some cases heavy grazing may have improved the rangeland. Sixth, range management that relies on increased sedentarization and decreased movement of animals will likely result in a decline, not an increase, in livestock production.

If these conclusions are correct, the most important implication for livestock projects in the pastoral zone is the need to relate projects to people rather than to land. In fact, the NRL project has begun doing this through the formation of herder associations involving 15 to 30 socially related families who travel together. Its successor, the Integrated Livestock Production (ILP) project, will carry on this effort. To the extent that techniques of range management can be successful, moreover, these will probably have to relate to the use of dry season pasture by these herders. The best way to achieve control over land in this case is probably to ensure control over water supplies. Traditionally, this control exists in Niger as long as the herders construct or pay for their own wells. This is not to say that there are no other issues involved. For example, what should be done in years of drought when there is insufficient normal dry season pasture? How is the problem of use of the same pasture by different groups during the wet and dry season to be handled? What is to be done about the encroachment of cultivators on pastoral land?

It is important, however, that livestock projects in the pastoral zone not be held up awaiting solution to overgrazing problems, or that the projects that are underway not be dominated by this issue. Instead, there is a need for better understanding of the problem and monitoring effects of population growth and project interventions on pasture utilization. This should not be done in isolation, but rather as part of a total system involving animal health and nutrition, livestock marketing, and the socioeconomics of pastoral society.

REFERENCES

Aronson, Dan R. Toward Development for Pastoralists in Central Niger; An Interim Synthesis Report of the Work of the Niger Range and Livestock Project. Discussion Paper Number 5, Niger Range and Livestock Project, February 1982.

Eicher, Carl K. and Doyle C. Baker. Research on Agricultural Development in Sub-Saharan Africa: A Critical Survey. MSU International Development Paper No. 1, Department of Agricutural Economics, Michigan State University, 1982.

Hardin, Garrett. "The Tragedy of the Commons." Science, 162 (1968): 1243-48.

Hopcraft, Peter N. "Economic Institutions and Pastoral Resource Management: Considerations for a Development Strategy." The Future of Pastoral Peoples, ed. John G. Galaty, Dan Aronson, Philip Carl Salzman, and Amy Chouinard, pp. 224-243. Ottawa: International Development Research Centre, 1981.

Penning deVries, F.W.T. and M.A. Djiteye. La Productivite des Paturages Saheliens. Wageningen: Center for Agricultural Publishing and Documentation, 1982.

Simpson, James R. and Gregory M. Sullivan. "On Institutional Change in Utilization of African Common Property Range Resources." Unpublished Paper, April 28, 1983.

USAID/Niamey. Niger Integrated Livestock Production Project Paper. Tufts University, May 31, 1983.

Wilson, R.T. and K. Wagenaar. An Introductory Survey of Livestock Population Demography and Reproductive Performance in the Area of the Niger Range and Livestock Project. Arid and Semi-Arid Zones Programme, International Livestock Centre for Africa, Bamako and Addis Ababa, 1982.

12
Livestock in African Culture and Society: A Historical Perspective

Harold K. Schneider

The purpose of this paper is to present an historical perspective on pastoralism in Africa, particularly East Africa, where the largest concentrations of cattle are (and which I know best). The emphasis is on how African people have viewed and used livestock based on what we know of them since first reports of their activities began to appear. The discussion will be limited by special attention to "development", which shall be interpreted as a shift of management of livestock from the classic mode which will be described, to integration with the goals of national and international economies. What this means is livestock become less important as repositories of value and more important as beef.

INTRODUCTION

Nearly 60 years ago Melville J. Herskovits, the founder of African studies in the United States, created a landmark in the history of anthropology by publication of an article entitled simply "The Cattle Complex in East Africa" [Herskovits 1926]. One quote from it will summarize the thesis Herskovits wished to express:

In East Africa, where currency in any form is absent, cattle constitute an almost exclusive hall-mark of wealth. The subsistence economy of these tribes is based on agriculture; but the number of cattle owned by a man correlates highly with his position. That is, among these people, as in most societies, position is related to wealth and cattle are the sole expression of wealth. It is of no consequence how much cultivated land or other goods a man possesses, for should he not have adequate resources in cattle, he can have no place of respect in society . . . Notwithstanding this, cattle can in no sense be

considered money; for nothing can be be acquired
with them except wives, and a long time has elaps-
ed since competent students have held to the
earlier naive concept that the giving of cattle by
the family of the suitor to a family of his bride
constitutes an act of wife-purchase. A cow is
eaten only on certain ceremonial occasions, or
when an animal dies; nor have cattle any other
subsistence utility aside from that of supplying
milk, since they are never employed as beasts of
burden. They are merely possessed and esteemed
for the prestige their possession brings. But
they are not money. [1926, pp. 264-265]

In the perspective of time, one must consider "The
Cattle Complex in East Africa" to be a remarkable
achievement, especially considering that it was done in
a library and was based on poor data. It seemed to
recognize, in a blurred fashion, a quality of cattle in
Africa which is only now being brought into focus,
namely that they are repositories of value, even
though he rejected the idea that they are money.
Subsequent research [Schneider 1957] raised serious questions about aspects of Herskovits' thesis. He
had said that cattle are eaten only on certain ceremonial occasions, or when they die, but the Pokot of
Kenya, one of the groups among whom the cattle complex
was said to be prominent, were shown to make up reasons
to eat their cattle. Investigation showed that they
regularly consumed male animals, except for a few bulls
saved for breeding stock, with a frequency which caused
the average herd to be composed of 2/3 females. Even
the females were eaten when no longer useful for breeding.
Ultimately the cattle complex thesis was undermin-
ed and replaced by a trend which now dominates the
interpretation of African uses of cattle, emphasis on
subsistence. Nearly everyone who writes on the subject
attempts to show that the way cattle are utilized is
consistent with maximizing their subsistence use,
principally through extraction of milk, but also by
consuming meat and blood. The other aspect of their
use which Herskovits stressed, their role as repositories of value, is only now beginning to be acknowledged
[Crotty 1980; World Bank 1982].
A balanced view of African pastoralist activity
requires attention to both points of view. Therefore,
the view of cattle as food will be taken up first after
which we can explore the theory of cattle as repositories of value. In the last part of this chapter some
examples of how change is occurring in pastoral societies will be discussed, followed by a concluding consideration of the implications for development.

THE SUBSISTENCE ROLE OF CATTLE

That cattle, along with sheep and goats, or camels in the more arid areas of Africa like Somalia, should be essential for subsistence is a thesis which on the surface looks unchallengeable. In Africa (Figure 12.1) there is a fairly close inverse relationship between rainfall and presence of livestock. This has been interpreted by many to mean that as opportunities to grow crops decline people shift to pastoralism to fulfill their subsistence needs. The distribution of high ratio groups follows a pattern, generally increasing as one moves from southwest to northeast into the more and more arid zone, until one reaches the semi-arid Somalia zone where camels are held in the largest numbers. There is a major exception to this generalization in much of Tanzania, which has relatively low rainfall but few cattle, an exception addressed later.

From these facts, some researchers, like Konczacki [1967] and Dyson-Hudson [1981], have concluded that the more arid a region, the larger the herds kept by pastoralists as insurance against the greater risk of famine. To Croze and Gwynne [1981], pastoralists are people "chasing protein across the landscape" [Dyson-Hudson 1981, p. 353].

Attempts to defend the primacy of subsistence in the valuation of livestock have run into some rather damaging arguments, especially in recent years. One of the most impressive came out of the work of Dahl and Hjort [1976], who used what data were available on East Africa to try to determine how herd size and composition would relate to the production of milk in sufficient quantities to support pastoral people dependent on it. Taking into account that the type of zebu utilized by Africans matures typically at about 250 kg, is a poor milk producer, and will not drop its milk unless there is a calf at the side, their simulations suggest that for a family of about six members to subsist on milk 60 head of cattle would be required during the wet season and ten times that amount during dry season. As Table 12.1 shows, the ratio of a minimum of 10 to 1 is very rare in East Africa and the dry season ratio of 100 to 1 is nearly unheard of. Therefore, milk cannot be a dominant staple food except on exceptional occasions, as when young men move the herds to better pastures during the dry season and subsist on milk during their movement.

As for beef, Aldington and Wilson [1968], calculating the amount of meat slaughtered from the sale of hides (which assumes that all hides are sold), showed that the herds of Kenyan pastoralists are not large enough to produce sufficient beef to maintain the populations, even if supplemented with milk. The 39 ethnic

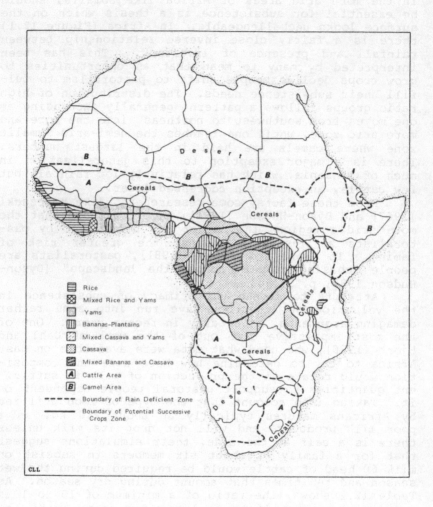

Figure 12.1. Distribution of staples and dominant livestock in Africa

Source: _The Africans_, by Harold K. Schneider, adapted from Bennett [1962].

Table 12.1.   Cattle per person in a sample of
East African pastoralists

Cattle per Person	Pastoral People

-Head-

Agriculture Lacking or Insignificant

18.0	Barabaig of central Tanzania
17.5	Samburu of Kenya
15.0	Maasai of Tanzania
9.0	Rendille of northern Kenya
8.0	Dorobo of Kenya; Uganda Pokot
6.5	Borana Galla of northern Kenya
6.0	Kenya Maasai

Agriculture Important in Varying Degrees

4.0	Karimojong of Uganda
3.7	Jie of Uganda
3.6	Dodoth of Uganda
3.0	Kenya Pokot
2.0	Kipsigis of Kenya
1.7	Meru of Kenya
1.4	Teso of Uganda
1.3	Giriama; Kitui Kamba
1.2	Taita
1.1	Turu of Tanzania
1.1	Machakos Kamba

Source: Schneider [1979].

groups these researchers studied consumed on average
only about 10 kg of meat per person per year. Even 'the
Turkana of northwest Kenya, one of the wealthiest
nomadic groups, consume only about 5 kg per person per
year despite having herds averaging more than 10 head
per person. This may be compared to the average American who, in 1976, ate about 60 kg [Ward, Knox, and
Hobson 1977].

These conclusions may be illustrated by reference
to the Turu of central Tanzania who had about 225,000
cattle for a population of 175,000 people in 1959-60,
giving a ratio of nearly 1.3:1. Various data suggest
that the maximum number of animals that can be taken
off the "national" herd for sale or slaughter cannot
exceed 13 percent in the traditional management system.
From this one can calculate an annual supply of animals
for the Turu as 29,500 head. Taking the average carcass weight of half the standing weight this yields an
average of 20 kg of meat and bone per person per year.
Of course the Turu do not eat all the animals they dispose of; in 1959-60 about 7 percent were sold, mainly
to pay taxes. In addition, as a person's herd size
increases, animals are loaned out to other people
creating a network of stock associates which enhances
the initial person's power. Even marriage, for which
livestock are nearly always used, is a kind of stock
associateship and a huge consumer of animals. It is
facts like these which also explain why people like the
Turkana slaughter only enough animals to provide 5 kg
of meat per person per year, assuming that Aldington
and Wilson's assumption that hide sales measure slaughter volume is correct, and taking into account that the
measure of meat consumed includes goats and sheep,
whose volume they show, only increased the amount of
meat available for consumption by a small amount.

Another damaging argument against the subsistence
thesis is that a substantial percentage of the comparatively wealthy pastoral people live in lush grasslands
and do not farm. A notable example is the Narok Maasai
in Kenya, who graze their animals on land which the
government of Kenya would like to convert to the wheat
cultivation. In Kenya 50 percent of the arable land is
occupied by people who possess 77 percent of all the
cattle in the country [Meyn 1970]. If subsistence were
foremost in the minds of these livestock owners one
would expect them to raise crops as well as animals or
even to abandon pastoralism for agriculture.

MONETARY ROLE OF CATTLE

Herskovits' cattle complex also stressed the role
of cattle in conferring prestige. It is this aspect
which is now often overlooked. Implicit in it are two

dimensions, that cattle are repositories of value and sources of power. Cattle hold value and their control confers power on their owner, who through patronage can control other people. The role of cattle as repositories of value can be demonstrated in various ways, beginning with their use as media of exchange.

The subsistence thesis was given inferential support by the anthropological practice of viewing each group of people as isolated from all others. If a "tribe" like the Turkana is indeed isolated, then they must obtain subsistence through the only means available to them, their herds. A regional perspective is increasingly emerging as the economic facts are exposed. Little [in press], in a recent dissertation on the agropastoral Il Chamus near Kenya's Lake Baringo, reveals an integrated system composed of both the valley Il Chamus and the Tugen of the nearby hills, between whom there is trade for grain and cattle using livestock as the medium of exchange. The Maasai, Turkana, Il Chamus and Somalis, to name only a few, all trade livestock of all types for grain and other goods. While milk and meat are important to the diets of these people, grain is also important and for many the principal staple food.

There is also competition between the monetary and subsistence roles as in bridewealth payments which are always relatively high and proportional to the rights obtained in a woman: most importantly rights to her children and rights to her labor, agricultural and otherwide. The average bridewealth among the Turu in 1959-60 was about 7 head of cattle. Since they have only about 1.3 head of cattle per person the average family would possess less than this. Since marriage is critical to the establishment of a production unit it must follow that the values residing in a women which a man will receive for cows will determine whether the cows will be used as money or for subsistence. The terms of trade, then, determine whether and to what extent livestock will be used for subsistence. In these exchanges, cows are valued over male animals because they reproduce and hence pay dividends. Hearty small animals are valued over larger ones because, generally speaking as monetary units, a heifer equals a heifer, whatever the size difference. Even today, with modern currencies competing with indigenous monetary systems, livestock continue to be used to mediate transactions, notably marriage.

Further evidence for this thesis can be derived from examination of the distribution of cattle and tsetse fly in Africa. Where there is fly there are no appreciable numbers of cattle, i.e., cattle ratios are less than one per human; where there are appreciable numbers of cattle there is no fly. And cattle are introduced everywhere that the fly is kept out. These fly-free areas are usually arid, but may be well-watered and fertile, such as the northeast quadrant of the Lake Victoria shore where people like the Kipsigis, Nandi and Teso raise cattle in high ratios on land which they also cultivate. Conversely, not all arid areas are good for cattle; much of Tanzania and Central Africa contains people who engage in swidden and other forms of labor intensive agriculture because the presence of tsetse fly makes pastoralism impossible.

One must naturally wonder why Africans place so much value on cattle. The temptation is to say because ultimately they provide subsistence. But this makes no more sense than to say we value gold because it can be used to make electronic components. In order to understand African behavior with respect to livestock we need only accept the fact that livestock are valued for themselves, just as understanding the behavior of the Western "economic man" does not require us to explain why he values gold.

DEVELOPMENT

A most pronounced characteristic of pastoralists has been their reluctance to engage in development. Sometimes they have accepted innovations, for example ox-drawn plows in southern Africa and parts of East Africa, but they have seldom agreed to manage their herds for beef production.

If one accepts the subsistence explanation for pastoral management practices, that the people are trying to maximize the utility of cattle for subsistence, then in the arid areas there is no reliable substitute for cattle or other livestock. As for the less arid areas one might argue that engaging in beef production in order to obtain money to buy food is too risky. This risk is in no small part a product of the propensity of African governments to administer prices.

If one accepts that the constraints to disposal of livestock relate to their monetary use then the reasons for failure to change become more obvious. In the first place, cattle as an investment seem more profitable than any other investment available to pastoralists. Dahl and Hjort's [1976] simulations suggest that some herders obtain returns in the form of calves up to 14 percent per annum, although average returns are probably lower. In addition, historically in East Africa the high-ratio (1:1 or better) pastoral people lived in "egalitarian" societies, societies without chiefs or other central authorities [Schneider 1979], in which there emerged a spirit of independence, even arrogance. This condition was apparently the result of the structure of pastoral economic operations, in which opportunity to become wealthy was widespread, the rate of growth of wealth very high, and in which the creation of large networks of stock associates gave political power to individuals, if only for short periods. Given these conditions it is unlikely that pastoralists turning to beef production can expect an annual predictable return on their investment of 14 percent, or that the proceeds can be used effectively to create the status of Big Man (there were occasionally powerful women, but no Big Women as such).

Despite the conservatism of many pastoralists with respect to herd management, change of one type or another has appeared in the past and is continuing today. Examination of some cases, such as the one referred to earlier by Little, are instructive in trying to overcome constraints to development. In that particular case, which might be labeled change without develop- ment, an agropastoral group has had to respond to modern conditions, but the response has taken place without development as it was defined at the beginning of this chapter. Traditionally, the Il Chamus have produced both livestock and grain but, because of opportunity costs they have not raised sufficient grain to satisfy their own needs and so have traded livestock for grain with Tugen in the hills to the west. Government policy has now disrupted this grain trade causing grain imported from this traditional source to be too expensive in relation to the maintenance of their herds. The result is that grain prices to the Il Chamus are now high in relation to livestock prices. Thus, they have begun to grow more grain especially in areas with irrigation. As the price of grain obtained from outside has gone up, the value of grain produced at home has also risen. However, the supply of irrigated land is severely limited. The result is that Il Chamus who managed to get control of irrigated land, mainly those with greater livestock holdings, were able to use the profits from grain sales to build their herds even more, while at the same time the people without irrigated land were forced to sell off their herds to purchase expensive grain. In short, while the Il Chamus still store wealth in livestock and still do not sell off cattle except for the purchase of grain when there is a shortage, class polarization has begun. In contrast to the past, agriculture and pastoralism now compete for first place. Pastoral people are responsive to changing conditions and are price sensitive, but such attitudes do not by themselves generate development.

A second type of response, seen among the agropastoral Teso of eastern Uganda, may be labeled mixed development, although from the point of view of livestock development specialists it might seem to be a failure. Just after the turn of the century the British colonial government in Uganda urged the Teso to grow cotton, to help pay the costs of government, and taught them to yoke their oxen to plows. Surprisingly, they took to cotton growing and the practice spread steadily so that today all Teso grow cotton with ox plows.

This happy outcome can be explained by studying the circumstances [Uchendu and Anthony 1975]. The Teso seem to have found cotton to be a profitable crop in several senses. Their staple food is finger millet (eleusine), whose production they discovered they could increase enormously by double cropping the already prepared ground used for the previous year's cotton. This increased profit from the sale of finger millet and also the production of beer, which they dearly love. They were able to do all this without reducing their cattle herds. In fact, they probably increased them with proceeds from cotton and millet sales, as all East African pastoralists are prone to do with profits from cash earning operations. But they are still unwilling to fence their fields, which would close down the system of common grazing, and they resist government attempts to get them to invest profits from cotton sales in agricultural improvements, such as fertilizer and pesticides. So, after about 80 years of development the Teso have ended up with a partially developed cotton sector but still have an undeveloped cattle sector.

A third kind of change process may be called decomposition, for want of a better word. Aldington and Wilson's study of slaughter offtake rates in Kenya showed a bimodal distribution. The offtake rate among more conservative pastoralists, like the Maasai, varied between 7 and 13 percent, but among people like Kamba, Meru, and Taita the rate varied from 17 to 24 percent. The higher offtake rate in the latter groups might be attributed to the fact that these people have shifted to beef production. But a study of the Mbeere of Mt. Kenya [Mwaniki 1982] shows that their high rate of offtake is due to the implementation of government-sponsored individualization of land tenure. The previously common lands have been broken up into closed, privately owned plots, thereby making herd management, as well as agricultural, in the traditional way impossible and forcing the sale of animals. Today Mbeere society is in a state of turmoil with a large increase in murder and assault as the people fight over possession of land. Mbeere society is being decomposed, its livestock and grain production withering away and its social structure coming apart.

A fourth process, which may be labeled develop-

ment, consists of the commercialization of pastoral
operations among a group of Kajiado Maasai, the
Kaputiei [Hedlund 1971]. In 1960-61 a major drought
struck in East Africa killing 65 percent of the total
herd of 300,000 animals in Kajiado District. At that
time the Kaputiei Development Committee was formed,
consisting of 28 families who split up 56,000 acres
into individual ranches. The members of these new
individual ranches withdrew from traditional stock
associate membership, closed their ranches to grazing
by other Maasai, and would not lend money or cattle to
Maasai who were not of the group. Most importantly,
they started to sell more cattle than others and opened
bank accounts. Evidently what happened is that after
the famine caused by the drought and the huge loss of
wealth these Maasai suffered they decided that the
potential for speedy recovery of wealth through beef
ranching was greater than through the traditional way.
This shift to private land ownership was, of course,
incompatible with communal ownership of land, stock
associateship and other forms of cooperation with
indigenous Maasai. The banding together of these
Maasai to help each other accomplish this change was
probably the beginning of hierarchy since grazing land,
by being shifted from open to closed range, becomes
amenable to exclusive control, like agricultural land,
and therefore a potentially scarce commodity, insuring
that in the end some Maasai will be without land, and
forced into relations of clientship with rich Maasai
patrons.

CONCLUSION

Anthropologists have often justifiably been perceived by others as conservative, defending the status
quo of the people they study. But this way of thinking
about pastoralists in Africa seems unjustified. They
are no more romantically wedded to one way of doing
things than are people of the industrial world. However, in discussing the constraints on expansion of the
beef industry among these people one should be as
explicit as possible about why pastoralists are unco-
operative in efforts to bring about change in their
herd management practices. The principal conclusion,
at least as regards East Africa, is that high-ratio
(one or more cattle per person) pastoral societies, in
which livestock are employed mainly as repositories of
value, are high opportunity societies with great
political freedom, at least among the men. Attempts to
induce change in such groups runs up against African
suspicions that if they shift to beef production they
will be losing money in the process and also losing
freedom. In this circumstance, the chief constraint to

change is refusal to limit the size of herds, any attempt to do so being interpreted as an attempt to limit their income.

On the other side, anyone who has studied the process of development extensively and on a worldwide scale recognizes that political polarization is very common, one might say universal, although, led by Julius Nyerere, Tanzania has rejected inequality in development as official policy. Furthermore, the situation described here for East Africa is not typical of all pastoral Africa. The large ratio livestock societies of southern Africa--Zulu, Tswana, Swazi, etc.--were all hierarchical for reasons that are still not understood. Therefore, methods of overcoming constraints may be classified in two categories--with or without inequality. If one should conclude that inequality must be accepted, at least in the short run, then the problem of overcoming constraints becomes relatively simple. African pastoralists, like other Africans, respond to price incentives. If they are not cooperative this may in large part be because the price of change is not sufficient. The pastoral people were always considered by themselves and other Africans as wealthy. The cost of change is higher for them than those Africans who had few livestock and who therefore resisted change less. If effective incentives can be found they will change their behavior, as did the Kaputiei Maasai.

Other than this, several courses of action seem to be suggested by the examples in this chapter. For one thing, further research is necessary to determine the extent to which the factors described still hold. In addition, research is needed on people such as the Kaputiei Maasai to discover how and why they made the transition to a commercial orientation. The Mbeere case suggests that many of the social constraints could be removed simply by individualizing land tenure, eliminating the old pastoral system, and replacing it with modern, beef oriented systems. Some would view this approach as a draconic tradeoff for increasing total national meat production.

REFERENCES

Aldington, T.J. and F.A. Wilson. "The Marketing of Beef in Kenya." University of Nairobi, Institute for Development Studies, Occasional Paper No. 3, 1968.

Bennett, M.K. "An Agroclimatic Mapping of Africa." Stanford: Food Research Institute Studies 3, 1962.

Crotty, R. Cattle, Economics and Development. Farnham Royal, Slough, England: Commonwealth Agricultural Bureaux, 1980.

Croze, H.J., and M.D. Gwynne. "A Methodology for the Inventory and Monitoring of Pastoral Ecosystem Processes." The Future of Pastoral Peoples, ed. J.G. Galaty, D. Aronson, P.C. Salzman and A. Chouinard, pp. 340-352. Ottawa: International Development Research Centre, 1981.

Dahl, G., and A. Hjort. Having Herds: Pastoral Herd Growth and Household Economy. Stockholm, Stockholm Studies in Social Anthropology 2, 1976.

Dyson-Hudson, R. "Indigenous Models of Time and Space as a Key to Ecological and Anthropological Monitoring." The Future of Pastoral Peoples, ed. J.G. Galaty, D. Aronson, P.C. Salzman and A. Chouinard, pp. 353-369. Ottawa: International Development Research Centre, 1981.

Hedlund, G.B. "The Impact of Group Ranches on a Pastoral Society." University of Nairobi, Institute for Development Studies, Staff Paper 100, 1971.

Herskovits, M.J. "The Cattle Complex in East Africa." American Anthropologist 28(1926): 230-272, 361-388, 494-528, 633-664.

Konczacki, Z.A. "Nomadism and Economic Development." Canadian J. of African Studies (1967): 63-175.

Little, P.D. "The Livestock-Grain Connection in Northern Kenya:" An Analysis of Pastoral Economic and Semi-Arid Land Development." Rural Africana, in press.

Meyn, K. Beef Production in Africa. Munich: Weltform Verlag, Institut fur Wirtschaftsforshung, 1970.

Mwaniki, N. "Social and Economic Impact of Land Reform in Mbeere." University of Nairobi, Institute for, Development Studies, Working Paper No. 391.

O'Conner, A.M. An Economic Geography of East Africa. New York: Praeger, 1966.

Schneider, H.K. Livestock and Equality in East Africa. Bloomington: Indiana U. Press, 1979.

_____. "The Subsistence Role of Cattle Among the Pakot and in East Africa." American Anthropologist, 59(1957): 278-300.

Uchendu, U.G., and K.R.M. Anthony. Agricultural Change in Teso District, Uganda. Nairobi: East African Literature Bureau, 1975.

Ward, G.M., P.L. Knox and B.W. Hobson. "Beef Production Options and Requirements for Fossil Fuel." Science (1977): 265-271.

World Bank. Market Forces and Livestock Development in Africa. The Center for Research on Economic Development, U. of Michigan, Sept. 1982.

13
Critical Socio-Economic Variables in African Pastoral Livestock Development: Toward a Comparative Framework

Peter D. Little

Analysis of the sociological dimensions of African pastoralism is a formidable task. This challenge is not so much due to the sociological complexities of African pastoral production systems, but rather to the difficulty of saying much which is comparatively useful, especially in establishing a general development strategy for Subsaharan Africa's livestock sector. Most sociological studies of African pastoralism are of a relativistic nature limited to analyses of one particular society or herding population. Often there is little comparative attention to pastoral systems outside of the specific study area. The regional differences in Africa's pastoral production systems—for example, between southern African pastoralism and Sahelian pastoralism—make general statements and models problematic. However, a general model or framework for development planning and policy needs to incorporate the socio-economic variables instrumental in the livestock development process.

The emphasis in this chapter is not on reiterating the general socio-economic problems and constraints in African livestock development, for they have been adequately summarized in several places [Horowitz 1980; Galaty et al. 1981]. Rather the focus is on eliciting certain key socio-economic indicators or "circumstances" [Sandford 1981] that can be identified in pastoral areas, and which can be used to develop a comparative analytical framework. This allows some comparison of pastoral development activities among the different regions of Subsaharan Africa and consequently permits easier incorporation of past development experiences and lessons into the design of present policy and planning concerns. A major problem at present is the lack of "transferability" of information and experiences pertinent to pastoral development among the different African regions [International Livestock Centre for Africa 1980]. In addition, the elaboration of such a framework can indicate where socio-economic information is lacking and type of data that must be collected prior to the design of any African livestock development program. It should be noted that several of the variables discussed below might logically come under the field of economics or agricultural economics. The term socio-economics is used here to explicity emphasize the interconnectedness of sociological and economic variables in the African pastoral context-- thus the blurry distinction between sociological and economic factors.

IDENTIFICATION OF KEY SOCIO-ECONOMIC INDICATORS

Pastoral production systems[1] in Africa range from the sedentary economies of highland Eastern Africa to the nomadic production systems of the northern Sahel, with several permutations between the two. The political and macro-economic environment that these systems operate within also vary considerably according to political and geographical boundaries. For example, the often cited case of inefficient livestock pricing policy, and consequent production disincentives [Little 1980], is not universally shared among African coun- tries. Thus, while in Kenya government pricing policy is disfavorable for many herders, in Botswana "the terms of trade in relation to other food commodities is very favorable for livestock producers" [Roe and Fort- mann 1981, p. xxv]. Given such variation of African pastoral systems and their macro-environment, a general socio-economic framework should be predicated on an identification of variables which are common in each context, although not necessarily equally important. Below are the most significant socio-economic indicators that should be examined at the planning or feasibility stage of a livestock development project. Where possible the socio-economic variables are illu- strated with specific reference to recent research and development experience.

Identification of Appropriate Production Units

Unlike certain farming communities, the production unit in the pastoral zone may be limited to a collec- tion of households or families, with some intra-annual variation based on climatic and ecological variables. In many of Africa's pastoral communities, aggregated production units are most common during the dry seasn when long migrations or transhumances are necessary. Similarly, independent family production units may split into two or more units during part of the year. For example, among the Karimojong of Uganda herding camps composed mainly of young men split off from the main domestic units during the dry season [Dyson-Hudson 1966]. Failure to recognize the appropriate production unit can lead to miscalculations regarding domestic labor availability, family investment patterns, and general production constraints.

Livestock Ownership and its Distribution

This is an important variable to consider when designing programs aimed at the poorer segments of the pastoral population. Contrary to common belief, livestock ownership in most pastoral areas of Africa is highly skewed, often with 10 percent of herders controlling up to 50 percent of livestock units and large ownership segments (often 20 to 30 percent of households) having little or no access to livestock. Figure 13.1 shows the distribution of livestock units among a pastoral group (Il Chamus) of Eastern Africa. Another exaggerated case of this is Botswana where in the livestock sector the poorest fifty percent of households obtain only 7 percent of their total income from livestock [Agency for International Development 1980]. Under such circumstances, livestock sector interventions such as veterinary services, water development and marketing inputs would have little benefit for the poorest segment of the population. The expansion of nonfarm employment opportunities would probably have a more favorable impact on the poorer stock owners.

The ownership/management distinction which is so common in the Sahel creates further problems in distinguishing ownership patterns. For example, in northern Upper Volta many of the Fulani herders are managing cattle that are owned by sedentary farmers [Delgado 1979]. This system, often called "cattle entrustment", creates a situation where the herding or management unit is not consistent with the ownership unit. A recent variation on this theme is the practice of pastoralists herding animals owned by civil servants or businessmen [Little 1983; White 1982]. This is particularly common in Sahelian countries such as Niger where the drought left many herders destitute. Evidence from post-drought Sahel suggests that many herders remain stockless and depend upon the herding of nonpastoral owned cattle for their livelihood.

Extent of Economic Diversification

The trend in many of Africa's animal-based systems is toward increased diversification of production and investment strategies. While many pastoral systems always included an important cropping component, the significance of this element as well as other nonpastoral activities, such as migration for wage employ-

204

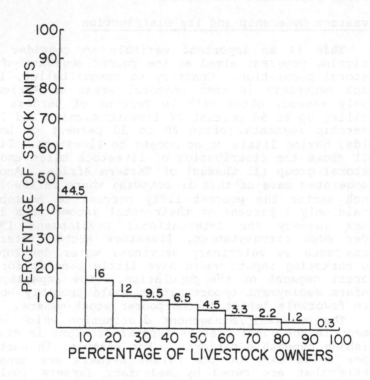

Figure 13.1. Ownership and distribution of livestock: Njemps
Location, Kenya

Source: Based on Little [1983].

Note: Stock Unit is equivalent to one bovine or six small
stock. This ratio approximates market equivalency in
Njemps Location, as well as the commonly used Tropical
Livestock Unit.

ment, is on the increase. In some cases, the strategy may be in response to pastoral "impoverishment"--often due to drought, disease, or deteriorating profitability; in other situations, it may be the "pull" factor of increased economic returns in a nonpastoral sector. For example, it recently has been noted for Somalia that "labor migration and permanent migration to towns from the pastoral sector appear to be a response to opportunities for higher incomes rather than to worsening environmental or economic conditions in pastoral areas" [Hoben et al. 1983, p.5]. The wage migration factor also is prominent in Botswana's pastoral areas, and is increasingly important in the rangelands of both Kenya and Niger [Horowitz 1983]. In Niger, it may create pastoral labor shortages at critical times in the production cycle [White 1982].

The diversification factor is important to identify regarding the appropriateness of mono-sector (livestock) versus integrated rural development programs. Where economic diversification is the norm, strict livestock sector investments may be misguided. Reference to a pastoral zone project in Kenya illustrates this point well. In this particular development project (Baringo Pilot Semi-Arid Area Project), the cropping component of the project proved more popular to local herders than assistance in livestock-related activities. The economic trend in Baringo over the last several years has been toward increased diversification, and the project's emphasis on cropping was well timed. Similarly, among central Niger's herders improved access to reasonably priced grain is considered a top priority for assistance [Tufts University 1982].

Local Resource Management and Pastoral Tenure

Several recent publications [Gilles and Jamtgaard 1982; Horowitz 1980; Draz 1978] point to the presence of local social devices (either formal or informal) for regulating access to grazing and water in pastoral areas. These institutions or regulations may be highly formalized, such as in the Dina (Mali) or Hema (Syria) systems, or they may be of a less structured nature. Nevertheless, the picture of "uncontrolled" exploitation of common grazing lands and water [see Hardin 1977] is not warranted by the empirical evidence. The importance of identifying these types of institutional devices is overwhelming, particularly when the program is focused on range management.

The extent to which political, economic and other factors have disrupted these indigenous mechanisms should be explored. For example, in northern Kenya it was found that the indigenous tenure system (called

Olokeri presently is ineffective at regulating grazing [Little 1983]. The change is related to the decline in the local elders' political power (they traditionally controlled the Olokeri system) and to the out-migration and increased nonpastoral activities among young men (in the past, they enforced the Olokeri system). While until the 1960s the Il Chamus under the Olokeri system restricted dry season grazing zones, little local effort is made today to enforce this pattern. However, it is important that before imposing a novel set of grazing regulations, efforts to restore the indigenous, more familiar system should be made. As Hoben et al. remark, "only when the indigenous institution forms can be shown to be ineffective and nonadaptable should new mutual grazing insurance mechanisms be created" [1983, p.4].

The Spatial and Temporal Dimensions of the Pastoral Economy

This indicator represents the "where" and "when" of pastoral livestock movements; that is, the areas pastoral livestock move to and the months such movements occur. The mobility and spatial dimensions of African pastoral production systems vary considerably among the different geographical regions. In general, the more mobile and nomadic a population, often the more difficult it is to apply standard development interventions, such as veterinary, marketing or range improvement services.

In many pastoral areas of Africa, the spatial aspects of herd movements have been transformed in recent times due to the loss of indigenous grazing lands. This type of encroachment perhaps is best illustrated in the Kenya Maasai case. In the past century, the Maasai of Kenya have lost access to grazing on the Laikipia Plateau, the Ngong Hills (outside of Nairobi), the Mara Plains, the Amboseli swamp, and the Mau Forest areas (in Narok District). Total losses are equivalent to more than 1,000 square miles of prime grazing area. The present trend in Massailand, as in most pastoral areas of Africa, is not toward recovery of lost grazing areas, but rather toward further depletion. Such situations are very difficult to redress given the political environment and anti-pastoral bias of many African governments. Loss of grazing land often is manifested in increased diversification (e.g., off-farm employment and crop production) and accelerated degradation of the rangelands.

The Terms of Trade for Livestock Producers

African livestock producers have been shown to be very responsive to changing conditions of profitability and trade [Kjaerby 1979; Little 1983; Sutter 1983]. The contemporary pastoralist reliance on imported foods and other consumption items makes terms of trade calculations very important. If time series data on prices exist for the particular area analyses should be based on as many entries as feasible, in order to distinguish short-term price fluctuations from long-term trends. Since agricultural commodity price changes in Africa are at times due more to government policy rather than to market forces, it is not surprising that livestock trade conditions reveal considerable regional variation. As already indicated, the terms of trade for livestock producers in Botswana is more favorable than in certain eastern African countries.

The terms of trade indicator should provide evidence of profitability changes over time in the pastoral production system. Since the main import item in most pastoral economies is grain, pastoral product price changes vis-a-vis grain price is the most important relationship to assess [Little 1983]. Further analysis of pastoral production profitability relative to other economic opportunities (i.e., the opportunity costs of pastoral production) should be supported by production cost data.

Articulation with the Regional Economy

What is seldom indicated in pastoral socio-economic studies is that pastoral systems often are subsystems of larger regional economies. The region itself might include an agricultural sector and an urban (nonfarm) sector. Failure to recognize this broader context leads to a misunderstanding of the causal elements in the pastoral subsystem. For example, in the regional setting changes in marketing patterns in a nonpastoral sector might have a more significant impact on the pastoral production system than varibles endogenous to the pastoral economy. Pastoral areas where the regional framework proves most useful are in areas of regional economic diversification. Predominant intra-regional linkages likely are to be in trade, particularly in livestock and grain.

The above indicators are all factors that can be identified, in may cases quantitatively, in each pastoral situation. They are not static indicators in that they can change drastically in a relatively short period of time. Also, the socio-economic variables discussed incorporate an element of change and, as such, the research effort required for each of the

variables differs. For the first five factors discussed, intensive household and community-level investigation is required. The broader issues of terms of trade
and regional linkages does not necessitate in-depth
production level research. It should be noted that the
socio-economic data base for many pastoral areas,
particularly in the Sahel and in parts of eastern
Africa, has improved dramatically over the past five
years. However, in many regions where livestock development programs are planned or underway, the socio-
economic data base is very inadequate. Ideally, the
socio-economic research noted above should be carried
out in conjunction with ecological and animal science
research.

It is important that the limitations on pastoral
research results, whether quantitative or qualitative,
be recognized, particularly when the research period is
less than three years. Such socio-economic investiga-
tions may be unduly influenced by the particular climatic conditions (drought or good rainfall years) at the
time of research. The investigator should account for
the impact of short-term climatic fluctuations.

PASTORAL RESEARCH METHODOLOGY

There is far greater consensus as to the critical
sociological issues in pastoral development [see Galaty
et al. 1981], than there is to how these parameters
should be examined. In fact, very little attention is
given to the methodological issues concerned with pastoral research. Even field-oriented methodologies in
agriculture (e.g., farming systems research) demonstrate little concern for livestock-related research.
Economic and social anthropologists, on the other hand,
who have dominated the arena of African pastoral
research have paid little attention to methodology,
although recent accumulation of quantitative data suggests that rigorous methodologies exist.

An adequate theory of pastoral production must
form a pastoral research framework. Here scholars
differ very much on exactly what is the nature or
theory of pastoral production (for contrasting views,
see Dahl and Hjort [1976] and Schneider [1979]). In
contrast to earlier research, recent literature
espouses a pastoral production model that views the
pastoral enterprise as both a production and investment
unit [Haaland 1977]. The rationale of pastoral producers is not merely to maximize food (milk, blood and
meat) production, but also to maximize investment
returns on animals, i.e. in the form of reproduction.
This latter perspective characterizes the pastoral
production unit as much more "outward-oriented" in its
production, marketing and investment strategies. This

paradigm influences the above selection of socio-economic indicators.

The systems approach typical of farming systems [Norman, Simmons, and Hays 1982; Shaner, Philipp, and Schmehl 1982; Gilbert, Norman, and Winch 1980] and ecological [Dyson-Hudson 1977] research is warranted in pastoral studies. While much can be learned from the farming systems approach, there are several difficulties with applying it to a pastoral production system. First, it assumes that the household is the appropriate unit of analysis and that a model of the production system should be at this level. As we have indicated above, other organizational forms (e.g., a collection of households) may be important production units, a point that others also have made for crop-based systems [Behnke and Kerven 1982; Guyer 1981]. Secondly, pastoral production systems often are best understood when examined from a regional perspective [Bates and Conant [1981; Hjort 1981]. The farming systems approach does not adequately address regional (macro/micro) linkages. In only a few African cases (Reeves and Frankenberger [1982], for example,) are regional marketing factors treated as an important component of the farming system. Quite often marketing and other regional variables are shown as exogenous to the farming systems model impending analysis of the entire food system (including both marketing and production).

An approach for examining the socio-economic indicators cited above should combine a production or farming systems orientation with regional analysis. Regional analysis has been demonstrated to be a useful tool for assessing agricultural production constraints (e.g., marketing or credit) at the nonhousehold level [Appleby 1976; Smith 1976]. In some cases, the region is the spatial unit that mediates between national policy and local production systems. As indicated earlier, many pastoral production constraints are best addressed within a regional framework.

The combination of regional analysis and production level research is best achieved when the research agenda is framed in the local production context. In such cases, it is best to identify a local production problem (e.g., critical labor shortages, or food bottlenecks in the production system) and then to address the regional dimension of the constraint(s). The constraints/problems identified at the production level should define the region--rather than relying on arbitrary regional/administrative boundaries. This problem-oriented approach is at the core of farming systems research [Norman, Simmons, and Hays 1982]. Our perspective differs from the latter, however, in that the production problem is traced to the reginal level.

LINKING SOCIO-ECONOMIC RESEARCH TO PASTORAL LIVESTOCK
DEVELOPMENT: SOME CONCLUDING REMARKS

It can be asserted that the prime beneficiaries of
pastoral livestock development programs should be the
herders themselves rather than the range, the animals
or the domestic and external consumers [Institute for
Development Anthropology 1980].
Obviously this priority is not universally shared.
Nevertheless, if the primary objective is to improve
the well-being of herders, it often follows that this
implies supporting the indigenous livestock production
system.   Other objectives--an improved livestock production system, increased offtake for markets, or
improved range productivity--should be consistent with
this goal. The way in which these objectives relate to
the design of specific development programs and are, in
turn, shaped by socio-economic variables, is discussed
below.
One of the deficiencies commonly cited in the
literature is that livestock project objectives and
strategies are rarely based on a sound understanding
of the pastoral system [Horowitz 1980].   In other
words, livestock development planners often have mis-
understood important variables in the pastoral context.[2] Certain information requirements need to be
fulfilled before project goals can be expressed
accurately. The relationship between minimum socio-
economic data requirements and some of the more
frequently stated objectives of livestock interventions
is exemplified as follows.

Objective (1) and Required Socio-Economic Information

Retard range degradation through the establishment
of a grazing block system and producer cooperative.
For this objective there must be an understanding of
present grazing patterns in the area, local institutions and grazing tenure laws, appropriate pastoral
production units and feasibility of combining these
into producer cooperatives, ownership distribution of
livestock and how this will be affected by proposed
project.

Objective (2) and Required Socio-Economic Information

Increase volume of marketed livestock through increased herd offtake and a better vertically integrated
market system.   This objective requires data on livestock marketing systems and their contribution to
herder income, wealth distribution and impact upon this
of marketing changes, cash investment patterns among

concerned herders, herd structures and potential for increased offtake of males.

Data collection at the planning stage should indicate the appropriateness of the above stated objectives, as well as point out where additional information is needed. It also should reveal where project objectives conflict with the goal of improving the "well-being of the herder". For example, measures to improve range conditions through restricted grazing often impede livestock movements required for combating seasonal fluctuations in rainfall and fodder production. This problem was associated with many of the colonial grazing block schemes in eastern Africa.

To conclude, several key sociological variables that should be investigated during the planning stages of a livestock development program have been discussed in this chapter. The emphasis has been on eliciting an analytical framework that allows regional comparisons of pastoral research and development experiences in Africa. It should be noted that socio-economic research often requires considerable field time. In many cases, initial socio-economic research should be followed up with the establishment of a monitoring and evaluation unit during project implementation [International Livestock Centre for Africa 1980]. While extensive socio-economic research often is unpalatable to donor organizations and African governments, when the research is focused on a limited set of variables it is more useful and cost effective for development needs.

NOTES

[1]The phrase "pastoral production system" does not exclude mixed production systems where cropping may account for as much as 50 percent of rural income. However, the phrase as used here does not include commercial ranching schemes or other capital intensive ventures. These livestock enterprises are not dealt with in this paper.

[2]There are exceptions to this and one recently cited is the World Bank's Eastern Senegal Livestock Development Project [Korten 1981].

REFERENCES

Agency for International Development. Country Development Strategy Statement: Botswana. Washington, DC, 1980.

Appleby, G., "Export Monoculture and Regional Social Structure in Puno, Peru." Regional Analysis: Volume I, Economic Systems, ed. C. Smith, pp. 147-178. New York: Academic Press, 1976.

Bates, D. and F. Conant. "Livestock and livelihood: A Handbook for the 1980s." The Future of Pastoral Peoples, ed. J. Galaty, D. Aronson, P. Salzman,and A. Chouinard, pp. 89-100. Ottawa: International Development Research Centre, 1981.

Behnke, R. and C. Kerven. "FSR and the Attempt to Understand the Goals and Motivations of Farmers." Culture and Agriculture, No. 19 (Spring 1983), pp. 9-16.

Dahl, G. and A. Hjort. Having Herds. Stockholm: Stockholm University Press, 1976.

Delgado, C. Livestock Versus Foodgrain Production in Southeast Upper Volta: A Resource Allocation Analysis. Ann Arbor, Michigan: Center for Research on Economic Development, 1979.

Draz, O., "Revival of the Hema System of Range Reserves as a Basis for the Syrian Range Development Program." Proceedings of the First International Rangeland Congress, ed. D. Hyder, pp. 100-103. Denver: Society for Range Management, 1978.

Dyson-Hudson, N. Karimojong Politics. Oxford: Claredon Press, 1966.

Dyson-Hudson, R. "An Ecosystems Approach to East African Livestock Production Systems." East African Pastoralism: Anthropological Perspectives and Development Needs. Addis Ababa, Ethiopia: ILCA, 1977.

Galaty, J., D. Aronson, P. Salzman, and A. Chouinard, eds. The Future of Pastoral Peoples. Ottawa: International Development Research Centre, 1981.

Gilbert, J., D. Norman, and F. Winch. "Farming Systems Research; A Critical Appraisal." Rural Development Paper No. 6., Michigan State University, 1980.

Gilles, J. and K. Jamtgaard. "Overgrazing in Pastoral Areas." Nomadic Peoples, No. 10 (April 1982), pp. 1-10.

Guyer, J. "Household and Community in Africa." Paper presented the Annual Meetings of the African Studies Association, Bloomington, Indiana, 1981.

Haaland, G., "Pastoral Systems of Production." Landuse and Development, ed. P. O'Keefe and B. Wisner, pp. 179-193. London: International African Institute, 1977.

Hardin, G., "The Tragedy of the Commons." Managing the Commons. ed. G. Hardin, San Francisco: Freeman and Co., 1977.

Hjort, A. "Herds, Trades, and Grain: Pastoralism in a Regional Perspective." The Future of Pastoral Peoples, ed. J. Galaty, D. Aronson, P. Salzman, and A. Chouinard, pp. 135-142. Ottawa: International Development Research Centre, 1981.

Hoben, A., T. Ahlers, D. Aronson, J. Harris, and S. Hoben. Somalia: Social and Institutional Profile. Boston: Boston University African Studies Center, 1983.

Horowitz, M. The Sociology of Pastoralism and African Livestock Projects. Program Evaluation Report No. 6, Agency for International Development, Washington, DC, 1980.

_____. Niger: A Social and Institutional Profile. Binghamton, New York: Institute of Development Anthropology, 1983.

Institute for Development Anthropology. "The Workshop on Pastoralism and African Livestock Development." Program Evaluation Report No. 4, Agency for International Development, Washington, DC, 1980.

International Livestock Centre for Africa. Pastoral Development Projects, ILCA Bulletin 9. Summary of paper and discussion abstracted from the Workshop and the Design and Implementation of Pastoral Development Projects for Tropical Africa, Addis Ababa, 25-29 February 1980.

Kjaerby, J. "The Development of Agro-Pastoralism Among the Barabaig in Hanang District." Bureau for Research and Land Use Planning Research Paper No. 56, University of Dar es Salaam, Tanzania, 1979.

Korten, D., "Social Development: Putting People First." Bureaucracy and the Poor: Closing the Gap, ed. D. Korten and F. Alfonso. New York: McGraw-Hill, 1981.

Little, P. "From Household to Region: The Marketing/Production Interface Among the Il Chamus of Northern Kenya." Doctoral Dissertation, Indiana University, 1983.

_____. "Pastoralism and Strategies: Socio-Economic Change in the Pastoral Sector of Baringo District, Kenya." Working Paper No. 368, Institute for Development Studies, University of Nairobi, Kenya, 1980.

Norman, D., E. Simmons and H. Hays. Farming Systems in the Nigerian Savanna: Research and Strategies for Development. Boulder: Westview Press, 1982.

Reeves, E. and T. Frankenberger. "Farming Systems Research in North Kordofan, Sudan." Report No. 2, College of Agriculture, University of Kentucky, 1982.

Roe, E. and L. Fortmann. "Water Use in Eastern Botswana: Policy Guide and Summary of the Water Points Survey." Ithaca: Cornell University Center for International Studies, 1981.

Sandford, S. "Organizing Government's Role in the Pastoral Sector." The Future of Pastoral Peoples, ed. J. Galaty, D. Aronson, P. Salzman, and A.

214

Chouinard, pp. 270-283. Ottawa: International
Development Research Centre, 1981.
Schneider, H.K. Livestock and Equality in East Africa:
The Economic Basis for Social Structure. Bloom-
ington: Indiana University Press, 1979.
Shaner, W., P. Philipp and W. Schmehl. Farming Systems
Research and Development: Guidelines for Devel-
oping Countries. Boulder: Westview Press, 1982.
Smith, C., ed. Regional Analysis: Volume I Economic
Systems. New York: Academic Press, 1976.
Sutter, J. "Commercial Strategies, Drought, and
Monetary Pressure: Wo'daa'be' Nomads of Tanout
Arrondisement, Niger," Nomadic Peoples, No. 11
(October 1982), pp. 26-60.
Tufts University. "Niger Integrated Livestock Produc-
tion Project." Unpublished Report, 1982.
White, C. "WoDaaBe Economic Subsectors." Niger Range
and Livestock Project Report, Niamey, Niger, 1982.

14

An Applied Approach to Range Management in Subsaharan Africa

Alex Dickie, IV and James T. O'Rourke

INTRODUCTION

Since African countries began to gain their independence, development agencies of the United States, European countries and other governments have changed social, economic and environmental conditions over vast areas of Africa. The changes were wrought by aid money, equipment and technical expertise. At least one generation of American and European foreign advisors has devoted careers to African development assistance projects.

Knowledge of African ecosystems has been advanced through years of basic and applied research. Ability to extend this knowledge to local livestock producers has increased. Many Africans have studied range management and livestock production and are now filling administrative and technical positions relating to range management. Development planners can now draw on the experience of range managers who have lived for years in remote areas of Africa and who work well with rural people. Ability to extend knowledge to local livestock producers has increased by African range technicians gaining a basic understanding of extension principles. A great deal of knowledge remains to be obtained from individuals with limited academic training but whose experience has taught them the fundamental aspects of making a living on rangeland.

This chapter discusses some constraints on range livestock development, the role of range management advisors and an approach to development assistance in range management. The approach integrates responses to technical and non-technical constraints to produce a comprehensive and viable rangeland development strategy. Discussion emphasizes work with African pastoralists. A pastoralist, as suggested by Gall [1981], is an agriculturalist whose primary means of subsistence is livestock, whether that person is among the small

number of pure nomads or among the various partially
settled groups.

CONSTRAINTS ON RANGE LIVESTOCK DEVELOPMENT

Several constraints on livestock development in
Subsaharan Africa that influence range management
efforts are listed below.

Ecological

1. Land production potential -- there is an
   absolute upper limit.
2. Disease, insects and parasites.
3. Climatic variability -- season, intensity and
   distribution of rainfall can have major
   effects on forage production, surface water
   availability and dependent livestock grazing
   strategies.
4. Fire -- its use can be a tool to benefit
   livestock production, but without control it
   limits the quality and quantity of forage
   that can be utilized by grazing animals over
   large areas of land.

Political/Economic

1. Boundary delineations -- community, regional
   and national.
2. Agricultural demands -- the expansion of subsistence agriculture often results in the
   exclusion of livestock from key grazing areas
   and/or the degradation of grazing land near
   agricultural areas due to concentration of
   livestock owned by sedentary agriculturalists.
3. Trade conditions -- market restrictions.
4. Bureaucracy -- policy constraints and absence
   of precedent regulatory mechanisms.

Social

1. Customs concerning animal husbandry.
2. Traditional land use.
3. Prejudices among tribal groups.
4. Technological development.

The list suggests the magnitude of the problems
confronted by development assistance projects. The
constraints in each category are discussed in detail in
other chapters of this book. Subsequent sections of
this chapter provide examples and procedures for planning and implementing projects. Examples are drawn

from the authors' personal experience as range management advisors in Africa.

THE JOB OF THE RANGE FIELD TECHNICIAN

For long-term success a management scheme must
meet the needs of plants, livestock and people involved. Range management is defined by Stoddart, Smith,
and Box as: "the science and art of optimizing the
returns from rangeland in those combinations most
desired by and suitable to society through the manipulation of range ecosystems" [1975, p. 3]. As suggested
by Workman,

> Range is a specific kind of land rather than
> necessarily being land devoted to a specific kind
> of use such as livestock grazing [p. 1] . . . .
> Range forage is a "flow" resource that with
> present technology can be converted on a large
> scale to products useful to man only through
> livestock grazing . . . . Rangelands are not graz-
> ed because they are well adapted to production of
> forage for harvest by domestic livestock. In-
> stead, rangelands are grazed because they are not
> well adapted to other land uses [p. 19] . . . .
> Many rangelands have become traditional grazing
> areas by default [1981, p. 20].

If more efficient uses can be found for rangeland, they
probably will be adopted.

The role of a range management advisor is to help
people make better decisions. The advice given must be
ecologically, economically and culturally sound.
Ideally a range advisor should help host country
nationals initiate a management strategy that will help
accelerate the adoption of improved management practices. To succeed as a range management advisor in a
developing country requires more than technical expertise. Like range extension specialists in the United
States, range management advisors overseas must communicate. Advisors must resolve conflicts among technical solutions and political and social constraints.
The required sensitivity to and understanding of non-
technical constraints to development are usually gained
through long-term living experience in less developed
countries. Adaptations of western technology that
enlarge management alternatives in the Subsahara are
invariably site specific and derive from technical
skill linked to indigenous people. A range management
advisor should encourage village-level participation in
planning, implementating and evaluating development
activities.

There is no easy way to improve management of rangeland resources. It is not possible to buy a prescription management package for range livestock production. Variation in ecological conditions and land use practices limit any attempt to generalize about appropriate technologies to improve pastoral economies. Justification for specific technological change should originate in the user group and be carried out by its members. Range planning and ongoing management must be site specific and requires continual backing of individuals who have both knowledge and authority to adjust daily operational activities. Authority in this case means authority in the local setting, in contrast to donor agency or departmental sanctions.

A lack of continuity has caused many range livestock development projects to be labeled failures. Occasionally improved adaptation of technology to the real needs of the field situation has been rejected by project evaluators because the procedure did not follow that prescribed in the original project paper. Regardless of good intentions it is unlikely that a newcomer can quickly and accurately determine the preferences, prejudices and needs of a group of people. It is for this reason that development projects must remain flexible. This flexibility requires that projects have the ability to integrate technical solutions and constraints with project goals.

LOOKING FOR AN ANSWER IN THE VILLAGE

From Tanzania to Senegal pastoral people are managing rangeland according to their perceived needs and traditional beliefs. On close inspection one usually finds that "traditional" use systems are highly refined. When environmental conditions are harsh, people adjust their life-style to minimize demands on labor and avoid exposure to the elements. People who must subsist on marginal land are concerned with labor efficiency and risk avoidance to insure survival more than they are with overall productivity.

Traditional mechanisms to control livestock utilization pressure on rangelands have been removed by the rapid and often inappropriate introduction of new technology (e.g. engineering and modern medicine) and the accompanying increased demand for land resources. Under those circumstances it may be necessary to modify the entire range livestock production system currently in place. Improved range management can benefit pastoralists, the countries they occupy and conserve land resources. In most cases economic and social constraints require that livestock production remain in the hands of pastoralists. Efforts to improve management practices are most effective when they establish a dialogue between technicians and pastoralists.

An example of a successful program complementing an indigenous land use system is the US AID Bakel Range Livestock Development Project for sedentary Toucouleur villages, covering an area of approximately 100,000 hectares in eastern Senegal. Small dirt tanks with water storage capacity of approximately 8 million liters facilitate the use of outlying grazing areas [Dickie 1981]. Project agents have collaborated with village leaders on the design, placement and utilization of dirt tanks since the project began in 1975. The success of the water development program results from technical support provided by US AID and the Government of Senegal, coupled with the rapport established among Senegalese field agents and Toucouleur villagers.

Pastoralists generally are willing to share knowledge about grazing land and are uniquely qualified to evalute management planning. Discussion of methods for providing livestock water are usually well received. As evidence of what modern technologies can produce, the Maasai, Fulani and other pastoral groups have access to large dams and boreholes. Pastoralists want to improve their living conditions and should be consulted by development planners as to their needs. Not all of the pastoralists' perceived needs will be genuine or ecologically or socially desirable. Nevertheless, such inforamtion can guide the determination of the type of assistance which may be appropriate for a given situation.

A Maasai elder, chairman of a livestock village in Tanzania's Maasailand, said,

> Good people (technicians) have done poorly here because they did not take their information to the people. A person who does your kind of work must know how to listen and respect the position of elders--or any age group. With real elders you must be very cautious. They are the deciding power. If you ignore these existent cultural age groups, you will never communicate or transmit your technology, and ultimately you will fail in your job. To be successful you must become accustomed to the area and have a good reputation. If your heart is good, the villagers will know it, and you will be accepted. People will speak of you, and others will hear. [Partalala Memeree, personal communication. Translated from Maasai by Gideon Soombe, Kijungu, Tanzania, 1978.]

Proposed changes in land use management strategies should be compatible with the strengths of the existing system but tailored to eliminate weaknesses. Leaders within herder groups must be given the opportunity and time necessary to incorporate innovations within their group's perception of management alternatives.

OPTIONS FOR EFFECTIVE DEVELOPMENT ASSISTANCE

Extension, formal education and research institutions must support one another. Building and strengthening these institutions provide options for development assistance. Physical development of rangeland resources is another option for assistance but is less effective.

Teaching and Research

Universities worldwide are increasing their expertise and preparedness to provide assistance to less developed countries in natural resource management. Education and research aid has improved international rangeland management. University training in-country and overseas has advanced the ability of Africans to properly manage their land resources. Research projects and surveys in developing countries are providing information about the productivity of rangeland and the livestock and people who are dependent upon them. Such information is badly needed by managers, who must understand traditional land use before helping livestock owners begin new resource utilization strategies.

The trained people who remain after a development project has ended determine the long-term success of the initiative. Training a sufficient number of people requires a long-term commitment to education anticipating high attrition rates. US AID initiated a long-term range management training program in Morocco in the early 1970s. Now, over ten years later, Moroccan researchers, teachers and extension personnel are available to work with foreign specialists to carry out specific projects.

The quality of training has often been inadequate but it is rapidly improving. Trainees from developing countries have found that many programs in the United States and Europe were inappropriate for them. Programs have been started by several university departments of range science to address this problem [Smith and Hays 1981]. Local technicians must be provided with the best training possible to facilitate the fit of technical innovations to cultural needs.

Some development assistance agencies insist that all training take place in the host country, or at least in a developing country. The establishment of degree and nondegree programs organized in developing countries is an important goal. However, until such programs and the necessary infrastructure exist, the personnel who will eventually staff them must be trained where there are adequate facilities. Countries that have recognized the importance of training have established effective range management institutions by insuring that they have personnel to develop and maintain them [O'Rourke 1982].

Experience has shown the one-to-one "counterpart" relationship to be both inadequate and expensive. It has been the rare exception where counterparts spend enough of their time together for the host country counterpart to obtain adequate training through day-to-day working activities. Furthermore, the counterpart relationship limits the source of training to one technician rather than a complete university faculty.

Extension

Application of research findings and technological innovations depends on a perception of value in such interventions. Information will be used if it seems to provide a realistic means of satisfying needs. Extension methodologies should be used to enhance peoples' understanding of new technology and accelerate its application in land management. An extension program taps the knowledge of experts and leaves responsibility for implementation in the hands of local people.

The application of range management expertise through extension education can amplify the effectiveness of an individual technician. Extension training has proven to be one of the most effective tools to transfer technology from one group to another and is perhaps the most efficient use of foreign advisors. Through the use of extension education the skills of a few individuals can be shared by many. Extension in this context has seldom been used in range management development assistance. Activities have been developed with no clear idea of audience or their need and thus no clear objectives. Advisors have diluted their potential training of large audiences by individually carrying out the physical development of isolated and, from a country-wide point of view, insignificant range construction projects.

One consistently missing element of extension type assistance has been the absence of a plan, not a general plan to satisfy donor curiosity of project purpose, but a plan as used in extension methodology. As stated by Johnson:

> To a very real extent, it may be said that an extension program will succeed, or conversely fail, in relation to the extent and quality of its plan . . . . Good planning is an extension imperative.

> There are . . . several approaches to planning
> extension activity . . . . (however) experience
> has shown that all such (approaches) share the
> concept of <u>audience</u> (who), and initial perception
> of their <u>need</u> (why), <u>objectives</u> which will help
> meet the overall need (what), and <u>activities</u> to
> accomplish each objective. [1981, p.12]

Hands-On Support

It is important to provide exposure to modern,
practical and accessible range management techniques.
However, such assistance may be ineffectual unless one
can also provide a helping hand in first-use of new
technology. For example, after years of training overseas it can be difficult for newly returned technicians
to put expertise to work in the field. This is the
result of long-term cultural isolation, as well as
being an agent of change in a traditional society.
Often, trained nationals are not from the traditional
society with which they are working. They too have
cultural barriers to bridge. As a complement to
sponsorship of overseas technical training, donor aid
agencies should provide additional in-country support
to newly returned range technicians. Host country
nationals usually are more effective than foreign
advisors in dealing with village people. This is
especially true where field agents are able to work in
their home areas. This aspect should be encouraged.
It is often the case that village leaders will be
impressed by the words and demonstrations of extension
personnel but will not have the confidence and/or
ability to communicate with distant neighbors to modify
the use of a shared grazing area. A lack of capable
manpower on the part of the host country may slow or
prevent the advance of coordinated land use. To accelerate range livestock development it is imperative that
skilled range technicians be available to assist
regional planners, newly trained host country range
technicians, field agents and villagers as they take
advantage of management alternatives.
Under most circumstances it is undesirable for
foreign advisors to manage or otherwise control host
country resources. Advisors must work <u>with</u> host country nationals to serve the cause of development assistance. It is usually better for a resource management
decision to go unmade than to be made by a foreign
advisor who will be present for a relatively short
time.
Technical assistance must be oriented toward
development of self-reliant skills to make good management decisions. Experience has shown that this is the
only effective way of fulfilling long-term needs.

Toward this end, rural development projects should include participation by the local economic community [Dickey 1982].

Local producer organizations should manage and control infrastructures such as water points and veterinary posts wherever possible. Participation of this type must be supported by community and government representatives. Through involvement of the target population in the process of community development, it is hoped that awareness gained, creativity released and activities initiated will continue to proliferate on their own.

In 1968 the President of Tanzania, Julius K. Nyerere, pointed the way for development assistance projects when he wrote:

Development brings freedom, provided it is development of people. But people cannot be developed; they can only develop themselves. For while it is possible for an outsider to build a man's house, an outsider cannot give the man pride and self-confidence in himself as a human being. Those things a man has to create in himself by his own actions.

CONCLUSIONS

Constraints to range livestock development in Subsaharan Africa are not insurmountable. There is a great deal of information available concerning the practical application of range management principles in pastoral areas. The importance of including user groups as full participants in the process of planning, implementing and evaluating projects cannot be over-emphasized. Improvement of management alternatives has already begun on a large scale. Range managers play an important role in the development of the African live-stock industry. Range management projects must be tailored to local conditions with allowance for flexi-bility in implementation.

Decision making is facilitated by profiting from the experience of others. In this respect the oppor-tunity to succeed is greater than ever. In the next decade, development assistance directed toward improv-ing the quality of rangeland management in Africa will be more efficient and more fruitful.

REFERENCES

Dickey, James R. "Livestock Development Assistance Strategy for Africa." Africa Bureau, Agency for International Development, Washington, D.C. 1982.

224

Dickie, Alex. "End of Tour Report." Range Management Advisor, US AID Bakel Range Livestock Development Project. Dakar, Senegal, 1981.

Gall, Pirie M. "Range Management and Livestock Development in the Sahel." Draft report prepared for Chemonics International Consulting Division. Washington, D.C., 1981.

Johnson, Kendall L. "A Guide to International Range Extension." Consultant Report to Morocco Range Management Improvement Project, Ministry of Agriculture and Agrarian Reform, Rabat, Morocco and Range Science Department, Utah State University, Logan, Utah, 1981.

Nyerere, Julius K. Freedom and Development. Policy handbook. Gov. of Tanzania, Dar es Salaam, 1968.

O'Rourke, James T. "Toward More Effective Range Management Training." Natural Resource Technical Bulletin No. 3. AID/NPS Natural Resources Project, 1982.

Smith, J. Allen, and V.W. Hays, ed. "Transfer of Grassland Research Findings." Proceedings of the XIV International Grassland Congress, Section XIII, pp. 813-824, 1981.

Stoddart, L.A., A.D. Smith, and T.W. Box. Range Management, McGraw Hill: New York, 1975.

Workman, J.P. Range Economics. Textbook Draft. Logan, Utah: Utah State University Press, 1982.

15
Development and Management of Livestock Projects in the Sahel Area of Africa

Thurston F. Teele

INTRODUCTION

In the current view of AID and other donors, livestock and range management projects in Subsaharan Africa are not held in very high regard. This is ostensibly because they have failed to achieve their objectives to the same degree as projects in other substantive areas. As a result, there exists a tendency to allocate resources away from the livestock sector to other sectors, a tendency which is increasing and is regrettable. The livestock sector remains of vital importance to the countries of the Sahel. It is the livelihood of significant portions of the Sahelian population, and it represents the best, if not the only, economic use of large areas of land.

The real or perceived failure of livestock projects can be attributed to many factors. Two stand out. First, there is virtually no agreement among experts as to what should be done to improve the livestock and range management sectors of these countries. Chemonics' recent paper on the subject [Gall 1982] makes this clear. Anyone who has attended meetings such as those sponsored by AID at Harpers' Ferry (1979), Mariottsville (1980), and Gainesville (1983), can observe the wide divergence of views expressed. There are differences, first, between those broadly classified as livestock and range management specialists, and the socio-anthropologists. The latter sometimes give the impression that any livestock or range management intervention which is proffered will cause more harm than good. And among livestock and range management specialists, there is the widest possible divergence of opinion about which interventions are appropriate and workable in the Sahel. This divergence is brought about not, as commonly stated, by the paucity of experience or research, but rather by the failure to make adequate use of the results of such experience and, more fundamentally, by the failure of project designers

225

to hypothesize a clear, causal relationship between interventions and impacts and to provide for the collection of specific data, pre- and post intervention, to prove that relationship.

A second reason for the perceived failure of livestock sector projects is that many have been poorly designed and improperly managed. Moreover, this second reason is a major contributor to the first (lack of consensus among experts), making the problem circular. Poor design often results in unattainable objectives, or objectives and interventions which are certain to be resisted by the target population. Poor design may also be characterized by the failure to provide a proper balance between research, implementation and monitoring of results, or a general tendency to focus on land and animals rather than the human populations. More often than not, project design provides inadequate resources, and time horizons which are far too short. Poor management, moreover, has brought about the misuse of resources, inadequate planning, planning without resource allocation, poor financial management, inadequate provision of logistic support and failure of communication between project personnel and the target population, to mention but a few results.

This paper concentrates on the second of the two broad problems introduced above, project design and management. While there is little dispute among practitioners about the effect of poor design on livestock projects, there is less agreement about the significance of poor management. Many evaluators tend to ignore management and implementation problems. Others, such as Allan Hoben, believe that management does not play as important a role in livestock projects as in other projects because, he observes, "livestock projects appear to suffer similar difficulties in...effectiveness, regardless of the quality of their management" [Hoben 1979]. Anyone who has attempted to implement livestock and range management projects on the ground, and who has experienced severe management problems in the process, as amply described in project reports [Chemonics 1983], would tend to believe that both design and management contribute to project effectiveness, and that poor management can doom even a well-designed project.

TYPES OF LIVESTOCK PROJECTS

Before discussing the design and management question, it would, perhaps, be useful to point out that there are many different types of livestock and range management projects, and that design and management requirements vary in each case. There can, for example, be a variety of technical interventions. However, more generically, five variables can be isolated: (1) the subject area; (2) scope of the project; (3) objectives and beneficiaries; (4) types of resources provided; and (5) geographic coverage. For example, in its subject area, a project may cover livestock alone, range management alone, or both in combination. Project scope may be limited to research, or it may involve a single intervention (such as animal health improvement) with or without supporting research. It may include multiple interventions, or may even be aimed at the total livestock/range management sector. Objectives may include improvement of pastoralists' livelihood, environmental preservation or improvement of the urban meat supply. Resources may consist mainly of technical assistance, or may include varying amounts of additional financial support for well drilling, construction, project operations, etc. Finally, geographic coverage may be limited to a relatively small region of a single country, or it may include the entire country or a multi-country region. A more complete listing is provided in Figure 15.1 under each of the five variables. Other variables could, of course, be included. It should be noted that the "research" shown under the category "scope" refers to formal research, rather than study and monitoring, which should be part of any project, even those which are otherwise considered as strictly implementational.

The focus for the balance of this chapter is on more complex projects, i.e., those which include multiple interventions, multiple objectives, technical assistance and significant financial resources for project operations and, perhaps, commodities and construction. This focus is justified because such projects are, (1) the most troublesome with respect to design and management, (2) potentially the most effective in bringing change to the recipient countries and (3) the most expensive.

LIVESTOCK PROJECT DESIGN

The following discussion of livestock and range management project design has two components, methodology and content. Only a few central ideas on design methodology are offered, since it is a large and specialized subject, impossible to cover fully in this brief analysis. More attention is given on the content of project design; that is, the elements that may be regarded as vital for a project's success.

Figure 15.1. Livestock/Range management project variables

(1) Subject	(2) Scope	(3) Objectives/ (Beneficiaries)	(4) Resources Provided	(5) Coverage
Livestock Production, including some or all of the following:	Research only	Improve livelihood or well-being of pastoralists (pastoralists)	Technical assistance (TA) only	Region of country
Animal health	Single Intervention (w/ or w/o research)	Improve environment (pastoralists, entire population)	TA and training (TNG) only	Entire country
Animal nutrition				
Feeding				
Milk production	Multiple Interventions (w/or w/o research)		TA w/wo TNG plus modest financial support for project activities	Multi-country region
Breeding				
Small ruminants				
Sociology/Anthropology				
Livestock/cropping systems	Total program (w or w/o research)	Improve meat supply (urban population)	TA w/wo TNG plus heavy or total financial support	
Livestock Marketing, including some or all of the following:		Institutional development (government, others)	Financial support w/o TA and/or TNG	
Infrastructure				
Price policy				
Institutions				
Range Management, including some/all of the following:				
Increasing Production				
Ecological protection				
Land use/land tenure				
Combinations of the above				

Source: Author.

Project Design Methodology

It should be recognized that, for donor-financed livestock and range management projects, the designers face severe limitations in that they are not dealing with a blank slate and much has already been decided by the time they intervene. Of the five variables constituting project type described above, the first, subject matter, and the fifth, geographic coverage, have already been determined. The other three; scope, objectives, and resources, have been partially decided. The designers must do their work within the broad parameters which have already been laid down.

Despite the above handicaps, in order to design a workable project within the given parameters (or, perhaps, to demonstrate that one cannot be designed), there is a need to study the situation to determine the real needs of the target population and identify interventions, in some detail, which will respond to those needs. The dilemma, mentioned by many authorities, is that a truly adequate series of studies which might yield definitive answers, when the subject is livestock and range management in the Sahel, would require a great deal of time and abundant resources. Two well-known project design authorities, Abercrombie and McElroy, produced in 1974 a 37-page outline of the points to be studied in designing a livestock and range managment project. Such lengthy efforts may be possible in some instances, but only in very few. Most situations provide neither the time nor the money for such exhaustive studies.

Any reasonable design requires that some field study must be done, yet often little is done. Projects have been designed with minimal discussion with the herders themselves, and with little effort to review prior studies and experiences which would be appropriate to the problem at hand. Examples abound. Chemonics once prepared a proposal to implement a range management project which called for a major, aerial data gathering phase in a situation where a well-known air survey firm had just completed a similar assignment. It was discovered that the voluminous reports were unopened in the host government's project office. The project had been designed without reference to this invaluable work and had even allocated time and resources to carry out the same task. With respect to the target population, it has been found in many cases that contact has been made, but in a very limited way, with the result that the views of the population are not well reflected in the project design. It is therefore important for project designers to ensure (1) that all existing data which bear on the areas and subject matter have been reviewed, and (2) that maximum contact has been made with the target population before the project design is finalized.

Contact with client populations is not always easy, especially with Sahelian herders who speak only their own language, are frequently on the move and who do not always provide the information requested. Nevertheless, numerous techniques have been developed for use in this situation. One which seems especially promising is the "rapid rural appraisal", a technique which consists of identifying and quantifying a series of key indicators appropriate to the problem, in this case range condition, livestock production, livestock marketing, the micro-economics of herders' lives, and resultant motivations [Chambers 1980]. This can be accomplished by a combination of methods: use of local people as sources and researchers, some direct observation of herder life, and guided individual interviews supplemented by group interviews.

The key in rapid rural appraisal is to establish some baseline data and indicators and, most important-ly, to provide the eventual implementation team with the information and methodology to be used for monitor-ing and modification as the project is implemented. This step is vital. If the primary study methods and the format and/or content of the baseline data are such that they cannot easily be used and repeated as part of the project implementation process, as sometimes occurs, they are useless. For many projects, this has been the case. The corollary, of course, is that the project must have resources to carry out the monitoring functions using these indicators and methods.

Project design sets out what will be done under the project and how it will be achieved based on the information obtained. Unfortunately, project design is often regarded as a "blueprint" which purports to set the project in concrete. This rigidity is a major problem, especially in livestock and range managment projects where there is great uncertainty about the effectiveness of selected interventions under specific conditions. Interventions which are ineffective because they are unexpectedly resisted by the herders or do not work functionally are difficult to change at a later date when no provision for modification exists in the design.

Due to the above problems, it is preferable that livestock and range management projects not be designed using the blueprint approach, but rather a "process approach." The process approach, described by a number of authorities, including Honadle [1980], and currently being used for a few projects financed by AID, permits much more flexibility in project implementation. This design methodology involves study of the situation and preparation of problem diagnosis. A series of possible interventions is then described and should be tied causally to specific impacts which constitute solutions to the problems. The programs and interventions for the first year or two are designed in some detail; they stress such initial activities as further appraisal by the implementation team, infrastructure development, establishment of management systems, and some quick impact activities.

The first phase of the work is, or should be, recognized by all as preliminary and experimental to avoid creating excessive vested interest in the specific interventions. Early on, the initial design is reviewed by the implementation team and an outside review team, necessary changes made, and the project continues. The review process should involve extensive discussion with the target population in order to obtain its views on the various interventions being tried or considered.

It should be stressed that the flexibility being advocated here refers to specific means used to implement the project, and not the existence or continuation of the project itself or the basic project objectives and beneficiaries. Although flexibility offers numerous advantages it puts strains on all participating parties, including the host government, donor and implementation team. All must be assured that the overall objectives will not be changed, and that the project itself will continue with adequate resources.

Much of this discussion on design methodology, centering on data gathering techniques and the process approach to design, impacts heavily on the content of project design because project design must provide the resources for frequent review and for modification or replacement of interventions. There is also a major impact on project management.

Project Design Content

Project scope. Project designers usually have some latitude with respect to project scope. If a project is mainly focused on implementation, there is a question whether there should be a research component. There is reason to believe that in most cases there should be despite the frequent claim by African government officials that ample research has been done and it is implementation which is needed. Adequate research is lacking in many areas, as is access to the results of prior research. Therefore, it is preferable that a research component be provided with the stipulation that prior research which may be applicable be collected and evaluated. All livestock/range management projects must include the mandate and resources to update the baseline study, and to monitor based on specific indicators and dialogue with the target population regardless of whether there is a formal research component.

Project resources. Resources must, of course, be appropriate to the project scope and objectives, but generally livestock/range management projects in Africa should include the full scope of resources. This means technical assistance and financial support for construction, commodities, supplies, equipment and, probably, project operating costs. In most cases, there must be a great deal of technical assistance and ample resources if the project is to have any chance for success. The fact that such resources are both expensive and indispensable must be recognized at the outset.

Resource-based planning. Perhaps more important than adequate resources is resource-based planning, a subject which is of sufficient importance that it is discussed both here and under "Project Management" below. All components (research, implementation and monitoring) require specific project resources which should be identified and assigned in the planning process. That is, if a program of water development is planned, the human, material, and financial resources necessary to carry it out must be acquired and assigned to the task. This may appear obvious, but experience shows it is rarely done effectively in African livestock and range management projects.

Duration of projects. It is widely recognized that effective livestock/range management projects, especially those directed toward pastoral herders of the Sahel, must be of long duration to be successful. Periods of ten years are not unreasonable, especially if projects are designed as process projects with the flexibility to modify interventions rapidly as lessons are learned. Many projects are still designed with excessively short time frames.

Project Components

Project components should include monitoring and participation in project review by both the evaluation team and target population, maximum use of existing herder organizations, and current range and livestock practices, as well as a broad range of activities not limited to those formally in the livestock/range management category. In addition, there should be quick-impact activities to gain confidence of the target population, and participation by them in a variety of ways including payment of fees for some services. Each of these desirable components is discussed briefly.

Monitoring and review by the implementation team. It has been noted that African officials are frequently concerned that projects will be oriented excessively toward research, suspecting that expatriate advisors are more comfortable with research than implementation. A frequent reaction is to reject any notion of research in implementation projects and to carry the prohibition against research to the extent that they reject monitoring and meaningful participation in project review as well. This is reinforced, in some cases, by the position descriptions for technical assistance personnel and by the selection process which concentrates exclusively on individuals who have practical, implementation experience to the exclusion of individuals with monitoring and project design skills.

Despite concerns by action oriented officials, it is important to include requirements and resources which allow the implementation team to carry out or update baseline data, and provide much or all of the project monitoring. This work by the implementation team, rather than relying exclusively on outside evaluators, reduces confrontational aspects of the monitoring process and encourages accurate and rapid response of results.

Maximum use of existing organizations and practices. It has been recognized in recent years that Sahelian pastoralists have adequate organizational arrangements and effective ways of managing their herds and ranges which have allowed them to survive in a very harsh environment. It is reasonable, therefore, that livestock and range improvement projects should be based, to the extent possible, on existing organizational arrangements and livestock and range management practices. This is not a new observation; but it is still sometimes ignored in practice at the design and/or the implementation stage when attempts are made to impose new, and often inappropriate organizations and ready-made technical solutions. As an example, Chemonics' own technical assistance team in the Dilly Zone of Mali experienced some aspects of this phenomenon, mostly because they were prevented, by the Malian project direction and lack of resources, from researching existing practices and devoting sufficient time to working closely with the herders.

Broad range of activities. Another desirable component of project design which is frequently mentioned but often ignored, is activities or interventions aimed at improving the life of the target population's lifestyle, a component which falls outside a strict livestock/range management rubric. Sahelian pastoralists are primarily herders, but they and especially their families, do engage in other economic activities such as growing millet. They also have other needs and concerns, especially drinking water and health. Projects which recognize this and provide assistance in some of these key areas may gain more acceptance by herders, and therefore have a better chance for success. There is, of course, a danger in designing and implementing projects with excessive interventions in too many areas since this can dilute effort and leave too few resources for the main livestock and range management activities. A balance is obviously needed, but a certain amount of such outside assistance is very helpful.

Quick-impact activities. Closely related to the desirability of supplementary activities is that of quick-impact activities. Livestock and, especially, range management interventions are slow to implement and even slower to bring about obvious benefits. They require sacrifices, sometimes very considerable ones on the part of herders, in the form of labor, rangeland left ungrazed, stocking limitations and general interference in their normal ways of life. Quick-impact activities, frequently in non-livestock/range management areas, can make the overall project more palatable and provide some early compensation for the initial sacrifices. Further, many activities, such as village wells, animal and human health measures, feed supplements etc., are not particularly controversial, and are successful virtually every time. They give the project an aura of success while the slower, more controversial, and problematic livestock and range management activities go forward.

Target population participation. It is important to involve the target population as much as possible in all aspects of the project. This can be achieved by frequent interaction between project personnel and herders, both individually and in small groups, and in larger meetings with herder village organizations. A second aspect of participation is through payment of fees for services rendered by, or through, the project. Experience shows that herders are willing to pay for services they recognize as beneficial, such as health care and sometimes stock water. Willingness to pay may be a good indication of true acceptance. It is also possible to have them pay for services not yet fully accepted, as part of a larger package of services, in order to demonstrate the value of them. This approach is valid provided it is not abused but, unfortunately, examples of serious abuse do exist.

LIVESTOCK PROJECT MANAGEMENT

Introduction

The requirements for successful management of livestock projects in Africa are essentially the same as those of any other project. There are, however, some special problems. First, with any donor-financed project, there are at least three entities involved in the project and its management: the host government agency(ies) charged with implementing the project, the donor agency and the implementing contractor. Often there are more: the head office of the donor agency and the contractor's home office. All of these have, or should have, some role in the management of the project.

There are also two or three sources or "modalities" of funding for project activities, especially for a project which is comprehensive and resource-heavy. There is the host government; the donor which makes some direct payments for project goods and services and also some direct reimbursements to the host government agency; and there is also the implementation contract, which may provide the vehicle for some and often a considerable share of the project funding. This situation of multiple participation in project management and financing naturally brings with it a potential for divided authority and weak overall management.

Division of Responsibility for Project Management

Since the major phenomenon which distinguishes donor-financed resource-heavy projects from other projects is the participation by three or more entities, the first requirement for successful management of such projects is a clear division of responsibility and authority. This seems obvious, but is not easily accomplished.

The first opportunity to set out a clear division of responsibility and authority is in the project design and approval documents, mainly the Project Paper and Project Agreement in the case of AID-financed projects. The second opportunity occurs in the technical assistance or implementation contract, and it is, in fact, vital that the technical assistance contract devote considerable attention to this issue. It appears that direct AID contracts rarely devote any serious attention to this division of responsibility and host country contracts only a little more, whereas it should be a major element in all contract negotiations.

Division of Responsibility Matrix

The proper division of responsibility and authority does vary somewhat with circumstances and specific projects, but certain basic guidelines can and should be set out as shown in Figure 15.2. Content of the matrix is open to discussion and, in any case, would vary with the specific project, and desires and capabilities of the three major participants. The important thing is that some version of such a matrix be developed as part of the project design and planning process and be negotiated and agreed upon. Even then, it is only a first step as the allocation of responsibility and authority must be tied to resources made available to each participant, i.e., resource-based planning.

Observations on Division of Responsibilities

Donor agency(ies) are, of course, the central element in any project and must be given as large a role as possible in planning and implementation. Further, one objective of almost any project should be strengthening capability of host government agency(ies) to implement such projects. However, it is important that one be aware of limitations under which host government agencies operate to avoid assigning more responsibility than they can possibly handle. A common error in livestock projects in the recent past has been to overestimate host government management capabilities.

To the extent that host government agency(ies) have administrative limitations, slack should be taken up mainly by the technical assistance contractor rather than the donor agency. In the case of the World Bank and other multinational donors, this cannot be disputed, since the donors have no resources and no mandate to provide project management. With respect to AID, the situation is less clear, because their missions are often rather large and there is a temptation for mission personnel to provide some project management. The temptation should be resisted, because missions are not well equipped to provide effective management, and attempts to do so interfere with the proper donor functions of planning, review and approval.

Key Management Responsibilities of the Contractor

The technical assistance contractor should be asked to provide any management services which the host government agency(ies) are unable to perform effectively. This could be a very wide range as suggested by Figure 15.2. The first area of particular

Figure 15.2. Division of authority in livestock projects

Project Activity	Contractor	Host Government	Donor
Planning			
Broad Planning, Goals	A	D	D
Annual Detailed Planning	A	D	OK
Allocation of Resources	A	D	OK
Establish Control Over Resources	A	D	OK
Management of Resources			
Technical Assistance Team	D	OK	OK
Counterparts	A	D	--
Training of Counterparts	P/A	D	D
Training of Beneficiaries	P/A	D	D
Financial Management	P	D	D
Logistic Support and Procurement	P	OK	OK
Maintenance including Vehicles	P	OK	OK
Management Information System	P or A	OK or P	OK
Evaluation and Re-Planning			
Continual Monitoring	P	P	OK
Modifications in Approach	A	D	OK
Major Evaluations	A	A	P
Major Replanning	A	D	D

(P = Provide, A = Advise On, D = Decide On, OK = Approve)

Source: Author.

importance is financial management. Resource-heavy projects in Africa require a great deal of financial management, a task which places a heavy strain on the host agency(ies). Generally, they cannot handle it, less because of inability of financial personnel than due to social and cultural pressures under which they must operate. Because of these pressures, financial management can become a disaster, one which might result in an entire project's failure. The contractor should be asked to actually provide the financial management of the project through furnishing and supporting appropriate personnel who hold adequate authority, or by providing advisory assistance with a strong mandate to ensure adequate management. The former is the preferred approach.

The second key area in contractor management responsibility is control and maintenance of vehicles and other key equipment. This may appear mundane, but those who have actual experience in project implementation in Africa (and elsewhere) are aware of the importance of a well-run vehicle park and adequate maintenance. This is another area where host agency personnel are faced with very severe cultural pressure and thus it is preferable to turn the problem over to the contractor.

A third, somewhat related area, is procurement and logistics. In livestock and range management projects, equipment requirements can be very heavy and African government agencies are generally not prepared to handle them. USAID missions can do so, but it is preferable to have the contractor in charge in order to improve coordination with project implementation. With respect to logistics, delivery and support of personnel and equipment to the project sites is best done jointly by the contractor and the host agency(ies) with the host agency (ies) developing logistical capability, through the assistance of the contractor, as rapidly as possible. Project designers should not automatically assume that the host agency(ies) can routinely handle this work, since they usually cannot.

Another key area of project management, and one which is difficult to deal with in this context, is construction as livestock and range management projects often involve considerable amounts of it. Given the management weaknesses of most African host-government agencies, construction work can often proceed very slowly and inefficiently. There is a temptation to recommend that the technical assistance contractor serve as the general contractor for construction. This solution can be successful, but it requires special attention to the need for adequate resources in the contract.

CONCLUSIONS

The perceived failure of livestock sector projects in recent years has occurred largely as a result of weaknesses in (1) project design and (2) project management. The design of livestock/range management projects presents major problems because of the large amounts of data needed and because of differences of opinion among experts on the potential effectiveness of any specific intervention. A "rapid rural appraisal" approach to data gathering and a "process" rather than a "blueprint" design is proposed to allow flexibility while assuring the continuation of a project for several years. Projects should include research as well as implementation, baseline data updates and monitoring by the implementation team, a stated causal relationship between interventions and results, a focus upon the people rather than land or animals, interaction between the herders and project personnel and maximum use of existing herder organizations and methods. For project management, a clear division of responsibility and authority for management among donor, host government agency and implementation contractors is needed. The contractor should be given a major role of project management. Through resource-based planning, each activity must be assigned to specific project participants and resources should be allocated and placed under control of that participant. This implies that the contractor be willing and able to undertake most management functions and provide personnel with the necessary talents.

REFERENCES

Chambers, R. Rapid Rural Appraisal: Rationale and Reportoire. Discussion Paper, Institute for Development Studies, University of Sussex, United Kingdom, 1980.

Chemonics International Consulting Division. Mali Livestock Development Project (Mali Livestock II 1977-1982, Mali Livestock I 1979-1982), Final Report. Washington, D.C., 1983.

Gall, Pirie. Range Management and Livestock Development in the Sahel. Chemonics International Consulting Division, Washington, D.C., 1982.

Hoben, A. Lessons From a Critical Examination of Livestock Projects in West Africa. Washington, D.C., AID Program Evaluation Working Paper 26, Agency for International Development, 1979.

Honadle, G., et al. Integrated Rural Development: Making it Work? Development Alternatives Inc., for the Agency for International Development, Washington, D.C., 1980.

PART IV

Approaches for the Future

The emphasis displayed throughout this book on Subsaharan Africa's livestock industry development being as much a people and land policy problem as a technical one is reinforced in this section. The first two chapters are devoted to evaluating approaches to land use policy--and the subject's complexity is clearly brought out by the authors' different conclusions about policy.

Steven Lawry, James Riddell and John Bennett review land tenure reforms and the rationale for establishment of individual rights to discrete grazing areas. They review Botswana data and conclude that, if anything, individualization of land rights can increase production costs and bring about a radical redistribution of assets and cattle wealth in favor of large holders. Thus, they feel that if equity for the ever increasing numbers of small and medium sized producers is a goal, changes in land use management will need to be other than through a privatization process. Their analyses show that while establishment of efficient administration of public, i.e. communal range will be a long and difficult undertaking, successful pastoral development will in large part depend upon the establishment of workable communal tenure systems.

Roy Behnke also draws heavily on Botswana as a case study area of subsistence pastoralism, but arrives at a very different conclusion--that improvement in the livestock industry implies greater commercialization of it. He argues that what is needed is not livestock development projects, regardless of how well they are designed, but rather livestock development programs and policies which place the pastoral sector within the context of the national economy.

Successful approaches for the future, regardless whether they take a project or program approach, will entail greater interaction between change agents and clients, and must be more interdisciplinary in nature. One area which is recognized as being indispensable to a successful future effort, but which has traditionally involved relative little participation of allied disciplines is animal health. Albert Sollod, Katherine Wolfgang, and James Knight attack the problem directly, arguing that veterinary medicine has often been a prima donna in development programs but, because diagnostic laboratories often have little integration with field and nonveterinary activities, research from it frequently gives scant or even misleading information. They describe how part of this problem can be overcome by interfacing research with the social sciences through veterinary anthropology, an approach which uses field methodologies from epidemiology and anthropology. Results from two West African countries, Niger and Upper Volta, are given which demonstrate, among other things, that classical top-down animal health schemes should be critically assessed as to the degree to which they ignore the role of women in veterinary care.

A different view of animal health, and yet one that is equally critical of contemporary practices, is provided by Tjaart Schillhorn van Veen, who argues that a trend is developing in veterinary epidemiology emphasizing impact studies using computerized technology without first obtaining sound field data. He feels that disease control must be an integrated process, which takes into account, for example, stress management rather than simply vaccination and drug treatments.

The variety of methods available to researchers and change agents is vast and continually growing. As Sollod, Wolfgang, and Knight as well as Schillhorn point out, the directions being taken in both research and transmission of veterinary services are changing rapidly. The same holds true in economic analysis now that computers are widely available. Edgar Ariza-Nino and Kenneth Shapiro demonstrate the way in which modeling can be used as an applied tool in analyzing African pastoral production and marketing decision making. Their capital theory model of herders' decisions concerning the age at which they sell animals, while general, does provide a common ground for comparing the variety of hypotheses, theories, and conclusions which have been proffered by the social and biological scientists involved in African livestock development. Ariza-Nino and Shapiro offer a surprising number of conclusions, some of which show that, contrary to widely accepted doctrine, pastoralists herds can recover very rapidly after a drought.

John DeBoer et al. provide results from a pioneering research effort in Africa to integrate livestock into the farming systems research and extension (FSR/E) methodology which, in itself, is in a developmental stage. Their refreshing approach in introducing an entirely new enterprise (dual-purpose goats) rather than attempting to improve existing farm enterprises is an important contribution to the literature on methods of technological change by which production of meat and milk can be expanded in a mixed farming system. Their experience also reinforces a conclusion set forth by many other authors in this book--that animal improvement programs are expensive, time-consuming, and long-term (at least ten years) in nature.

16
Land Tenure Policy in African Livestock Development

Steven W. Lawry, James C. Riddell
and John W. Bennett

With very few exceptions, livestock development in Subsaharan African has had two broad policy objectives: increased animal output and range conservation. Land tenure reform in some guise has often been seen as instrumental to the pursuit of these objectives. On the simplest (but most widely accepted) level, it is communal land tenure that has been pointed to as a major constraint. Thus, it is not surprising that many programs and projects have tried to introduce tenure reforms which involve, in one way or another, a reduction of multiple claims to, and uses of, specific grazing areas.

This tendency towards "individualization" is especially apparent in projects which emphasize range conservation. The rationale for establishment of individual rights to discrete grazing territories is often provided by (and attributed to) the "tragedy of the commons" paradigm popularized by Hardin [1968] whose rather simplified parable of what are in fact highly complex processes has frequently been taken much too literally by project planners.[1] This criticism especially applies to an uncritical adoption of Hardin's policy solution. Only under individualized tenure, Hardin argues, would the individual herder be assured that self-restraint in balancing herd size with range carrying capacity will not be exploited by the actions of other range users.

The "tragedy of the commons" paradigm found its way into African land tenure policy in remarkably explicit ways. Seretse Khama, the late President of Botswana, used the following variant of the "tragedy of the commons" in introducing the Tribal Grazing Lands Policy to Botswana's parliament in 1975:

> Under our communal grazing system it is in no one individual's interest to limit the number of his animals. If one man takes his cattle off, someone else moves his own cattle in. Unless livestock

numbers are somehow tied to specific grazing areas
no one has an incentive to control grazing . . . .
[Khama 1975]

Individual land rights have been held to promote
conservation for other reasons.[2] Since a first principle of managing animal production on natural range is
the establishment of appropriate herd size, some
analysts see limiting the available grazing territory
as an essential preliminary step to limiting animal
numbers. Only then will the herder be able to comprehend the implications of running excessive numbers on
what would presumably be that person's only possible
range. Under open access, not only is the responsibility for range abuse shared, and thereby diluted among
the community of herders, but the individual herder
does not suffer in a proportionate or unique way from
his or her contribution to range degradation. Also,
under individual tenure, it is held, herders will
become disabused of the notion that there are available
pastures elsewhere when the local range is depleted.[3]

Assignment of leasehold rights to individuals or
small groups is the more common approach to tenure
reform. A leasehold agreement is often seen as an
appropriate instrument for specifying legally binding
stock limitations, usually under the rubric of the
"good husbandry" conditions typical to leases for
state-owned agricultural land. Stock limitations
specified in leases are almost never enforced nor are
they, for that matter, practically enforceable.
Reluctance or inability to invoke penalties against
violations of lease agreements is attributable to the
same sorts of political realities that militate against
implementation of more general statutory prohibitions
against resources abuse.

Individualized tenure has also been advanced as a
reform that will accommodate growth policies. Two arguments are typically offered. First, circumstances that
favor conservation will also favor growth, as sustained
development and growth in market offtake depend in part
upon the steady introduction of improved production
techniques and, perhaps most importantly, a stable
production environment. Both of these conditions are
facilitated, it is argued, by the increased control
that individual producers will have over grazing land.
Second, individual rights will provide greater assurance to investors that land holders are in sufficient
control of ranching assets to warrant confident extension of greater loan financing. Even though reposses-
sion of leased state land is usually not an option
available to private loan insitutions, a legally recognized exclusive land right by the ranching enterprise
is a signal to banks and other lending agencies that

the rancher has made certain entrepreneurial management commitments to commercial production.

While individualization of tenure rights has been seen as the solution for most effectively handling large herd owners in Botswana, for example, governments and projects have recognized that it is inapplicable to many livestock management situations elsewhere on the continent, and for smallholders in Botswana. There has been a growing tendency for tenure reform to specify the exclusive rights of a particular group to a definite grazing territory. The best known examples of this approach are the group ranches of Kenya and Tanzania, but the principle in one form or another is found in most Sahelian and East African project designs. See for example Riddell [1982] and Bennett [1983].

Government and project planners have cast group rights in terms that provide a legal context for corporate range investment. The data, however, indicate that many herders welcome group ranches in countries like Kenya, not because they are anxious to limit stock numbers, or curtail traditional strategies, but rather because the new legal machinery gives them a less ambiguous route to follow in protecting their range from invasion by cultivators [Galaty 1980].

In point of fact, experience has shown that tenure reform has often not been an effective instrument in the pursuit of either growth or conservation policy objectives. It can be argued that the tenure reforms offered have not taken adequate account of the broad economic and ecological environment of pastoral systems, or of the nature of the changes that are underway in the organization of livestock production. Some of the more salient structural aspects of pastoral production and their implications to policy are examined below, but for purposes of the present discussion of conventional tenure policy, the following observations are offered.

While tenure policies have tended to emphasize assignment of exclusive rights to discrete land areas, the circumstances of livestock production for the vast majority of cattle producers require maintenance of some form of communal tenure. In fact, in most pastoral economies, livestock production and use of grazing commons are still inseparable for two main reasons, the first of which is related to problems of herd size. The great majority of livestock holdings in Africa are small, fewer than 100 head of cattle [FAO 1975]. No single production unit could capitalize a ranching operation including water supply, with such small holdings, especially given the noncommercial orientation of many producers. Of course, the group ranch concept offers the economies of scale necessary to finance ranch development, but in most cases critical issues of asset management and herd disposition have not been successfully resolved.

Second are ecological reasons that militate against imposition of systems of individual land rights to replace communal tenure. Livestock production in semi-arid savannah areas is a land extensive enterprise, typically requiring quick response to highly variable rainfall patterns. Land tenure must take into account the variable environmental base. Hence, we should not be surprised that transience of resource use is a near universal condition as specific land resources can normally only be expected to have use value for limited amounts of time each season. The timing of this use will depend on type of animal, seasonal variation and so forth, which in the Sahel, for example, results in different groups utilizing the same resource base at different times during the year (See Gallais and Boudet [1980] for a project design that explicitly tries to deal with this factor). Transiency will remain de facto an essential component of most tenure systems, if not de jure.

The transiency component means the intensity of use on any given landed resources will vary by time, space and social group. Planning will have to come to grips with the time-thing-person relationships that make life possible in these arid rangelands. Individual tenure is not easily made compatible with regular, transhumant movements between seasonally available water supplies, especially where dry season pasture conditions are not predictable. Exclusive tenure requires, in most cases, a technical infrastructure that is not economically feasible given present and foreseeable market conditions.

The conclusion is that while the number of options for making production more efficient are severely limited, existing circumstances virtually dictate that some form of communal tenure will have to continue at the present time regardless of the tenure reforms proposed. But, we hasten to add that the existing situation, characterized by a virtual absence of grazing controls, widespread land degradation, growing impoverishment and inequality among producers, does not provide the elements of a long-term communal tenure model of great inherent promise. Furthermore, the changes affecting African pastoralism are not well dealt with by the institutional resources of traditional society. In fact, the atrophy of traditional management rules is but another symptom of the changes that are overtaking the pastoral sector. Thus, new models of communal tenure must be designed to meet emergent circumstances of pastoral production and resources use. In the following section, several relevant aspects of the changing pastoral environment in relation to tenure policy are examined.

TRANSITIONAL ECONOMIES AND TENURE POLICY

The economic organization of livestock production and resource management practices are changing in response to a general reorientation of household economic interests away from subsistence production toward engagement with more cosmopolitan economic institutions. This process has two important implications for pastoral production.

First, resource management tends to become abusive. Especially today, herders have even less incentive to maintain or initiate agreements pertaining to resource allocation and control. The local-level institutions that traditionally have performed that function have yielded to supra-local market institutions as an important new factor in gauging production decisions.[4] This dissolution of local-level controls is further accommodated by other phenomena that accompany rapid economic change, such as population growth, income diversification, technological changes, and, of course, development projects. The latter, including those that aim solely to re-establish ecologically sound management practices, are cast with reference to the emergent, market-oriented economic institutions.

The second key aspect of economic change is the emergence of enterpreneurship, a term used in the broadest possible sense. Simply stated, as herd ownership becomes less constrained by collective economic and managerial controls, private rather than collective benefits are maximized. Or, put another way, the economic interests of the household or herd ownership unit are pursued with increasing reference to external market institutions, and commensurately less so to local social obligations. This process of increasingly autonomous decision making reinforces the breakdown of local-level management controls.

There are three major attributes of the economic change process that are relevant to the development of tenure policy. First, the process of adjustment to the new economic reality has been a tremendously uneven one, not only among pastoral groups, but within groups as well. In fact, the highly differential character of producer adaptation and response to economic change is perhaps the single most important attribute of the change process from the tenure reform viewpoint. Greater decision making autonomy coupled with a wider choice of technologies and product outlets has given rise to what we choose to call differential production orientations and management styles [Bennett 1982]. On

the most general level, "production orientation"
divides along the lines of market and nonmarket production, but the actual situation is one of a broad continuum between these two extremes. "Management style"
refers to the kinds of herd management and enterprise
investment practices typically characteristic of each
production orientation. For example, a "commercial"
production orientation would normally indicate a
management style characterized by relatively high
capital investment in water supply and ranch infrastructure, hired labor, and fairly large herd size. A
small subsistence producer, on the other hand, would
probably act to minimize expenditure on the herd, given
that household cash requirements might be more
efficiently secured by applying limited assets and
labor to other activities, perhaps involving labor
migration. These distinctions are important for tenure
policy because production orientation and management
style indicate general tenure models appropriate to the
prevalent production systems.

A second major attribute in the process of economic and structural change is its implications for
local-level resource control practices, including
formal and informal regulatory institutions. Recent
research has led to an approach that has many appealing
implications to institutional development for range
conservation, buttressing traditional institutional
controls over the range use practices of local herders
[Horowitz 1979; Gulbrandsen 1980]. Traditional institutions hold promise as broad organizational frameworks
for extension and planning programs, but it is doubtful
that they alone retain the essential attributes and
authority necessary for achieving conservation objectives, for several reasons. First, the authority of
traditional institutions (as vested in chiefs, ward
heads, and lineage heads) is mainly derived from the
exercise of political and economic functions that have
atrophied as institutions external to the traditional
order have gained ascendance. As stated above, household production and labor allocation decisions are
increasingly less confined by local conventions.
Market conditons, external employment opportunities and
new technologies have all resulted in a fundamental
reorientation of economic interest and herd management
almost everywhere on the continent.

In some parts of Subsaharan Africa, such as Botswana, the process of change from traditional subsistence oriented production toward more commercialization
is well advanced, while in others, such as among the
Dinka and the Nuer in the southern Sudan, it has barely
begun. The Maasai and the Fulani are probably at an
intermediate stage in the process. The decline of
traditional authority has often been promoted by

modern political elites as part of the program for
nation building, and often as a means of consolidating
their own positions. Reinvesting traditional authorities with control over important land matters would be
considered a step backward by most modern political
leaders as well as by many herders. Finally, there has
even been a tendency by some analysts to exaggerate the
extent of control formerly exercised by traditional
authorities over community resource use. Those controls that were in place were tailored to the requirements and circumstances of relative resource abundance,
and were largely concerned with assuring equitable
access to resources by group members.

Range use has truly become a chaotic situation in
many areas, and the prospects for local institutions
alone maintaining control of the situation are not very
good. This is happening because the processes of
structural change described above imply that the
relevant economic institutions affecting the production
and resource use decisions of pastoralists are increasingly situated beyond the level of local exchange and
redistribution networks. To be effective, resource
control institutions must somehow be scaled to these
new influence "jurisdictions". Typically, some measure
of state level control is necessary for the effective
regulation of economic activity integrated by national
markets. This is not to deny, in the least, a role for
local-level institutions in the management of resources, but it does suggest that the power and authority of such bodies will probably have to be supported
by, and integrated into, higher levels of state
authority.

Institutions, only part of the equation, must be
seen as arbiters of what is currently absent in most
communal tenure situations today: a body of consistent
and accepted common property law that defines the
terms, conditions and rights of access to common resources. Arriving at effective common property law is
a matter of interpretating customs and practice, combined with considerations of desirable public policy
toward economic development and land use. In effect,
taking into consideration both national and individual
goals, common property law must be restated at the
level of the nation, taking cognizance of local variations in custom and practice. The evolution and formal
restatement of common property law will in most cases
be a long-term process.

A third major attribute of the changes affecting
pastoral production is the transitional character of
the new economic and ecological relationships facing
the producer at any given time which makes for an
inherently unstable policy making environment. Producers assume fundamentally new economic and social

attitudes while simultaneously attempting to retain old ones. Official institutional resources are weak and poorly defined. Rules of behavior and definitions of rights tend to be vague and uncertain. Projects themselves push production and conservation objectives that appear contradictory to the producer. Signals are mixed, detracting from the already weak credibility producers grant modern sector authorities.

Such problems are endemic to situations of rapid economic and social change. But the implications of inherent institutional weakness and widespread public uncertainty over resource rights regarding the efficacy of proposed tenure reforms are rarely considered. Economic change is a dynamic process, putting severe limits on the ability of usually static legal rules to maintain relevancy. This is a problem not easily dealt with under any circumstances, especially by policy planners who are faced with a multitude of tradeoffs.

A MODEL OF TENURE POLICY FOR PASTORAL SYSTEMS

The changes presently underway are characterized by divergent responses of animal producers to a changing economic environment, especially in the area of commercialization of the herd, and by increasing individualization of decisions about resource use, accommodated in part by a decline in the efficacy of local-level range use controls. For reasons discussed above, grazing land is still primarily communal, as necessitated by the intrinsic requirement of smallholder animal management on low productivity range of seasonally variable carrying capacity. These characteristics of production with respect to land use require that communal tenure be retained, in one form or another, as an essential feature of most pastoral production systems. Once the necessity of communal tenure is accepted, the key policy issues center upon the design of communal tenure rules and institutions appropriate to the needs and potentialities of producers of varying production orientations and management capabilities.

A policy model which holds promise for Subsaharan Africa is summarized in Figure 16.1. It should be emphasized that as a general model it is meant to be illustrative of the principles that underlie the policy relationships that are discussed below. That is, we attempt a theoretical framework for approaching the specific details of any number of tenure policy problems. The model appears to assume a large measure of spatial separation between large, commercial holdings, and smaller, non-commercial enterprises. This, of course, is typically not the case, and a key question in most tenure reform programs will be how to tailor

Figure 16.1. A general model of tenure policy variables for African pastoral systems

Role of Livestock in the Household Economy	Production Orientation	Management Style	Tenure	Type of Institutional Controls
1. Large Holdings				
High reliance upon livestock sales to meet large cash needs.	Commercial production for market.	Fairly high investment in ranching operations.	Exclusive: ranging from private property rights to some form of leasehold.	State issues specific right via legal instrument (freehold, leasehold, etc.).
2. Small to Medium Holdings				
High dependence upon cattle for cash and subsistence needs; and as input into other aspects of farming enterprise.	Broad continuum from essentially tradi- tional to mainly commercial; typically cattle still important for subsistence but small levels of planned commercial offtake achieved.	Ranges from "tradi- tional" strategy of minimizing expenses to a "commercial" willingness to under- take investments.	Modified communal, formal allotment to extensive group including management provisos; also indirect control over land exercised via private water rights. Group ranch model.	Supra-local board or authority allots grazing rights to local grazing committee, group ranch, etc. Negotiation, not strict regulation, of range use preferable.
3. Small to Very Small Holdings				
Low reliance upon cattle as source of current income; used as form of investment and savings, but generally aspire to build up herds.	Marginal "itinerant" production; only occasional, and then unplanned cattle sales possible.	Minimal expenditure on farm operation; asset and labor short.	Communal use of public water supplies; cattle keeping in mixed farming areas.	Local level agree- ments; extent of overgrazing limited by water availability and perhaps by land use zoning.

Source: Authors.

specific reforms for specific groups utilizing shared
range. This will be difficult under the best of circumstances, and the evolution of greater spatial
separation may in the long-run be necessary. Also, the
model applies to semi-arid and arid production environments.

Tenure is treated in the model essentially as a
dependent policy variable. Tenure rules and institutions normally should be scaled to the circumstances of
livestock production, as indicated by the role of livestock in the household economy, and the production
orientations and management styles of the producing
units. The first measure is the role of livestock in
contributing to the overall income requirements of the
producing unit. This provides an indirect measure of
the relative economic interest of the household in
livestock, and the willingness (and ability) of the
household to make available labor and other productive
assets necessary for the adoption of certain types of
tenure-dependent management practices.

"Production orientation" refers to attitude of the
livestock enterprise to the market. Most herders
produce both for subsistence consumption and for the
market, so it is the proportional mix that is really
important. A potentially useful measure for classifying mixed production units as either predominantly
subsistence oriented or predominantly commercial
oriented is whether sales are undertaken on a regular
and planned basis. This would not, of course, be fail-
safe, but it exemplifies the qualitative considerations
that are involved in assessing changes in production
orientation.

"Production orientation" is important to tenure
policy for two reason. First, the degree of production
for sale indicates the general potential for undertaking private investments in water development and other
range improvements. Second, production orientation
provides an indirect measure of producer integration in
national economic (and public) institutions, including
marketing networks. These institutions provide a structure, or medium, for the conveyance of production and
resource management incentives. In the absence of a
reasonably high measure of producer integration, in
terms of overall political and economic interdepen-
dency, it is unlikely that the supra-local land authorities necessary for the negotiation and administration
of tenure rules will be effective. "Management style"
is a derivative of "production orientation," and is
used here as a measure of the willingness and ability
of producers to undertake expenditures on herding
operations. It is a supplementary measure of producer
reliance upon livestock, and susceptibility to public
incentives.

IMPLICATIONS FOR LAND TENURE POLICY

The large-scale commercial operations described in the first row of the model may often warrant granting of exclusive leasehold rights to qualified producers, although implementation of such a radical tenure reform should be approached with great caution as competing rights must be thoroughly adjudicated. Rights of stock movement should normally be preserved. Planning for the Tribal Grazing Lands Program (TGLP) in Botswana incorporated an overestimation of the commercial orientation and management capabilities of many large holders originally believed qualified for the special rights and privileges involved in leasehold agreements. Instead of assuring a production environment conducive to the investment and improved management practices characteristic of commercial ranches, the program instead provided an opportunity for wealthy and influential large holders to claim exclusive rights to land without being obliged to make the improvements appropriate to commercial enterprises. Granting of exclusive rights to individual stockholders should be undertaken only when there is reasonable expectation that the benefits that will accrue to society, in terms of increased output, income and improved resource guardianship, outweigh the loss of societal welfare involved in the displacement of other producers utilizing the land. Most livestock producers fall within the category of small to medium sized herders. Communal tenure is an essential aspect of this sector's production environment. Policy development must accept communal tenure as a given, and undertake to develop rules and promote institutions capable of making livestock production on common range work in the interests of producer welfare and environmental conservation. Policy emphases to date have not given sufficient direct attention to the problems of communal tenure.

Two elements have been suggested in the preceding section as essential elements of a workable communal tenure. First is a specific body of law governing rights and limits of access to communal resources, while second is an institutional framework for alloting land rights and policing land use. What is needed is the creation of institutions at both local and supra-local levels, the first under the control of stock-holders, the latter responsible for implementing range use standards and assuring equitable participation. Communal range policies would evolve out of a process of negotiation, compromise and regulation which in the long-term may lead to the reasonable satisfaction of most interests. The group ranch model is illustrative of a local level organization broadly representative of

herder interests. Though it has typically, and appropriately, been promoted by planners for its advantages
as a production unit, greater attention should be given
to its potential as an organization for engaging regulatory institutions in negotiations over range use
standards. Supra-local bodies must be backed up by
suitable administrative resources, regulatory authority, and, of course, political commitment. To be
effective, any supra-local institution must enjoy a
wider political legitimacy, achievable only from a
general public appreciation of the need for a formal
institutional role in regulating resource use. This
latter requirement has probably not been adequately met
anywhere in Subsaharan Africa. Establishing institutional legitimacy on matters involving the regulation
of resources is perhaps the single most difficult
resource development constraint.

The third group in the model presents very different policy problems. These small to very small holders
typically secure only a small portion of total household income from cattle in the form of milk, blood, and
only very occasional cash sale. For them, the small
family herd may be an important input to other aspects
of the farming enterprise and may also serve as the
household's only significant form of savings.

It is just because the smallholder is so often
unable to provide either the labor or capital to effectively manage the few animals owned that special
difficulties are presented. Often the very animals that
cause the greatest damage and are unattended or only
casually cared for belong to this category of owner.
Yet at the same time, the owner is frequently incapable
of providing more animal supervision. In addition,
these small holdings are the only secure form of
"wealth" possessed by this lower stratum of the pastoral community. In the aggregate, the number of
animals on the African range belonging to this category
is substantial, and unless we address the property
rights involved, there is little hope of effective
management. The land rights of small holders are
probably best provided in relatively sedentary mixed
farming areas. These areas need to be identified and
secured for smallholders as a first step in any tenure
reform program.

CONCLUSIONS

In most pastoral production areas of Subsaharan
African, communal tenure makes economic and ecological
sense. Though communal tenure systems throughout the
continent are undergoing severe stress in the face of
rapid economic and institutional change, individualization of range land will only in the rarest cases solve

the problems characteristic of communal tenure systems today. At the same time, establishment of communal tenure systems that accommodate growth, conservation and equity objectives presents formidable challenges. In any given situation, analysts must be prepared to rigorously assess the environment of livestock production and producer decision making in terms of what it implies for land tenure, producer cooperation and forms of administrative regulation. Though traditional institutions may in some circumstances retain sufficient legitimacy to play a role in range management, the economic and political bases for traditional authority are becoming increasingly tenuous across Africa. The contemporary production environment presents several unique problems unfamiliar to traditional institutional experience.

The continuing importance of communal land use to pastoral production indicates that, over the long-run, increasing attention should be given to the development of policies in the areas of common property law (including the relationship between individual and corporate rights and responsibilities as well as arrangements such as group ranching) and regulatory and community management institutions for communal land usage. These two institutional realms will provide the working rules for communal tenure. The latter area, regulatory and community management institutions, has some implications for technical assistance, for it suggests greater emphasis on approaches to resource management similar to the tradition of public lands management as known and practiced in North America [Calef 1960]. This tradition, with its predominant emphasis upon the negotiation, assignment and regulation of grazing rights to common pastures, has been remarkably absent in providing even the most general background to pasture management in Africa.

Achieving efficient adminstration of public, communal range will be a long and difficult undertaking. Land managment agencies will become factors to be reckoned with at a rate roughly commensurate with two important developments in Africa's political economy: the economic integration of pastoralists and their livestock production into the national economies; and the public recognition of the state's legitimate interest in matters affecting the use of natural resources. The former is proceeding rapidly, the latter will be granted only grudgingly.

NOTES

[1]Hardin recognized the danger and his subsequent work edited with Borden [1977] more fully elaborates the multitude of intervening variables.

[2]We use the terms individual, private, and exclusive rights more or less interchangably.

[3]This issue has recently been applied to the Botswana case by Paul Devitt [Carl Bro 1982]. That there are in fact "greener" pastures elsewhere has been the basis of traditional range use strategy. Loss of land to competing users, demographic growth, etc., have made such solutions to range degradation increasingly unviable.

REFERENCES

Bennett, John W. Of Time and the Enterprise. Minneapolis: University of Minnesota Press, 1982.

_____. "The Political Ecology and Economic Development of Pastoralist Societies in East Africa." Research Paper No. 80, Land Tenure Center, University of Wisconsin-Madison, 1983.

Calef, Wesley. Private Grazing and Public Lands. Chicago: University of Chicago Press, 1960.

Carl Bro International. An Evaluation of Livestock Management and Production in Botswana: Main Report, Volume 2, Gaborone: Ministry of Agriculture; and Brussels: European Development Fund, 1982.

Food and Agriculture Organization/PNUE. Conference FAO/PNUE sur l'amenagement ecologique de parcours arides et semiarides d'Afrique et du Proche-Orient, Rome, 3-8 fevrier 1975.

Galaty, John. "The Maasai Group-Ranch: Politics and Development in an African Pastoral Society." When Nomads Settle, ed. Philip Carl Salzman, pp. 157-172. New York: Praeger, 1980.

Gallais, Jean, and G. Boudet. "Projet de code pastoral concernant plus specialement la region du delta central du Niger au Mali," Maison-Alfort: IEMVT, 1980.

Gulbrandsen, Ornulf. Agro-Pastoral Production and Communal Land Use: A Socio-Economic Study of the Bangwaketse. Gaborone: Ministry of Agriculture, 1980.

Hardin, Garret. "The Tragedy of the Commons," Science 1962 (1968):1243-1248.

Hardin, Garret, and John Baden, eds. Managing the Commons. San Francisco: W.H. Freeman and Co., 1977.

Horowitz, Michael M. "The Sociology of Pastoralism and African Livestock Projects." Background paper for USAID Workshop on Pastoralism and African Livestock Projects, Harper's Ferry, West Virgina, 24-26 September 1979. Binghamton, New York: Institute for Development Anthropology.

Khama, Seretse. <u>National Policy on Tribal Grazing Land</u>. Government White Paper No. 2 of 1975. Gaborone: Government Printer.

Lawry, Steven W. "Land Tenure, Land Policy, and Small-holder Livestock Development in Botswana." Research Paper No. 78, Land Tenure Center, University of Wisconsin-Madison, 1983.

Riddell, James C. "Land Tenure Issues in West African Livestock and Range Development Projects." Research Paper No. 77, Land Tenure Center, University of Wisconsin-Madison, 1982.

17
Fenced and Open-Range Ranching: The Commercialization of Pastoral Land and Livestock in Africa

Roy H. Behnke, Jr.

In widely scattered parts of the world, subsistence forms of pastoral nomadism are being displaced by the spread of commercial systems of livestock production. Documentation of this trend comes from both the European and North American Arctic, from southwest Asia, and the various regions of pastoral Africa.[2] To date, however, these emerging forms of commercial livestock management have been the primary focus of very few field studies and have excited even less theoretical interest, aside from Ingold's [1980] major work on commercial reindeer pastoralism in the Arctic.

As a result of this neglect, there has accumulated a formidable list of questions regarding the nature of African systems of commercial livestock production which make use of natural rangeland. Included in this list are questions of interest to both donor agencies and national governments involved in agricultural development and administration, questions regarding the impact of commercialization on range degradation, offtake rates, producer price responsiveness, and economic stratification within pastoral communities (see Galaty et al. [1981] for a recent review of applied pastoral studies). While each of these issues merits individual attention, it is equally true that specialized studies of this kind will be of greater practical utility if they can be incorporated into a general model of commercial change in pastoral Africa.

Two alternative approaches immediately present themselves as possibilities for building such a model. On the one hand, we can attempt to penetrate the workings of pastoral economies by examining the systems of exchange in which they are involved [Schneider 1979]. In this way we directly confront the phenomenon in question, commercial exchange, and make that phenomenon (or analogous processes in noncapitalistic economies) the independent variable around which we arrange all other factors. This approach has the added advantage

261

of being consistent with the interests and methods of
formal economists working with aggregated statistical
data on offtake and sales rates [Crotty 1980; Jarvis
1974; Reutlinger 1966; Ariza-Nino and Shapiro 1984].
On the other hand, we can view the supply of marketable
livestock as a dependent phenomenon controlled by the
capacity of different kinds of production systems to
sustain such output. This approach is consistent with
the assumption that humans must provision themselves in
order to exist and that, as a consequence, economic
production is historically, logically, and causally
prior to economic exchange [Sahlins 1972; Gudeman
1978]. It is this latter perspective which is adopted
in this analysis.

Central to this approach is the notion of the
production rationale. Briefly, a production rationale
can be defined as a distinctive set of production goals
which are consistent with a particular configuration of
human, technical, and natural resources. Based on the
concept of production rationales, it is possible to
distinguish different pastoral regimes which are
designed to operate efficiently under different sets of
conditions, are especially suited to solving different
kinds of problems, and will yield different volumes and
kinds of pastoral produce. The history of range
livestock production reveals three such regimes:
subsistence pastoralism, open-range ranching, and
fenced ranching.

Figure 17.1 provides an overview of the
distinctive features of these three regimes, and a more
detailed description of each regime has been provided
elsewhere [Behnke 1983]. The following remarks describe in capsule form the rationale which underlies
each of these systems of livestock management:

i. Subsistence production is often treated as a
null category, as the kind of primitive production
orientation that develops when markets are not available and when people cannot sell their produce. With
respect to subsistence pastoralism, this point of view
has recently become very difficult to sustain in the
face of extensive evidence of pastoral participation in
systems of economic exchange [Barth 1964; Bates and
Lees 1977; Schneider 1979]. A more useful conception
of the purpose of pastoralism may instead focus on the
relatively low ratios of livestock to dependent humans
in these systems of production [Allan 1965; Helland
1980]. Precisely because they are subsistence oriented, pastoral economies can employ and directly feed
more people than alternative forms of commercial livestock management [Dahl and Hjort 1976; Dyson-Hudson
1980; Ingold 1980].

ii. Conceived in these terms, the contrast between
subsistence pastoralism and open-range ranching could

Figure 17.1. Three pastoral production regimes

Production Regime	Techniques of Animal Management	Type of Animal Produce	Land Tenure
Subsistence Pastoralism ("Pastoral Nomadism")	Labor-intensive shepherding and nomadism; animals are docile and easily controlled	Subsistence production: repeated harvesting of live-animal products, especially milk	"Communal", i.e., tenure systems tend to be localized and exceedingly complex arrangements which reflect political and military as well as legal considerations
Open-Range Ranching	Labor-extensive roundups and drives; stock is feral and largely cares for itself	Commodity production: animals for slaughter	"Range rights", i.e., de jure communal and de facto private control of critical resources
Fenced Ranching	Capital-intensive methods of "modern" management which supplement natural pastures (fodder, tame pasture, feed and mineral supplements, etc.); based on highly productive improved breeds	Commodity production: animals for slaughter	Private leasehold or freehold tenure

Source: Author.

not be clearer. Pastoralism requires intense contact between humans and stock, and is labor intensive; commodity production under range conditions requires almost no contact between man and animal, and is labor extensive [Ingold 1980]. These different systems of labor use are based, in turn, on different balances between numbers of livestock owners and livestock owned. Herders are relatively poor in stock and must exploit their animals frugally; ranchers are relatively rich in animals and can afford to take only those products that are most readily obtainable or marketable.

iii. Like the other production regimes, fenced ranching makes sense as an integrated whole in which one aspect of the production system sustains and is sustained by the other aspects. Relative to open-range ranching, enclosed ranching involves heavy capital investment in landed improvements which are attractive only if rangeland is privately owned. These capital investments make possible technical innovations (such as fodder production, pasture rotation, improved watering facilities, etc.) which are required to meet the higher standards of husbandry demanded by high-yield breeds of livestock. In return, the higher productivity of these breeds is able to pay for the expense of their nursing. We have, therefore, come full circle-- back to an intensive form of animal exploitation. But fenced ranching has substituted commodity production for in-kind production, and heavy capital investments for pastoral investments in human labor.

THE COMMERCIALIZATION OF LIVESTOCK AND LAND: AN OVERVIEW

The three production regimes presented here are ideal types, abstractions which are not meant to represent the diversity of livestock production in Africa. It would be a mistake, therefore, to construe the typology as a classificatory scheme, or to attempt to modify or elaborate it to fulfill this role. What the typology does provide is a simple, logical framework which we can use to discipline our thinking about change in pastoral economies. By comparing different production regimes, the typology immediately suggests the existence of different kinds of commercialization processes; it also helps us to isolate a limited set of factors which may account for these processes, and to conceptualize the causal linkages among the factors. In short, the typology forces us to make certain simplifying assumptions. Our next step is to examime the adequacy of these assumptions through an analysis of case material on pastoral economies undergoing commercialization. According to the scheme advanced here, we should be able to distinguish two fundamentally

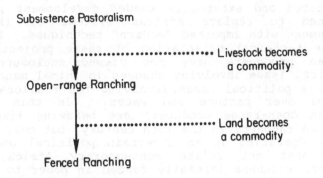

Figure 17.2. Commercialization processes in contemporary Africa

different kinds of commercial producers--open-range and fenced ranchers--and two distinct processes of commercialization--one involving livestock and the other involving land (see Figure 17.2). The remainder of this analysis will argue that the commercialization of African pastoralism does indeed make remarkable sense within the confines of this simple explanatory scheme.

In pastoral Africa animals themselves are the first factors in production which lose their local definition of value, their subsistence use, or in-kind value, and are transformed into commodities. This redefinition of the value of livestock has, in turn, a ramifying impact on almost all other aspects of the production process, for pastoralists who wish to maximize the long-term production of saleable animals must undertake fairly standard modifications in the way they manage their herds and deal with their fellow pastoralists. If macro-economic conditions will permit them to undertake these changes, then their system of livestock management will tend to shift from intensive subsistence pastoralism to extensive open-range ranching. In contemporary Africa this kind of commercial change can be characterized as "spontaneous"; it will be affected by government policies, but it is usually based on capital accumulated by individual producers and reflects the changing needs and attitudes of these producers. As a form of commercial production adapted to the margins of a market-based economy, the development of this indigenous form of open-range ranching may be one of the most important innovations in the organization of production in the arid areas of modern Africa.

A second, less pervasive kind of commercial change is embodied in the shift from open-range to fenced

ranching in Africa. Almost all examples of range
enclosure in Africa are associated with formally
instituted and externally funded development projects
designed to replace African systems of livestock
management with imported "modern" techniques. In spite
of the technical orientation of these projects, most
African producers have not viewed enclosure as a
technical issue involving changes in animal management,
but as a political issue involving the reallocation of
control over pasture and water. In this respect
African open-range producers are behaving like North
American ranchers in the 19th century, but only up to a
point. Operating in an uncertain political and legal
environment not unlike contemporary Africa, North
American ranchers initially fenced in order to control
land, not in order to better manage their animals
[Anderson and Hill 1977]. Once they were trapped
inside their own fences, however, American ranchers
adjusted their herding practices to compensate for
their loss of mobility [Osgood 1929]. For a variety of
reasons, it is this last and most fundamental step in
the enclosure process which has not yet occurred on any
widespread basis in pastoral Africa.

The succeeding sections of this chapter describe
in greater detail the different phases in the commercialization process, from pastoralism to open-range
ranching, and from open-range to fenced ranching.

FROM INTENSIVE PASTORALISM TO EXTENSIVE RANCHING

The spread of range ranching in Africa poses a
major analytical problem. In its classic form open-
range ranching made perfect sense for European colo-
nists living on an economic frontier under conditions
of free resources and few people [Behnke 1983]. But
the nature of the industrial frontier has changed in
the last century. Instead of pushing indigenous producers aside, industrial expansion now proceeds by
incorporationg them and transforming their economies.
African pastoralists responded to this process of
incorporation by increasing animal sales, thus trigger-
ing a series of changes which ramify throughout pastoral economies. Beginning with a shift in production
rationale, commercialization entails predictable
changes in husbandry techniques, herd size, economic
stratification, and land tenure, all of which are
discussed below.

Changes in Production Rationale

Pastoralists can respond to market incentives in
two different ways. At a superficial level of commercial involvement, they may simply sell livestock

products that they would otherwise have used themselves. At a more fundamental level, they must change
their style of herd management in order to maximize
commodity production.

Confronted with new markets and increased demand
for animal produce, an initial response by pastoralists
may be to exchange the by-products or occasional surpluses generated by a subsistence oriented system of
production. Since milk is the principal pastoral product, commercial dairying may experience a temporary
florescence under conditions of initial commercial
penetration. For example, the production of cream for
sale was a flourishing industry in Botswana in the
1930s [Pim 1933]; clarified butter was an important
Somali export in the 1920s and 1930s [Swift 1977]; and
Bedouin families regularly produced clarified butter
for local sale in the years immediately following
Libya's oil boom [Behnke 1980]. Thus, in an unpredict-
able commercial environment, pastoralists manage to
exploit new markets without irrevocably committing
themselves to producing primarily for those markets.

Dairy production for local exchange may constitute
a stable pastoral adaptation under certain ecological
and social conditions, as among the Fulani of the Sahel
[Stenning 1959; Dupire 1962] or among pastoralists of
Anatolia and Iran [Bates 1973; Barth 1964]. In Botswana, however, the commercial dairy industry had
collapsed by the 1950s [Ryan 1958], and by the same
date ghee exports from Somalia had declined and been
replaced by a vast expansion in the export trade in
live animals [Swift 1977]. In Libya by the late 1970s,
the same Bedouin families who had once sold clarified
butter now produced it solely for home consumption, and
many were shifting out of goat production (a mixed milk
and meat operation) and into commercial sheep ranching
[Behnke n.d.]. At the same time, in both Botswana and
Libya there was a decline among commercial producers in
the production of milk for home consumption. Libyans
entirely abandoned the milking of marginal dairy
animals such as sheep and camels; large commercial
operators in Botswana stated that they now milked their
herds less, and statistical evidence supports this
generalization [Solway 1979; CARL BRO 1982].

Herein lies one of the essential dilemmas of the
commercialization process. In attempting to exploit a
lucrative market for live animals, herd owners are
tempted to redeploy the productive capacities of their
most valuable asset, their herd. In so doing, however,
they cannot retain the essence of their old dairying
operation, for there is direct competition between
humans and nursing animals for milk. A commercial
pastoralist can produce a healthier and larger crop of

slaughter animals in a shorter time only if he allows young animals to suckle all their mothers' milk.

An almost identical calculation of costs and benefits will obtain in the case of other kinds of pastoral products. Pastoralists are specialists in prolonging the useful life of an animal in order to extract the maximum amount of replenishable, live animal products. While this pattern of animal use may not kill the animal, it is usually carried out at some cost to the animal's health, vigor, or suitability for slaughter. The use of oxen for plowing in Botswana, for example, alters their conformation and produces weight losses which diminish their commercial slaughter value [Solway 1979]. Much the same can be said about the use of camels for transport, or the bleeding of cattle among East African pastoralists. The abandonment of these subsistence forms of animal use will constitute an integral part of the commercialization process because such uses interfere with the increased production of a single profitable commodity.

The ultimate result of these shifts is a radical revaluation of the purpose for livestock keeping. Among pastoralists an animal's eventual death or slaughter (often in anticipation of death) is simply the last stage in its productive career. This pattern of animal use will tend to postpone the age of sale well beyond the optimal age of slaughter, thereby producing the skinny, old, toothless, cull animals found in many African markets [Doran et al. 1979; Quam 1978]. A commercial producer, on the other hand, cannot realize a profit from his animal until he disposes of it. Whereas the pastoralist will want to hold onto his animal as long as possible, the commercial producer will want to cut his costs by selling it at the earliest profitable opportunity. At this most fundamental level, therefore, subsistence and commercial approaches to animal management are incommensurable production strategies.

Changes in Husbandry Techniques

Changes in the kinds of produce extracted from herds are inextricably linked to changes in herding techniques. Subsistence pastoralism is a laborious undertaking predicated on almost continual contact between humans and animals. The characteristic pastoral herd management methods are all labor intensive: shepherding, household nomadism, and milking. These are also tasks which commonly require the attention of the entire household, including young and old, male and female workers.

Commercialization changes this picture. If the household can be provisioned through animal sales and food purchases, it will no longer be necessary for the family to migrate in order to be near its food supply. At the same time, the elimination of milking will mean that an abundance of household labor is no longer required in order to effectively exploit the herd. Thus, herds may continue to move on a seasonal basis to avoid disease or gain access to water or good pastures, but household nomadism can be discontinued. Under favorable circumstances, even shepherding may be curtailed since the herd must no longer be rendered either docile or accessible for milking or draft purposes. The larger herd species that can look after themselves may be left to run free, and periodically rounded up for branding, inspection, transport or sale. As a result, the management techniques of African commercial producers will resemble in an attenuated form the husbandry practices of open-range ranchers of European descent, particularly if these ranchers are operating on a small scale and in remote areas where livestock marketing is difficult [Riviere 1972; Dobie 1941].

Herd Size and the Effect of Scale

The increased levels of livestock marketing and the changes in management style described here are not, in the phraseology of economics, neutral to scale. Commercialization will affect large and small subsistence operations in very different ways.

A critical consideration for the small herd owner will be the effective decline in herd performance and productivity under commercial conditions. Part of this decline is a direct result of the shift from milk to meat production:

> Since each step in the chain of conversions from pasture to milk, and from milk to meat, involves a net loss of energy, the interception of this chain in such a way as to deduct a proportion of milk yield "at source" permits the support of a far greater human population than if the same proportion were deducted only after its conversion into meat. Given that man is not biologically equipped to digest the plant food of ruminants, milch pastoralism represents the most efficient possible use of uncultivable grazing land, if measured in terms of population carrying capacity. [Ingold 1980, p.176]

In essence, commercial producers have traded a biologically efficient production system for an economically profitable one. This decline in biological output per animal will be exacerbated if the shift from dairy to meat production encourages the discontinuation of close herding and increases animal losses due to straying, theft, or predation [Ingold 1976]. Under these circumstances, small herd owners who shift from intensive to extensive forms of management will probably need to increase their herd sizes simply to maintain their prior standard of living, unless they are lucky enough to experience a simultaneous improvement in the pastoral/nonpastoral terms of trade [Behnke 1980; Swift 1979].

There are several factors, however, which impede any effort by small producers to simultaneously commercialize and increase their herd size. In the first place, by shifting to commercial production they have acquired new cash needs since they must substitute purchased consumer items for goods and services formally supplied by the herd--the provision of shelter, clothing, traction and transport, and food. At the same time, they must come to terms with the classic dilemma of the commercial livestock producer: they can either sell an animal for immediate profit, or hold it as a capital investment in future profits, but they cannot do both at once. They must therefore possess a herd which is large enough to fulfill two separate functions, the biological reproduction of the herd itself and the maintenance of an offtake rate which will meet household expenses. Lacking a herd which meets these minimal requirements, small commercial producers may find themselves selling off their breeding stock simply to feed the family [Sutter 1982; Quam 1978].

A different set of considerations will affect the response of large herd owners to commercial incentives. Whereas small operators must view commercialization in the light of declining animal productivity, large operators may view it in terms of potential increases in the productivity of human labor. Maximum pastoral herd sizes are highly variable and depend on practical considerations regarding the herd species, the terrain, the presence of predators, etc. [Bates 1972; Dahl and Hjort 1976; Khazanov 1980; Irons 1972; Swidler 1972]. Paramount among these considerations, however, must be what Horowitz has called "the carrying capacity to labor" under different techniques of management [1979, p. 45]. Since large inputs of pastoral labor are rendered redundant in a commercializing operation, restrictions on herd growth which resulted from high labor demands are relaxed. At the same time, herd growth is encouraged by economies of size in which larger herds tend to have lower per unit operating costs for labor and other critical inputs [Strickon 1965]. Thus, technical innovations permit and economic pressures encourage a marked increase in the maximum size of large subsistence herds undergoing commercialization.

Economic Stratification

Livestock is a potentially unstable form of wealth susceptible to both rapid increase and sudden loss from epizootic disease, drought, theft and raiding. Subject to these risks, wealthy pastoralists have frequently sought to transform surplus livestock into more secure forms of wealth. The nature of these conversions has varied from society to society and according to the nature of the wider political order and investment options open to wealth producers, giving rise to the debate on economic inequality in pastoral societies [Asad 1979; Burnham 1979; Irons 1979; Salzman 1979]. In some settings--particularly common in Subsahara Africa--wealthly pastoralists have attempted to transform surplus animal holdings into political prestige through the medium of clientship and in-kind payments for herding labor, long-term stock loans, or emergency gifts to rebuild the herds of impoverished pastoralists [Baxter 1975; Lewis 1961; Peters 1965; Doughty 1979 first published 1888; Rigby 1979]. In redistributive economies of this sort there may emerge elaborate systems of political ranking based, almost paradoxically, on a minimum degree of permanent economic inequality [Gluckman 1971].

The shift to commercial production is likely to call into question the existence of these inward-turning, relatively unstratified economies. Because of the effect of commercialization on herd sizes, animals that pastoralists once thought of as surplus to their needs will be perceived under commercial conditions as essential to their survival, while any genuine surpluses will be sold. There will, in short, be both a real and perceived shrinkage in the pool of animals available for redistribution within the pastoral economy, a development which will further undermine the viability of small herds and promote the growth of economic inequality in climatically unstable environments [Hertel 1977; Swift 1979]. At the same time, commercialization will open new opportunities for pastoralists to invest outside the livestock sector. The building of permanent houses or new water points, the purchase of cars, trucks, mechanized agricultural implements, retail businesses or agricultural land are all potential mechanisms for rendering transient differences in livestock wealth more permanent.

Land Tenure

In communal systems of land tenure, effective control of rangeland is contingent on the ownership of livestock which can exploit this common resource. It is also contingent upon having access to scarce water sources which make pastures available for use by animals in certain seasons. Thus, even if the traditional land tenure system remains unchanged, commercial herd owners take effective possession of the range as their herds grow in size. This position of superiority will be reinforced if large commercial operators invest in the development of private water sources which they do legally own. The result may be the emergence of an unstable system of de facto land control reminiscent of old North American systems of "range rights" in which restricted access to pasture is based on herd wealth, local political influence, and the legal ownership of scarce water [Osgood 1929; Webb 1931]. Thus, commercialization may lead to both increased economic inequality within pastoral society and to the solidification of these differences into class distinctions based on permanent, differential access to productive resources.

Summary

The preceding analysis allows us to dispose of two persistent misconceptions regarding the initial shift to commercial production among African pastoralists. First, western-trained development planners are accustomed to think of economic development in ahistorical terms as progress involving increased levels of commodity production and improved levels of biological productivity. This is not normally the case in the open-range phase of commercial pastoral development. To the contrary, the initial increases in commodity production may occur at the expense of a decline in total animal output. Commercialization involves a redefinition of production goals, not an absolute increase in the level of productivity.

A second aspect of the commercialization process follows from this decline in herd performance. Commercial production will appeal most strongly and will confer a competitive advantage on large herd owners who can sustain reduced productivity while profiting from labor-saving economies of size. Herds will therefore tend to grow and commercialize at unequal rates, and some herd owners will successfully make the shift to commercial production and others will not. The commercially successful will, then, attempt to reinforce their new superiority by acquiring private use rights to land that in theory is still communally owned. Thus, the process of commercialization entails the exclusion of some former subsistence pastoralists from full participation in the emerging commercial system. Again, this process of exclusion is a necessary feature of commercialization if rangeland is to escape over-grazing and if ranchers are to nonetheless succeed

in creating the larger units of production which are
necessary in a commercial setting.

RANCHING SCHEMES AND THE DEVELOPMENTS OF
PRIVATE TENURE

Outside Africa the enclosure of the commonage has
been a comprehensive process of social and economic
change comparable in scope to the shift from subsistence pastoralism to open-range ranching. The centrally planned and externally funded ranching schemes of
the 1960s and 1970s were designed to precipitate this
kind of radical change in pastoral Africa. That the
projects did not transform African pastoralists into
"modern" fenced ranchers is now generally accepted
[Sandford 1983]. In condemning these projects, however, we sometimes forget that ranching schemes did
occasionally elicit an enthusiastic response from pastoralists who were prepared to manipulate the projects
for their own purposes. The projects became, therefore, integral parts of a process of social change,
although not the process that project designers had
foreseen. In essence, what the projects did was put
open-range ranching on a firm legal footing by
formalizing an ambiguous land tenure situation. Far
from transforming the African livestock industry, the
projects themselves were transformed by evolutionary
processes ongoing within that industry.

This analysis will briefly examine the history of
three ranching schemes--the Ankole Ranching Scheme in
Uganda, the Tribal Grazing Land Policy and Livestock
Project II in Botswana, and the creation of individual
and group ranches among the Maasai of Kenya. All these
projects offered the standard ranching package in vogue
in the 1960s and 1970s in which the privatization of
land tenure was to serve as an enabling factor in a
multifaceted process of social, economic and technical
change. Initiated in the mid-1960s by U.S.A.I.D and
the Government of Uganda, the Ankole Ranching Scheme
was designed to install ranchers in an area of south-
western Uganda which had been cleared of tsetse fly and
reopened to livestock. The ranches were designed to
prevent the reoccupation of the area both by fly and by
the traditional pastoralists whose deficient husbandry
practices were deemed to encourage spread of the fly
and were "rooted in the attitude that cattle were a
factor of social status, were wasteful of grazing land
and did not make a contribution to the economy of the
country [Doornbos and Lofchie 1971, p. 170]. In this
context planners conceived of the ranches as ecologically sound and economically profitable ventures which
would "function as a nucleus of modernization" by
exemplifying "more up-to-date methods of animal husban-

dry [Doornbos and Lofchie 1971, pp. 173, 171].

Like the Ankole Scheme, the Tribal Grazing Land Policy of Botswana was an amalgam of land reform and economic development designed to prevent the spread of indigenous forms of cattle keeping, in this case onto "unused" land in the Kalahari. The reasoning adduced in favor of this package was also familiar. Cattle production could be "more than doubled" it was asserted, through the institution of simple management innovations such as fencing, a move that would also protect pasture resources, for "unless livestock numbers are somehow tied to specific grazing areas no one has an incentive to control grazing" [Government of Botswana 1975, p. 1]. Finally , there is the case of both individual and group ranch development among the Maasai of Kenya. The second phase of the Kenya Livestock Development Project, in which the creation of group ranches was a major component, cites the usual panoply of objectives:

i.   to increase marketed offtake and improve pastoral incomes,
ii. to rationalize land use by limiting livestock numbers to carrying capacity,
iii. to improve formal education for pastoralists,
iv. and to increase the biological productivity of animal husbandry through loan-financed investment by pastoralists in "modern" infrastructural improvements [Oxby 1981].

In sum, it is clear that planners envisaged each of these projects as an integrated "package deal" made up of interdependent components; the pastoral perception of these projects, it can be shown, was somewhat different, and was dictated by the needs and interests of various segments of the pastoral population.

In Ankole, western Botswana, and among Maasai the local proponents of individual ranch development represented a very small segment within the larger pastoral community. In Ankole these interested parties were members of an ethnically distinct cattle-owning elite who successfully lobbied for the right to own ranches in absentia. Thus, of the first forty ranches allocated, fifteen were owned by absentee landlords who were also members of the political elite, with four of the six Ankole members of the Uganda parliament receiving ranches. As Doornbus and Lofchie emphasize,

The distinctive and possibly unique feature of the political elite that has been able to exploit the Ankole Ranching Scheme is its local character . . . . It is predominantly composed of individuals whose basis of political status is within the Ankole area. Even those who occupied formal positions at the national level are members of

Ankole society and have risen to national prominence by virtue of their ability to exercise influence within the Ankole community. [1971, p.168]

Ranch allocation under Botswana's TGLP closely replicates the Ankole pattern. The descendants of large cattle owners, the political elite of modern Botswana have come to town to work for government, but they have not abandoned their interests in cattle. They have used their urban incomes to finance the expansion of their rural cattle enterprises, and used their political influence to create an official climate favorable to the interests of large commercial cattle-men [Parson 1981]. If private ranches were to be created, then these individuals were--like the Ankole elite--in an excellent position to compete for their ownership. In terms of the formal allocation criteria, wealthy cattle owners had the equity needed to secure government loans, the outside income which would permit the private financing of ranch development, and the large herds necessary to effectively populate a ranch [Government of Botswana 1981]. It also quickly became clear that they had a vested interest in the land being taken over for ranch development. In addition to sup-porting hunter-gatherer populations, almost all the supposedly "empty" areas designated for ranch develop-ment were already inhabited by cattle which were being watered at private boreholes and were grazing on land that was under the de facto control of whoever owned the water. In some areas nearly a third of all these water source owners were civil servants and were in a position to influence government policy [Hitchcock 1980]. Thus, prior occupancy of rangeland, political self-interest, conservative lending practices, and the formal criteria for ranch allocation all favored the large cattle owner.

It remains to be shown why these individuals should seek formal ownership of land that they already controlled. On the basis of research conducted prior to ranch allocation, Hitchcock [1980] concluded that these absentee land and cattle owners did not intend to put up fences, reduce stocking rates, or hire trained ranch managers should they receive a ranch:

The question arises,then, why people might want exclusive rights to land, if they have no inten-tion of developing it beyond the way it is at present....I found that the most commonly voiced reason for wanting exclusive right was so that the "squatters" could be forced off the land. In this way the Grazing Policy is playing into the hand of those who wish to remove unwanted people who live

276

around the boreholes but who, under Traditional Tswana law, cannot be forced to leave. [1980, p. 21]

The subsequent history of TGLP largely substantiates Hitchcock's contention that prospective ranch owners did not intend to immediately develop their property. A comparative survey carried out in 1980-81 demonstrated that there was no significant difference in performance between ranch herds and other large cattle herds using communal land in the vicinity of the ranches [CARL BRO 1982]. Even more indicative of the attitude of large cattle owners is the way they have responded to government efforts to demarcate, allocate, and lease the ranches. Interest has been high and competition between prospective owners has been keen up to the point when the ranches are allocated and it becomes clear who will eventually receive private title. Once this issue is settled, however, the new ranch owners become difficult to locate, and when located are reluctant to actually sign leases which would commit them to even a minimal rent after a three year grace period [Government of Botswana 1981; Bekure and Dyson-Hudson 1982]. The release of government loans for ranch development is nevertheless contingent upon the signing of a formal lease, and as a result, none of the money set aside by the World Bank for ranch loans had been committed by the end of 1981 [Government of Botswana 1981]

A similar picture emerges from an examination of ranch development in Kenya's Maasailand. The situation here is more complicated only because pastoralists have participated in two different kinds of new tenure arrangements--group ranches and individual ranches--and each of these tenure systems has developed its own local constituency. The allocation of individual ranches began in certain areas of Maasailand almost immediately after independence, and was organized by local County Councils acting without clear legal authority. On the familiar pattern of Botswana and Ankole, these ranches went to a group of individuals who constituted, in Hedlund's words, "a newly created distinct social and economic class" [1971, p. 16]. The individuals who made up this class differed from the mass of the pastoral Maasai on the following counts:

i.   They owned more vehicles.
ii. They were better educated.
iii. They were relatively wealthy, but reluctant to loan money and cattle, or to exchange grazing rights with other pastoralists.
iv. They sold large numbers of cattle and held a local monopoly on cattle trading.

v. They were influential and politically active,
    although they held few political offices in
    the old age-set system in which there was
    little opportunity for economic advancement
    by an office holder [Hedlund 1971].

Hedlund's profile leaves little doubt as to the socio-
economic characteristics of the individual Maasai
rancher. As in Ankole and Botswana, the advocate of
private ranch ownership is the large, commercially
oriented pastoralist who has already made the shift to
open-range ranching.

Compared to individual ranching, group ranching
has met the needs of a more broadly-based segment of
Maasai society. According to Galaty [1980] and Hedlund, group ranching has appealed to large numbers of
Maasai for two related reasons. Within Maasai society,
group ranching was an attempt both to emulate the success of individual ranchers and forestall the alienation of Maasai land to this small group of large commercial producers [Hedlund 1971]. On the wider political scene, group ranching was a tool used by the Maasai
to renegotiate with the independent Kenyan government
the colonial Maasai Treaty of 1911. What the Maasai
needed was a suitably modern tenure format that would
prevent the piece-meal alienation of Maasai pastures to
neighboring, ethnically distinct groups of cultivators
such as the Kikuyu. As Hedlund succinctly phrases it,
"registration itself was considered more important than
other kinds of development" [1971, p.4], or in Galaty's
terms, "The major significance of the Group-Ranch
structure lies not in the field of economic innovation,
which was occurring previously through individual and
neighborhood channels, but in the essential area of
political security" [1980, p. 165, emphasis not in
original].[3] Consistent with these conclusions,
numerous authors note that there has been little
progress made by group ranches in meeting any of the
economic and technical objectives of fenced ranch
development such as destocking, increased cattle sales,
the curtailment of cattle movement among ranches, or
the borrowing of money for capital improvements
[Halderman 1972; Oxby 1981; Galaty 1980; Helland 1980].

This analysis of fenced ranching began with an
examination of the many components which made up the
typical ranching scheme of the last couple decades. It
is now clear, however, that African pastoralists and
open-range ranchers rejected all components of these
projects which did not meet their immediate needs, and
persistently rejected the use of fencing. Batswana and
Maasai livestock producers cite a consistent set of
reasons for this rejection. Fences, say the Batswana,
would trap herds on ranches that were periodically
untenable due to borehole breakdown, veldt fires, and

localized drought [Hitchcock 1980]; Maasai, on the other hand, stress the problems of erratic rainfall and insufficient resources on particular ·ranches [Halderman 1972]. Like subsistence pastoralists, open-range ranchers rely on mobility as a technique for balancing localized deficiencies in resources needed by the herd. In this way they maintain within a wide geographical region a total livestock population far greater than that that which could be sustained, ceteris paribus, by independent herds operated separately on small plots of land. From the point of view of open-range ranchers, fencing will immediately precipitate declines in total herd productivity (through forced destocking), while simultaneously increasing risk from environmental variability, and requiring capital investments that inflate the costs of herd management. As with the previous shift from in-kind milk production to commodity meat production, the techniques of open-range herd management are inconsistent with those of fenced ranching, and obtaining the benefits of one system of management is contingent upon giving up the benefits of the other. Thus far commercial African producers have used ranching projects as a means of reinforcing their legal position vis-a-vis competitors, but few have found it in their interest to incur the considerable short-term costs that accompany the shift to a fenced system of animal management.

CONCLUSION

In comparison with subsistence pastoralism, modern fenced ranching increases the volume of marketable off-take, improves the cash income of livestock owners, and (if Hardin's [1968] reasoning is to be believed) provides a built-in check against overgrazing. Here, then, is an improved technology with something for everyone--more meat for the urban consumer, more money for the pastoral producer, and improved resource management for the national planner. All that remains, it would appear, is to coerce or cajole African pastoralists into transforming themselves into a reasonable facsimile of North American ranchers.

But there is one problem. The direct comparison of pastoralism and fenced ranching is irrelevant to the situation at hand, for commercializing pastoralists do not immediately adopt all the components of a fenced system of production. For most African pastoralists, the real choice is between open-range subsistence and open-range commercial production. In this case the switch to commercial production will confer considerable benefits on certain producers but will also entail considerable costs in the form of larger herd sizes, potential increases in range degradation, the abandonment of subsistence techniques in intensive animal use, and the promotion of economic inequality. These are not costs which are borne equally by all segments a commercializing pastoral society. It follows, therefore, that the creation of open-range ranching is not simply a technical process requiring changes in the way herders manage their animals and land; it is an eminently political process involving changes in the distribution and control of vital productive resources.

In the past, development planners subscribed to a normative theory of commercial change, a theory which ignored dimensions of the commercialization process which did not conform to project targets and which attributed lack of change to pastoral conservatism rather than to self interest. Livestock development planners can no longer afford the luxury of this illusion. Commercialization is not a uniformly attractive process, either for displaced pastoralists or for administrators who must find ways of accommodating and employing displaced people. But however unacceptable the process may be, we can profit from the simple fact that it occurs in predicatable ways and results in foreseeable consequences. If we can achieve some understanding of the principles involved, then it may become feasible to marginally speed up, redirect, or selectively suppress aspects of a process of change that already exists but would otherwise be unplanned and unassisted. Only in this way can planners hope to exercise some beneficial effect on a spontaneous process that they did not initiate and cannot fully control.

NOTES

[1]This analysis is concerned with Africa as a whole and does not attempt to distinguish between Africa north and south of the Sahara. There is insufficient case material at the present time to undertake a regional breakdown of African commercial production systems, and it is unlikely that a simple North Africa/ Subsahara dichotomy would yield results of much interest. The Congo River Basin, not the Sahara, has probably been the greatest barrier to the movement of livestock and pastoral people in Africa, and while pastoral economies and societies in North Africa are certainly different from those in East, Central, Southern and Sahelian Africa, it is doubtful whether these differences are significantly greater than those which obtain between any other two subregions of the continent.

[2]Evidence of commercialization comes from the following sources in the Arctic [Ingold 1974, 1976; Paine 1972; Dolan 1983], from Southwest Asia [Cole

1975; Bates and Conant 1981; Bates 1980], and within
Africa from the Sahel [Delgado 1979; Bourgeot 1981;
Sutter 1982], the Horn [Swift 1976, 1979; Aronson 1980;
Box 1982], East [Hedlund 1971; Meadows and White 1981;
Quam 1978], Southern [Doran et al. 1979; Cruz de
Carvalho 1971; Behnke 1983] and North Africa [Behnke
1980].

[3]"In Northern Kenya Borana pastoralists support the concept of grazing blocks to the extent (and
only to this extent) that they provide a device for
keeping Somali pastoralists out of certain areas, and
the Somali pastoralists dislike the proposals for the
same reason." [Sandford 1983, p. 93]

REFERENCES

Allan, William. The African Husbandman. Westport,
    Connecticut: Greenwood Press, 1965.
Anderson, Terry L. and P.J. Hill. "From Free Grass to
    Fences: Transforming the Commons of the American
    West." Managing the Commons, eds., Garret Hardin
    and John Baden. San Francisco: W.H. Freeman and
    Company, 1977.
Ariza-Nino, Edgar, and Kenneth Shapiro. "Cattle as
    Capital, Consumables and Cash: Modeling Age-of-
    Sale Decision in African Pastoral Production."
    1984, in this volume.
Aronson, Dan. "Kinsmen and Comrades: Towards a Class
    Analysis of the Somali Pastoral Sector." Nomadic
    Peoples 7(1980):14-23.
Asad, Talal. "Equality in Nomadic Social Systems?
    Notes towards the Dissolution of an Anthropologi-
    cal Category." Pastoral Production and Society.
    Cambridge: Cambridge University Press, 1979.
Barth, Fredrik. Nomads of South Persia. London: Allen
    and Unwin, 1964.
Bates, Daniel. "Differential Access to Pasture in a
    Nomadic society: The Yoruk of Southwestern
    Turkey." Perspectives in Nomadism, eds., W. Irons
    and N. Dyson-Hudson. Leiden: Brill, 1972.
    _____ . Nomads and Farmers: A Study of the Yoruk of
    Southeastern Turkey. Ann Arbor, Michigan:
    University of Michigan Press, 1973.
    _____ . "Yoruk Settlement in Southeast Turkey."
    When Nomads Settle. ed., P.C. Salzman. Brooklyn,
    New York: Praeger, 1980.
Bates, Daniel and Susan Lees. "The Role of Exchange in
    Pastoral Specialization." American Anthropologist
    79(1977):824-841.
Bates, Daniel and Francis Paine Conant. "Livestock and
    Livelihood: A Handbook for the 1980s." The
    Future of Pastoral Peoples. eds. J. Galaty et al.

Ottawa: International Development Research Centre, 1981.

Baxter, P.T.W. "Some Consequences of Sedentarization for Social Relationships." Pastoralism in Tropical Africa, ed., T. Monod. London: OUP, 1975.

Behnke, Roy. "Production Rationales: The Commercialization of Subsistence Pastoralism." Nomadic Peoples, October 1983.

_____. The Herders of Cyrenaica. Urbana: University of Illinois Press, 1980.

_____. Unpublished field notes on Eastern Libya 1978-79, n.d.

Bekure, Solomon and N. Dyson-Hudson. The Operation and Viability of the Second Livestock Development Project. Gaborone, Botswana: Government Printer, 1982.

Bourgeot, A. "Nomadic Pastoral Society and the Market." Journal of Asian and African Studies 16 (1981): 116-127.

Box, Thadis. "Nomadism and Land Use in Somalia." Economic Development and Cultural Change 19 (1971): 222-228.

Burnham, Philip. "Spatial Mobility and Political Centralization in Pastoral Societies." Pastoral Production and Society. Cambridge: Cambridge University Press, 1979.

CARL BRO International A/S. Livestock Management and Production in Botswana. Gaborone, Botswana: Ministry of Agriculture, 1982.

Cole, Donald. "The Enmeshment of Nomads in Sa'udi Arabian Society: The Case of Al Murrah." The Desert and the Sown, ed., Cynthia Nelson. Berkeley: Institute of International Studies, 1973.

Crotty, R. Cattle, Economics and Development. Farnham Royal, Slough, England: Commonwealth Agricultural Bureaux, 1980.

Cruz de Carvalho, Eduardo. "Traditional and Modern Patterns of Cattle Raising in South-Western Angola: A Critical Evaluation of Change from Pastoralism to Ranching." Journal of Developing Areas 8(1974):199-226.

Dahl, Gudrun and Anders Hjort. Having Herds: Pastoral Growth and Household Economy. Stockholm: Department of Social Anthropology, 1976.

Delgado, C. Livestock and Grain Production in Upper Volta. Ann Arbor, Michigan: Center for Research on Economic Development, 1979.

Dobie, J. Frank. The Longhorns. Austin, Texas: University of Texas Press, 1980 (first published in 1941).

Dolan, Carrie. "Reindeer Roundup on Frozen Tundra Is No Sleigh Ride." Wall Street Journal, February 25, 1983 pp.1.

Doornbos, Martin and Michael Lofchie. "Ranching and Scheming: A Case Study of the Ankole Ranching Scheme." The State of the Nations: Constraints on Development in Independent Africa, ed., Michael Lofchie. Berkeley: University of California Press, 1971.

Doughty, Charles M. Travels in Arabia Deserta. New York: Dover Publications, 1979 (first published in 1888).

Dupire, M. Peuls Nomades: Etudes Descriptive Des Wodaabe Du Sahel Nigerien. Paris: Institut d'Ethnologie, 1962.

Dyson-Hudson, Neville. "The Human Factor." Pastoral Development Projects. Addis Ababa: International Livestock Centre for Africa, 1980.

Galaty, John. "The Maasai Group-Ranch: Politics and Development in an African Pastoral Society." When Nomads Settle, ed., P.C. Salzman. Brooklyn, New York: Praeger, 1980.

Galaty, John, D. Aronson, P.C. Salzman and A. Chouinard, eds. The Future of Pastoral Peoples. Ottawa: International Development Research Centre, 1981.

Gluckman, Max. Politics, Law and Ritual in Tribal Society. London: Basil Blackwell, 1971.

Government of Botswana. National Policy on Tribal Grazing Land. Gaborone, Botswana: Government Printer, 1975.

_____. Livestock Development Project Progress Report 1980/1981. Gaborone, Botswana: Government Printer, 1981.

Gudeman, Stephen. The Demise of a Rural Economy. London: Routledge and Kegan Paul, 1978.

Halderman, J.M. "Analysis of Continued Nomadism on the Kaputiei Maasai Group Ranches: Social and Ecological Factors." University of Nairobi, Institute for Development Studies Discussion Paper 152, 1972.

Hardin, Garrett. "The Tragedy of the Commons." Science 162 (1968): 1234-1248.

Hedlund, Hans G.B. "The Impact of Group Ranches on a Pastoral Society." University of Nairobi, Institute for Development Studies Staff Paper No. 100, 1971.

Hertel, Thomas. The System of Mafisa and the Highly Dependent Agricultural Sector. Gaborone, Botswana: Rural Sociology Unit Report Series No. 11, Ministry of Agriculture, 1977.

Hitchcock, Robert. "Tradition, Social Justice and Land Reform in Central Botswana." Journal of African Law 24 (1980): 1-34.

Horowitz, Michael. The Sociology of Pastoralism and African Livestock Projects. Washington: Bureau for Program and Policy Coordination, U.S.A.I.D., 1979.

Ingold, Tim. Hunters, Pastoralists and Ranchers. Cambridge: Cambridge University Press, 1980.
_____. The Skolt Lapps Today. Cambridge: Cambridge University Press, 1976.
_____. "On Reindeer and Men." Man 9 (1974): 523-538.

Irons, Williams. "Political Stratification Among Pastoral Nomads." Pastoral Production and Society. Cambridge: Cambridge University Press, 1979.
_____. "Variation in Economic Organization: A Comparison of the Pastoral Yomut and the Basseri." Perspectives on Nomadism, eds., Williams Irons and Neville Dyson-Hudson. Leiden: Brill, 1972.

Jarvis, Lovell S. "Cattle as Capital Goods and Ranchers as Portfolio Managers: An Application to the Argentine Cattle Sector." Journal of Political Economy 82 (1974): 489-520.

Khazanov, A.M. "The Size of Herds among Pastoral Nomads." Nomadic Peoples 7 (1980): 8-13.

Lewis, I.M. A Pastoral Democracy. London: OUP, 1961.

Meadows, S. and J. White. "Structure of Herd and Determinants of Offtake Rates in Kajiado District in Kenya." ODI Pastoral Network Paper, 7d, 1979.

Osgood, Ernest. The Day of the Cattleman. Chicago: University of Chicago Press, 1970 (first published in 1929).

Oxby, Clare. Group Ranches in Africa. Rome: F.A.O. [W/P3098], 1981.

Parson, Jack. "Cattle, Class and the State in Rural Botswana." Journal of Southern African Studies 7 (1981): 236-255.

Peters, Emrys. "Aspects of the Family among Bedouin of Cyrenaica." Comparative Family Structures , ed., M.F. Nimkoff. Boston: Houghton Mifflin, 1965.

Pim, A.W. Financial and Economic Position of the Bechuanaland Protectorate. London: His Majesty's Stationery Office, 1933.

Quam, Michael. "Cattle Marketing and Pastoral Conservatism: Karamoja District, Uganda, 1948-1970." African Studies Review 21 (1978): 49-71.

Reutlinger, S. "Short-Run Beef Supply Response." American Journal of Agricultural Economics 48 (1966): 909-919.

Rigby, Peter. "Olpul and Entoroj: The Economy of Sharing among the Pastoral Baraguyu of Tanzania."

      Pastoral Production and Society. Cambridge:
Cambridge University Press, 1979.

Riviere, Peter. The Forgotten Frontier: Ranchers of
North Brazil. New York: Holt, Rinehart and Win-
ston, 1972.

Ryan, G.C. Report to the Bechuanaland Protectorate
Government on the Livestock Industry. Gaborone:
National Archives, 1958.

Sahlins, Marshall. Stone Age Economics. New York:
Aldine, 1972.

Salzman, P.C. "Inequality and Oppression in Nomadic
Society." Pastoral Production and Society.
Cambridge: Cambridge University Press, 1979.

Schneider, Harold. Livestock and Equality in East
Africa. Bloomington: Indiana University Press,
1979.

Solway, Jacqueline S. People, Cattle and Drought in
the Western Kweneng District. Gaborone: Rural
Sociology Report Series No. 16, Ministry of
Agriculture, 1980.

Stenning, Derrick. Savannah Nomads. London: OUP,
1959.

Strickon, Arnold. "The Euro-American Ranching Com-
plex." Man, Culture and Animals, eds., Anthony
Leeds and Andrew P. Vayda. Wasington, D.C.:
American Association for the Advancement of
Science. 1965.

Sutter, John. "Commercial Strategies, Drought, and
Monetary Pressure: Wo'daa'be Nomads of Tanout
Arrondissement, Niger." Nomadic Peoples 11 (1982):
26-60.

Swidler, W.W. "Some Demographic Factors Regulating the
Formation of Flocks and Camps among the Brahui of
Baluchistan." Perspectives on Nomadism, eds., W.
Irons and N. Dyson-Hudson, Leiden: Brill, 1972

Swift, Jeremy. "The Development of Livestock Trading
in a Nomad Pastoral Economy: The Somali Case."
Pastoral Production and Society. Cambridge:
Cambridge University Press. 1979.

      . "Pastoral Development in Somalia: Herding
Cooperatives as a Strategy against Desertification
and Famine." Desertification: Environmental
Degradation In and Around Arid Lands, ed., Michael
H. Glantz. Boulder Colorado: Westview Press,
1977.

Webb, W.P. The Great Plains. Waltham, Mass.: Blais-
dell Publishing Co., 1931.

18
Veterinary Anthropology: Interdisciplinary Methods in Pastoral Systems Research

Albert E. Sollod, Katherine Wolfgang and James A. Knight

In development programs, veterinary medicine has often been a prima donna. Diagnostic laboratories are operated with little integration with non veterinary activities and without measurable impact on pastoral production. Field activities such as the rinderpest vaccination campaigns are often carried out in sweeping military fashion which pastoralists barely comprehend, and governmental extension systems require facilities and logistical support which are uneconomic to use and maintain. Because it is out of context, research undertaken in laboratories on samples delivered from the field usually gives scant, and often misleading, information on the actual economic and social impact of diseases from the pastoralists' point of view.

Veterinary activities need not be divorced from the realities of pastoralism. The problem is not a lack of relevance for as Aronson has pointed out, "[veterinary interventions] are the most direct way to the herders' hearts" [1982, p. 17], but is the result of disciplinary and institutional limitations which define veterinary medicine in developed countries. The way out of this dilemma is for veterinary research and application to move out into the field where pastoralists live and to transcend disciplinary boundaries, to undertake a holistic systems approach. "Veterinary anthropology" has been used in this context to define an interdisciplinary domain of study:

> Veterinary anthropology uses the basic repertoire of anthropology's research skills and techniques, which include: 1. the observation of human behavior and its interpretation in the context of a constantly evolving understanding of cultural, social, economic and historical matrices, 2. interview, 3. participation in daily activities, and 4. some manipulation of situations. All activities are directed toward a progressively more precise understanding of the range of systematic behavior. [Sollod and Knight 1982]

Veterinary anthropology is a form of systems research
which is aimed at the integration of component findings
from different analytical frameworks such as ecology,
biology and sociology. It is also concerned with
policies and existing political structures which control these components. In this paper, we draw on field
work at the interface of two disciplines, anthropology
and epidemiology, to confront issues of herd health and
pastoral productivity. In veterinary anthropology,
findings are analyzed in a socioeconomic context that
is relevant to the herders' needs; thus, veterinary
medicine is used to further an anthropological approach
to development in which the primary responsibility for
sustained improvement in animal health belongs to the
pastoralists themselves [Aronson 1981; Goldschmidt
1981a]. Gaining an understanding of the pastoralists'
own knowledge of animal health problems and their solutions is seen as a key to introducing gradual change.
In pastoral societies, where herds are economically and
culturally focal, anthropological methods can be used
to gather a wealth of traditional veterinary knowledge.
Epidemiological methods are also applicable since there
are herds of animals which can be studied as units.
Here, the patterns in which diseases present themselves
are keys to reliable diagnosis.

One region which contains the necessary elements,
herders and their herds, for the application of veterinary anthropology, is the West African Sahel and northern Sudan, which sweep across Subsaharan Africa for
4,000 km between the Atlantic Ocean and Chad. The
region is pastoral in the north where rainfall is less
than 350-400 mm, and agropastoral in the south where
rainfall may be up to 900 mm. Extensive husbandry is
practiced with herds which are often mixed - cattle,
camels, goats, sheep and donkeys - but it may be
dominated by a single species in some production systems. These systems have evolved over centuries and
have adapted to the exigent ecologies of the region,
but some have been increasingly modified by economic
and political events of the past century.

This chapter presents descriptions of veterinary
anthropological research from two areas in the region:
a pastoral area of central Niger which has been the
site of the USAID-financed Niger Range and Livestock
Project, and an agropastoral area of south-central
Upper Volta in the White Volta basin. Our efforts were
directed at identifying feasible animal health
interventions for future development activities. While
particular findings may be site-specific, the overall
tenor and methodologies of the research are applicable
to other pastoral communities.

CENTRAL NIGER

The study was undertaken within an area of 80,000 square km of arid rangeland in the departments of Tahoua, Maradi and Agadez. The northern sector receives 150 mm of rainfall annually while 400 mm falls in the south. Although there are large areas of open grassland, both hilly and level, the zone is not a uniform grassland ecosystem; a significant portion is occupied by scattered brush and open forest, tree-lined rivers, and low drainage areas and ponds.

Very extensive mixed species herding is practiced under purely pastoral conditions in most of the region, but along the southern fringes there is millet farming and agropastoral herding. Twareg pastoralists, who usually raise small stock and camels, are the most numerous; they tend to inhabit lowlands or the margins of seasonal water courses which have ample browse. Wodaabe are the second most numerous group but represent only about 15 precent of the total population; they are zebu cattle herders but often keep sheep and goats as well. Most Wodaabe utilize a swath of land which cuts across the zone and has the highest density of trees which are more or less evenly dispersed. Other pastoral groups which use the land include the agropastoral Fulbe from the south who transhume into the zone during the rainy season, and several groups of Arabs which live in the north.

Field surveys were carried out as collaborations between veterinarian and anthropologists, in which the epidemiological studies were woven into ongoing anthropological research. The objectives of the study were to: - define the diseases which occurred in the herds,
- determine the incidence, prevalence, seasonality, and geographic distribution of the diseases,
- establish the economic importance of the diseases both as production constraints and as they were perceived by the pastoralists themselves,
- determine and test practical means of combatting the diseases and,
- develop strategies for animal health service.

In order to determine the nature of the diseases in the herds, the following study methods were used:

1. An all-terrain vehicle made it possible to systematically survey most of the the project zone. Interviews with herders were carried out to obtain detailed accounts of current or past herd health problems. The length of the discussion ranged from several minutes at a well site to several days of a camp.

2. Examinations were made of herds on a group basis in order to observe their general condition and nutritional status, their responsiveness to their surroundings, their carriage and mobility, and signs of disease. Observations were made on variations within and between herds.
3. Individual sick animals were removed from the herd and examined clinically.
4. Sick sheep and goats were purchased from herders and autopsied in the field.
5. Continuous detailed assessment was made of natural and man-made components of the environment.
6. Trials were carried out with vaccines and medications which could be incorporated into an overall herd health program. Special efforts were made to determine the herders' perceptions about the medications and their benefits.
7. A cornerstone of the field work was the keeping of a detailed record of the observations and interviews which were made each day. The day's events were analyzed each evening by a veterinarian and an anthropologist and were recorded for subsequent cross-sectional analysis.

The studies were carried out in two phases: a dry season phase was undertaken in April and May and a wet season phase was undertaken between late July and early September, 1981 [Sollod 1981]. In the dry season phase herds were studied under the stresses of inadequate feed and water, whereas in the wet season phase, herds were observed under different environmental circumstances, when their geographic distribution was altered by migration and their breeding capacity was enhanced by improved nutritional intake. Many observations were made on the attitude of herders towards states of health and disease in their herds. At times, trial sales of drugs and vaccines were carried out, and herders were asked to treat their own animals, often on a herd basis. It was considered that the importance which a herder attaches to herd health would influence his willingness to expend money, labor and time to resolve health-related problems. Since good rapport between pastoralists and government veterinary personnel was considered essential for the efficient provision of herd health services, observations were also made on the ability of government veterinary personnel to interact with herders and to allow the latter to participate in the treatment of their own animals.

Table 18.1. Herd health problems in central Niger

	Degree of Importance		
Species	Greater	Lesser	Uncertain
Cattle	Anaplasmosis Blackleg and malignant edema Brucellosis Foot-and-Mouth disease Vitamin A deficiency	Babesiosis Streptothriocosis	Anthrax Septicemic pasteurellosis
Sheep	Caseous lymphadenitis Enterotoxemia Nodular worm disease Stress pneumonias Protein-caloric inanition Sheeppox	Contagious ecthyma Pododermatitis	Anthrax Psoroptic mange
Goats	Enterotoxemia Nodular worm disease Peste des petits ruminants	Contagious ecthyma Goatpox Labial papillomata Stress pneumonias	
Camels	Salmonellosis		Sarcoptic mange Mandibular abscesses
Donkeys		Strangles	

Source: Authors.

Herd Health

During the period in which the study was undertaken, at least 26 endemic diseases were observed in
various classes of livestock (Table 18.1). The diagnoses were made by examining herds and individual animals and by obtaining thorough accounts from herders of
the patterns of disease in their herds. No diagnosis
was made in the few cases where single animals were ill
and there was no relevant history. The case for using
this methodology, without laboratory support, has been
argued by Schwabe et al. [1977]. Under pastoral conditions, where herders are acutely aware of health
related problems, both the quantity and quality of
information can be superior to that obtained from
laboratory studies [Maliki 1981; Sidi 1982].

The importance of each disease was estimated
according to three criteria. First, the pastoralists'
own perceptions were taken into account. For example,
blackleg is considered very important by Wodaabe and it
occurs in sporadic fashion in one- to two-year-old

calves with an annual incidence of 2-5 percent in this age group. Although there are no losses in milk production or of market age or breeding animals, we deferred to the herders' own judgment. Second, we considered the herders' subsistence imperatives. Diseases which cause late abortion or neonatal death, such as brucellosis or vitamin A deficiency, result in lost milk production for the entire season. Therefore, these diseases were judged important, even though the value of the calf at birth is miminal. Finally, diseases which affect marketable animals and cash income, such as enterotoxemia and caseous lymphadenitis, were considered important. We are convinced that all three criteria must be used in evaluating the importance of disease in pastoral systems.

It has been written frequently that, under pastoral conditions, veterinary interventions have led to increases in herd size beyond the capacity of the ecosystem to support [e.g., Ferguson 1976; Goldschmidt 1981b]. In fact, there is little evidence to support the validity of this notion for the West African Sahel where rinderpest vaccination has been the only animal health intervention to be widely applied. Statistics from various sources [Franke and Chasin 1980] show no effect on Niger's cattle population from the JP-15 vaccination campaign which was implemented in the 1960s. At the present time, in the region in which our study was undertaken, there is a balance between animals and vegetative biomass except during consecutive drought years [Greenwood and de Leeuw 1983].

If veterinary interventions do affect herd size, a corollary must also be true: that the state of animal health plays a determining role in the size of a herd. In fact, there is little precise information on any factors which may affect herd size in pastoral systems [Horowitz 1981]. In the present study, we did attempt to determine whether herd size and health status were related but could find no evidence for this. There was no noticeable difference in the health status of large herds when compared to small ones. However, the productivity of the herds was adversely affected by disease, and endemic disease may increase the number of animals which a herder requires to maintain economic security, thereby forcing up animal numbers.

Finally, we determined that the annual calf mortality in most cattle herds was only about 5 percent and did not exceed 25 percent, even in herds that were kept in disadvantageous habitats, such as at saline wells. This is consistent with the findings of Maliki [1981] in the same region where, over a six year period, Wodaabe calf mortality averaged 10.8 percent. These values are far lower than the average of 30-40 percent which is often given for the Sahel, and although mortality may be higher during sporadic droughts, optimistic expectations for declines in mortality as a result of veterinary or nutritional interventions may not be achievable in normal years.

Production Systems Factors

It is possible to associate some diseases with specific production systems, that is, Twareg or Wodaabe, and to recognize the underlying management practices which lead to the expression of disease. For example, stress-related pneumonias and protein-caloric inanition in sheep are severe problems in Wodaabe herds but not in herds belonging to Twareg. Twareg herders control the breeding season of their sheep to produce a single lamb crop each year. After the first breeding season is over, the penile sheath of the breeding rams is ligated and connected with a cord to the base of the scrotum to cause penile deviation and prevent intromission. This practice is not unique to Twareg production systems and is common to many other small stock pastoral groups. Its advantage is that lambing is prevented at the end of the dry season when nutrition is poor so that ewes are spared the additional stress of pregnancy and lactation.

The Wodaabe consider themselves to be cattle herders, and although many sheep are raised, the amount of labor expended on sheep management is low and, with few exceptions, the rams are not ligated for breeding control. However, rams may be sold from the herd after the first breeding season, but whether this is economic depends on an individual herder's access to markets and the seasonal fluctuations in market prices. The market prices are influenced by the Id festivities, at which time there is a strong demand for mutton. If the festivities come after the breeding season, as they will in the next few years, the Wodaabe will be favorably placed to control sheep breeding. However, after that, they will again be in a position to suffer losses for lack of a marketing mechanism which accommodates their production system.

In contrast, there are some diseases which are prevalent in Twareg sheep herds and uncommon in the Wodaabe production system; these include nodular worm disease and caseous lymphadenitis. These chronic and debilitating diseases reach a high prevalence rate when the environment becomes contaminated over a long period of time and, consequently, affect herds of low mobility which exploit the same habitat for most of the year. The diseases affect Twareg sheep because the Twareg reside with their herds in relatively fixed locations, while the Wodaabe move with their herds throughout the year.

These examples illustrate the importance of exogenous determinants of disease, that is, factors which are external to the etiologic agents and the hosts that determine which herds will be affected. Exogenous determinants include herd management and environmental factors which are characteristic of specific production systems, and their identification under pastoral conditions has obvious implications in the development of any herd health program. It makes possible the use of nonmedical approaches to animal health which include marketing and management interventions, and allows the use of a simplified package of veterinary commodities for each production system.

Animal Health Delivery

The provision of animal health services to pastoral communities is extraordinarily difficult to achieve. High herd mobility and demanding management practices preclude extensive governmental intervention. However, productivity could be improved if more responsibility for veterinary care were given to the pastoralists themselves. A herder-based extension network could be supplied from private or governmental sources and herders, during very short training sessions, could be taught to use verterinary commodities. During our field trials in which herders treated their own animals, it became obvious that they would have to be taught about the specificity of medications and vaccines and how to store and administer them and time their administration. However, the pastoralists' comprehensive knowledge of animal diseases facilitates their dialogue with competent trainers.

The idea of herders as veterinary auxiliaries is not new. It has been attempted in Ethiopia [Sandford 1981] and also, to some extent, within the zone in Niger where our study was carried out. These experiences indicate that six principles should be observed to establish a sustainable extension system:

1. The commodities package must be restricted to a few items which are easy to store and use.
2. Each intervention should bring about an economic increase in production or a reduction in risk.
3. Herd problems rather than individual animal care should be stressed.
4. Herders should pay full price for all commodities.
5. Auxiliaries should be volunteers rather than paid governmental agents.
6. Auxiliaries should fit into an organizational structure such as herders associations, so demand can be effectively transferred from the herders to the auxiliaries and then to the suppliers.

Even if these guidelines are followed, the system may break down for reasons which affect supplies outside the pastoral community; however, in the long run such efforts to increase pastoralists' participation in veterinary care should be rewarding.

THE WHITE VOLTA BASIN

This study was undertaken in and around a 2000 square km area of Sudan tallgrass savannah west of the Nouhao tributary of the White Volta River. The area, which is open to common grazing, receives about 800 mm of rainfall annually and is lightly infested with riverine tsetse flies. It is part of a region where blackflies had been eradicated for onchocerciasis control, and the area is undergoing resettlement, largely unorganized, by both crop farmers and herders. Most of the herders, in fact all within the study sample, are agropastoral Fulbe who herd zebu cattle. The broad goal of the study was to determine ways to improve the articulation between the Fulbe herders and the governmental veterinary service and, in support of this, there were three operational objectives:

- to determine and analyze Fulbe concepts of veterinary medicine,
- to examine the particular role of women in disease diagnosis and treatment, and
- to assess the responsiveness of the governmental veterinary service to the needs of the pastoralists.

We have already stated a concern for the pastoralists' own knowledge, and in this case the specific concern is the use of that knowledge to create a more effective dialogue between pastoralists and veterinary agents, and to identify veterinary interventions which support the pastoralists' own goals. The field work was performed between September, 1982, and January, 1983, that is, from the late rainy through the mid dry season. Open-ended interviews were conducted with Fulbe herders, with women from herding families, and with trained veterinary agents, both private and governmental. Also, scheduled interviews were administered to a random sample of 38 herders from the approximately 250 who live in the area. The survey instrument consisted of 37 major questions, with clarifying subordinate questions, which dealt with modern and traditional treatment and vaccination practices, the herders' perceptions of the governmental veterinary services, production and mortality losses from disease, the relationships between disease and transhumance, the importance of ectoparasites including ticks and tsetse files, pregnancy detection, and trypanotolerance. In

Table 18.2 Some characteristics of wilsere

	The Most Serious Disease	Caused Death This Year	Caused Abortion This Year	Problem During Trans- humance	Water Is Contami- nated	Pasture Is Contami- nated	Associa- ted with Tsetse Flies
Percent of herders responding positively	36	34	24	71	50	21	71

keeping with the local Fulbe production system, most of the questions referred to cattle. The last source of information which contributed to the analysis was from direct observation of Fulbe husbandry practices and the field work of a veterinary agent. A number of interesting and pertinent concepts developed from this study, and a few illustrative examples underscore the usefulness of veterinary anthropology in research and analysis for development.

Fulbe Conceptual Framework

A central concern of cattle health for Fulbe of the White Volta is the concept of wilsere. Wilsere is a deviation from normal health which occurs most fre- quently as the beginning of the rainy season, in April or May - this period is always difficult in single- rainy-season, pastoral ecosystems - and during the cold season, in December and January. It is caused by infectious agents, often described as minute worms, which are associated with river where cattle are watered. Occasionally pastures are said to be contami- nated. Many herders mention flies as disease vectors although water is usually thought to be the primary reservoir. Some herders thought that the fly bites resulted in transmission, but others thought that the disease agents were carried on the flies' bodies or excreted into the water; nevertheless, only 30 percent of respondents said that they avoided rivers infested by dangerous flies. Fulbe herders believe that ante- lopes and elephants can also act as reservoirs of wilsere. There are other names which describe wilsere: ladde nyau and ferlo nyau. Ladde and ferlo both mean "the bush", and nyau is translated as "disease"; thus, wilsere is thought of by Fulbe as a bush disease. Other characteristics of wilsere appear in Table 18.2

Wilsere always causes systemic illness; localized diseases such as streptothricosis are excluded. This bush disease has endless symptomatic forms; as one knowledgable herder explained the multifaceted char- acter of wilsere: "We all sit here, some black and some white, but we are all human. That is how wilsere is." A wasting disease is described as the most common form. Cattle become dirty, and the hair coat is rough and matted. The animals may be covered with flies and stand with lowered heads. The manure is dry and hard, causing painful defecation. Appetite is poor and the cattle slowly become thin and dehydrated. The disease lingers for up to several months, until the animals die. Other syndromes may affect the gastrointestinal or respiratory tracts, or may cause reproductive problems and abortion. Each syndrome is given a Fulfulde name which designates it as a specific form of wilsere.

Older animals are more resistant to wilsere unless they are weak or otherwise ill. Fulbe also recognize that herds can become habituated to local infection, and that cattle which come from a distance will have a higher mortality rate. As one herder explained, "If you see an area with very fresh grass from the first rains, you know it is infected, but you also rejoice because your cattle can eat and be content. If the herd is accustomed to the disease, the animals will grow fat; if not some may die before they get used to it." The herders must face the choice between finding green pasturage which is in, or near, wet infected areas or not getting forage for their cattle.

What can be concluded from these descriptions? Wilsere is a concept which embraces all fatal systemic, infectious diseases which are endemic, those which are autochthonous and occur every year. In contrast, localized infections and the epidemic diseases are recognized as specific entities. For example, there are Fulfulde names for pulmonary tuberculosis (douyrou), streptothricosis (gugna), rinderpest (carra), blackleg (baleeyel), anthrax (pidoli), pleuropneumonia (bola) and others. These epidemic diseases are thought to be blown in by the wind or carried by herds passing through, but there is a clear understanding that some are soil-borne, and that after they are brought into the region, the soil may become the main reservoir and source for indigenous herds.

Women's Role

This is a most difficult area to explore, although as Horowitz points out:

Participation of women [in livestock sector projects] has been nonexistent. Whereas the role of women in agriculture has begun to be explored, the economic role of women in pastoral societies has been simply ignored. . . . This is most unfortunate, because the position and the status of women in pastoral societies are threatened by the very objective of many of the productivity-oriented interventions: converting the economy from one that emphasizes dairy production and the feeding of the herding population directly on the produce of the herd to one that emphasizes meat production and the feeding of urban populations. [1981, p. 85]

The Fulbe women's role in maintaining herd health is unquestionably important. Women have a large stake in keeping their milk cows healthy. They provide essential protein for their families, and the sale of milk is usually their primary sources of income. But obtaining extensive information on women, either directly or from their husbands, is difficult. Women have less contact with people from outside their local communities, and our use of a male interpreter was inhibiting; nevertheless, certain questions on the scheduled interviews and several open-ended interviews were devoted to understanding the women's role. Questions on the survey instrument which directly asked herders about their wives' involvement in herd health were the following: a series of questions on milking cows which coughed or had tuberculosis, questions which asked the wives' ability to diagnose certain diseases, and a question on tick control.

Through interviews with women, informal questioning and direct observation of milking, the women's role was seen as integral to the herd's well-being. Removal of ticks was mentioned by the herders as a woman's job. That particular activity, shared by the children, is essential to avoid streptothricosis and blowfly attack of the udder. Tick removal may also prevent other tick-borne diseases such as babesiosis, theileriosis and anaplasmosis.

General attention to an animal's daily appearance and disposition is probably the most important way to avoid or to detect the early signs of health problems. During milking, the wives and older girls had close contact with the cows twice daily. Wives that were interviewed said they took notice of many aspects of the cow's general appearance and the milk production level. If there were any changes, they would be reported to the herders.

Only a few herders responded that their wives decided themselves when not to milk a sick cow, and virtually all answered that their wives could not recognize any disease symptoms. The herders' responses reflected how they, as men, wanted to present themselves to the interviewer, but may not have clearly reflected their wives' actual role. One woman stated that frequently the herder will point out a cow that has not been grazing well and will ask her to tell him if milk production has decreased. Obviously, the

herders realized that this is important information.
Interviews with the women also indicated that they did
know how to diagnose many diseases. When symptoms of
important diseases were described, they almost always
knew the name of the diseases. One woman said she
often noticed the first signs of neonatal enteric
infection (duce) when the calf could not suckle and his
mouth smelled rotten.

Women also play a role in maintaining the general
health of young calves. The day a calf is born, a woman
boils some water and feeds it to the calf with a small
gourd. The Fulbe say this is essential to avoid
dangerous constipation in young calves. Wives also
mentioned that if a calf of one of their milk cows
appeared weak, they would shelter it in a hut or a
courtyard for the day, while the cow was out grazing.
The women must also decide how much milk the calves
will get and how much the family gets. After a cow
gives birth, the woman waits a certain amount of time
before she begins to take some of the milk for family
consumption. Usually the woman waits longer before
beginning to milk in the dry season when there is less
forage than in the lusher wet season. Although average
waiting periods were reported - 60 percent waited one
to two weeks, others up to two months - the length
depended on the cow's milk production and the strength
of the calf. Initially, the woman takes only a small
amount of milk and increases the quantity as the calf
grows and can forage a little. When milking was
observed, it was clear that the woman adjusted how much
milk she took to the condition of the calf. With a
very young calf the wife might milk one quarter, and
with an older calf she might milk three quarters. If a
wife is a poor judge, the calf may grow weak and get
sick; therefore, the woman's expertise is important to
the calves' well being.

Finally, women play a role in traditional herbal
treatments of animals. They are responsible for the
preparation of medications from the leaves, bark, stem
and roots of plants which are gathered by the herders.

Responding to Pastoralists' Needs

In making suggestions for development strategies
we underscore the need to understand the pastoralists'
technical rationale inherent in their herding practices, but at the same time, we are constantly searching for means by which governmental and private sector
personnel can assist the pastoralists to obtain their
own development objectives in an equitable fashion.
All too often projects align themselves with one or the
other constituent almost exclusively, and there may be
little regard to one of the principal contributions

which a project can make: to assist all of the
indigenous actors in constructing a dialogue to define
the meaning of development. Development must come from
within; it cannot be imposed by technical assistance
projects, and there must be a convergence of ideas
within the developing country or region.

To determine whether the Fulbe were interested in
having assistance within this framework, the scheduled
interview asked a number of questions relating to
"traditional" and "modern" concepts and practices.
When asked about traditional treatments, many herders
responded that they had in the past used Fulbe treatments, but that they did not work well any more. Some
said that they did not know the traditional treatments,
and many older herders said that their children did not
want to learn about old medicines. Eventually, most
herders would say that the veterinarian's medicine is
more effective and much easier to use, as traditional
medicines involve the extensive collection of materials
and tedious preparation. Each animal must be dosed
individually and the whole process is very difficult
and time consuming. When the veterinarian comes, he
can treat many herds in a single morning. Thus, there
is an underlying desire by pastoralists to incorporate
new technology into their production systems as long as
this offers substantial improvements over traditional
methods.

The disease syndrome wilsere presents a particularly good example of difficulties that can arise in
attempting a constructive dialogue. When working with
a veterinary agent, Fulbe herders usually speak Moore,
rather than Fulfulde, as most of the agents in the
region speak Moore as a first language. Wilsere is
translated into Moore as ouaogo which, because of the
agent's formal technical training,is taken to mean
trypanosomiasis. Thus, a demand is created for a
trypanocidal drug, and one is administered to large
numbers of animals, regardless of the etiological
nature of the incipient threat. This situation is
inefficient, as the drug is highly specific in its
activity against trypanosomes, and excessive and
ill-timed usage is a waste of money and may lead to the
development of drug resistant strains.

Veterinary agents often describe this situation as
proof that Fulbe are poor diagnosticians and do not
understand cattle diseases, but in reality, the situation arises from lack of common ground for communication. To correct this, it is necessary to understand
the backgrounds of both the herders and the veterinary
agents. The herders' taxonomy of disease rests on
epidemiological and pathological grounds. On the other
hand, the veterinary agents have had formal training in
medicine which emphasizes specific etiologies. The

answer would seem to come from educating both parties
in what may be viewed as their complementary deficiencies. Verterinary agents should receive more training
in the epidemiology and pathology of animal diseases,
beginning with a strong orientation in ethnoscience in
which concepts such as wilsere are explained. In fact,
ways should be found in which herders could participate
in the instruction, perhaps through prearranged visits
to the field.

At the same time herders could be taught about
specific agents of disease, and could be assisted in
sorting out the causes of wilsere. Trypanosomiasis is
certainly a smokescreen which obscures the diagnosis of
many diseases [Sollod et al. 1982] and herders could
use help in unweaving this tangled web. A laboratory-
supported epidemiological study could shed some light
on the situation, but the information cannot be rele-
gated solely to reports and the formal classroom; it
must be passed on to the herders in its entirety so
they may be able to make more specific diagnoses based
on geographic, temporal and herd management determinants. Who should teach the herders? The veterinary
agents themselves should be responsible. Very short,
but recurrent, teaching sessions could be held in the
field with the herders. Veterinary agents, in order to
be effective teachers, would have to demonstrate a
comprehensive understanding of wilsere.

This strategy would entail a change in the responsibilities of both herders and veterinary agents.
Herders could be given more responsibility for the
treatment of their herds, with the veterinary agents
more assured that an adequate job would be done, and
veterinary agents would be required to take on more
responsibility as teachers. If the veterinary agents
were taught to give outstanding performances as teachers, they would feel less threatened in relinquishing
their perceived prerogative to treat animals with
modern medications and may be persuaded to allow
herders more freedom to administer vaccine and medica-
tion to their own animals. If this strategy is followed, the complementary deficiencies of herders and
veterinary agents would diminish and full use would be
made of their complementary strengths.

The role of women in caring for cattle herds may
also be strengthened by this approach to development.
The study suggests several specific activities which
deserve attention. First, tick control is the domain
of women. Any scheme which involves the systematic
dipping of animals, as attempted with Fulbe herds in
northern Ivory Coast, would automatically pre-empt this
function. Under pastoral conditions, removal by hand
may be the most economic practice, but if a program of
acaricide application is attempted, either by hand

spraying or dressing, it should be done with women's participation and with adequate training to ensure proper usage and to avoid the contamination of milk.

Secondly, women are relied upon in the detection of neonatal calf enteritis. Since management practices allow the calves adequate consumption of colostrum, it should be possible to save most cases if oral hydration therapy were practiced. Women could easily be taught to prepare a solution of salt, sugar and bicarbonate to administer to the calf as soon as its suckling ability is affected, before the onset of diarrhea and dehydration. Thirdly, women are responsible for the preparation of traditional medications, and the introduction of modern drugs displaces this activity if they are administered by veterinary agents or herders. Perhaps this trend could be reversed if medications were sold to herders in bulk powder form which the women could dissolve or suspend in boiled water at an appropriate dosage for the age and size of the animal. Finally, women may one day be responsible for small diagnostic laboratories for local herds. A microscope, if it proved valuable for the diagnosis of trypanosomiasis under pastoral conditions, would undoubtedly be maintained by women in a much better state of repair than those found at most official veterinary posts. We have seen pastoral women who manage to maintain a sewing machine in working order under the most difficult and isolated conditions.

CONCLUSION

This paper shows how veterinary medicine can play a greater role in pastoral systems studies by interfacing its research with social science. Veterinary anthropology, which uses field methodology from epidemiology and anthropology, was applied in two areas of West Africa to demonstrate the importance of endemic diseases in pastoral systems. There was a systems specificity to herd health problems and also the conceptual framework in which diseases were classified and analyzed by pastoralists; thus, veterinary anthropology could be a useful tool in theoretical studies on comparative ethnography or cultural ecology. However, the main concern was to identify diseases and to suggest practical ways to improve animal health. By tapping the common knowledge of herders, it was possible to identify the most important problems and to show how the herders themselves could assume the greatest responsibility for improving animal health.

For development agencies and planners, veterinary anthropology has several policy implications. First, veterinary medical field studies should be supported more fully if they are integrated into problem-oriented

systems research. There should be a better balance
between field work and more costly laboratory activities. Second, pastoralists should always participate
in animal health extension systems. This may require
innovative education of both herders and governmental
veterinary agents in order to develop a dialogue
between the two groups. Finally, classical, top-down
animal health schemes should be implemented only after
careful consideration of their possible effects on the
role of women in veterinary care.

REFERENCES

Aronson, Dan R. "Development of Nomadic Pastoralists:
    Who Benefits?" The Future of Pastoral Peoples, ed.
    J.G. Galaty, D. Aronson, P.C. Salzman and A.
    Chouinard, pp. 42-51. Ottawa: International
    Development Research Centre, 1981.
_____. Toward Development for Pastoralists in
    Central Niger. Niamey: United States Agency for
    International Development, Feb. 1982.
Ferguson, D.S. A Conceptual Framework for the Evalua-
    tion of Livestock Production Development Projects
    and Programs in Sub-Saharan West Africa. Ann
    Arbor: Center for Research on Economic Develop-
    ment, 1976.
Franke, Richard W. and Barbara H. Chasin. Seeds of
    Famine. Montclair: Allanheld, Osmun & Co.
    Publishers, Inc., 1980.
Greenwood, Gregory and Peter de Leeuw. "Annex 10.
    Natural Resources Management". Integrated Live-
    stock Production Project Paper. Boston: Tufts
    University, 1983.
Goldschmidt, Walter. "An Antropological Approach to
    Economic Development". The Future of Pastoral
    Peoples, ed. J.G. Galaty, D. Aronson, P.C. Salzman
    and A. Chouinard, pp. 52-60. Ottawa: Inter-
    national Development Research Centre, 1981a.
_____. "The Failure of Pastoral Economic Develop-
    ment Programs in Africa" The Future of Pastoral
    Peoples, ed. J.G. Galaty, D. Aronson, P.C. Salz-
    man and A. Chouinard, pp. 101-118. Ottawa:
    International Development Research Centre, 1981b.
Horowitz, Michael M. "Research Priorities in Pastoral
    Studies: An Agenda for the 1980's" The Future of
    Pastoral Peoples, J.G. Galaty, D. Aronson, P.C.
    Salzman and A. Chouinard, pp. 61-88. Ottawa:
    International Development Research Centre, 1981.
Maliki, Angelo B., Ngaynaaka: Herding According to the
    Wodaabe. Tahoua: United States Agency for Inter-
    national Development, Niger Range and Livestock
    Project Discussion Paper No. 2, Nov. 1981.

302

Sandford, Dick. Pastoralists as Animal Health Workers:
The Range Development Project in Ethiopia. London:
Overseas Development Institute Pastoral Network
Paper 12c, July 1981.
Schwabe, C.B., H.P. Riemann, and C.E. Franti.
Epidemiology in Veterinary Practice. Philadelphia:
Lee and Febiger, 1977.
Sidi, Abou. L'Art Veterinaire des Pasteurs Saheliens.
Dakar: ENDA, 1982.
Sollod, Albert E. Patterns of Disease in Sylvopastoral
Herds of Central Niger. Niamey: United States
Agency for International Development, Niger Range
and Livestock Production Discussion Paper No. 4,
Nov. 1981.
Sollod, Albert E. and James A. Knight, "Veterinary
Anthropology: A Herd Health Study in Central
Niger". Proceedings of the Third International
Symposium on Veterinary Epidemiology and Eco-
nomics, forthcoming.
Sollod, Albert E., Roger Poulin, Elon Gilbert, David
Gow, Warren Leatham, James Ogborn, Fred Sowers,
Roger Steinkamp and Thomas Stickley. Agricultural
Sector Assistance Strategy for Upper Volta. Wash-
ington, DC: Development Alternatives, Inc., 1982.

19
Observations on Animal Health, Especially on Approaches to Identify and Overcome Constraints in the Subhumid Zone of West Africa

T.W. Schillhorn van Veen

INTRODUCTION

Although the importance of animal health as a constraint in the development of African livestock is well recognized, the issue has only obtained limited attention in livestock or agricultural development projects during the last decade. In most cases the animal health component of a project was left to the local authorities and was rarely considered a variable or researchable subject.

Also until recently, disease control measures were only aimed at the major, and often fatal, diseases in Africa, e.g. rinderpest, anthrax, blackleg, contagious bovine pleuropneumonia (CBPP), East Coast Fever, peste des petits ruminants (PPR), and trypanosomiasis (Table 19.1). Limited attention has been given to the day-to-day problems of livestock owners. At present, still little is known about these problems and, unfortunately, there are many in and outside Africa who believe that the African livestock industry will thrive as soon as rinderpest, CBPP, trypanosomiasis and tickborne diseases are brought under control, forgetting that even in the western world disease is still a major impediment in the livestock industry. Table 19.2, depicting some major reasons for slaughter in a rural slaughterslab in the Nigerian Guinea zone, demonstrates that the unthrifty cow is as much a problem in Africa as it is in Europe or the U.S. Unthriftiness is associated with various diseases including internal parasites and trypanosomiasis; still the problems encountered in this village differ considerably from those listed in Table 19.1.

CONSTRAINTS IN LIVESTOCK PRODUCTION

The major problems of the livestock industry in the savanna, e.g. the 3-D group: Drought, Disease and

Table 19.1. Major animal health problems in the humid tropics of Africa

Cattle	Sheep and Goats	Pigs	Poultry
Rinderpest	Helminthiasis	Pneumonia	Newcastle disease
Trypanosomiasis	PPR	Helminthiasis	Coccidiosis
Tickborne disease	Pneumonia	Mange	Gumboro disease
Dermatophilosis	Mange	Trypanosomiasis	Leucosis

Source: Unpublished research.

Table 19.2. Prices and health status of cattle slaughtered at a rural slaughter-slab in northern Nigeria during 1974 and 1975

Condition	Number of Animals	Mean Price[a]
	—head—	—Naira—
Healthy	539	112.74
Unthrifty[b]	127	48.63[c]
"Daji"[d]	49	55.28[c]
"Hanta"[e]	100	93.90
Broken leg	6	63.33
FMD[f]	22	42.73[c]
"Kirci"[g]	33	61.51[c]
"Diarrhea"	5	56.00
Old age	3	60.00
Reproductive problems	3	88.66
Others	6	56.00

Source: Schillhorn van Veen and Buntjer [1983].

[a]Exchange rate 1.00 Naira = U.S. $1.50.

[b]Includes trypanosomiasis.

[c]Statistically different from price of healthy animals (p<0.05).

[d]Condition with nervous signs, mainly due to heartwater disease.

[e]Condition with signs of depression and liver changes often associated with liverfluke disease.

[f]Condition with secondary lesions associated with foot and mouth disease (mainly young animals).

[g]Condition with skin lesions, mainly due to Dermatophilus infections.

Disorganization are, except for drought, also found in the humid zone. Although it is intended in this presentation to concentrate on health, some remarks have to be made about the last "D", disorganization. This issue is important as disease control becomes more and more dependent on an organizational infrastructure. The disorganization does not so much relate to the situation at the farm level, but to the "outside world". Livestock production is a high risk business in the tropics where markets are highly volatile and where basic supplies may not always be available. In the humid tropics where farmers are more and more dependent on the availability of supplies such as commercial feed, the survival of the industry depends to a great extent on the organization of supply and livestock product marketing. Other support services, including veterinary care are often too unreliable to ensure the producer the security he needs to make long-term production plans. On the other hand, most livestock owners produce more than one end product, e.g. milk, eggs, meat, manure, skins etc., which tends to ease some of their risks.

Although in general the animal health constraints are known in Africa, there are considerable local differences in prevalence and epidemiology of various diseases. A complete understanding of the role of disease in a given area may require efforts beyond routine investigation. Various methods have been developed to provide an overview of the major disease problems. Slaughterhouse surveys, especially for the smaller slaughterslabs, provide a general picture of the health status of slaughter animals. As many of the animals slaughtered in smaller slaughterslabs or at home are considered emergency-slaughter cases it may be useful to determine the reason for slaughter by interview and clinical diagnosis. An example of the changes in status of slaughter animals in a small village in northern Nigeria is given in Figure 19.1. These data, collected during 1972-1973, were obtained by interviewing the butcher and owner; further study of 51 cases revealed that the suspected cause of disease was correct in 35 cases. More sophisticated diagnostic systems, including mobile clinics and diagnostic laboratories are very useful but are rarely available on a regular basis in rural areas.

Without proper diagnostic facilities it is difficult to determine the economic importance of disease. In some cases the economic impact of disease is briefly discussed [Akerejola, Schillhorn van Veen, and Njoku 1979; Ilemobade and Balogun 1981], but in general the importance is rarely determined in economic terms. This is probably related to the fact that until recently there was no doubt about the economic impact of

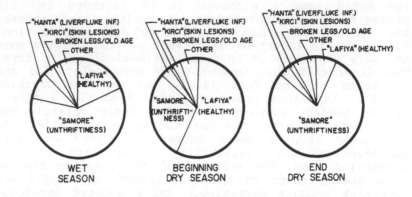

Figure 19.1. Graphic representation of the seasonal changes in
reasons for slaughter of 449 cattle slaughtered in
a rural village in northern Nigeria, 1972-1973

fatal epidemics, and recent cost-benefit analyses of
some of the control efforts prove this point (Table
19.3). Very little is known about the economic impact
of chronic diseases such as internal parasitism, mange,
mastitis and foot rot. Models to study the economic
importance are mainly based on European and American
impact studies and are rarely relevant in the tropics
where diseased animals are often salvaged and consumed,
where inputs and initial investments are limited, and
where animals are rarely kept for short-term economic
reasons only. Other studies look at macroeconomic
data, e.g. export of live animals or skins, or slaughter data from large abattoirs, without sound data on
the indigenous losses.

Intensification of the livestock industry changes
the pattern of disease problems, but unfortunately does
not decrease the prevalence. Diseases like brucellosis, tuberculosis and pneumonia generally increase in
settled livestock systems and appear to be emerging in
many parts of humid Africa. Keeping animals in fenced
pasture may prevent losses to predators or certain
plant poisonings but increases the risk of parasitic
diseases and nutritional imbalances. In general, settled livestock systems require an increased knowledge
and responsibility with respect to management and
health.

Management is often recognized as a crucial factor
in livestock production but is rarely defined. Management interventions such as providing colostrum to new-

Table 19.3. Losses and benefit cost ratio of treatment or prevention of some selected animal health problems in Africa

Diseases	Losses	B/C Ratio	Reference
	—Percent—		
Cattle			
Rinderpest	10-50	1.77[a]	Felton and Ellis [1978]
Trypanosomiasis	?	1.14[b]	Habtemariam et al. [1983]
Small Ruminants			
Internal Parasitism	11	2.4[c]	Akerejola et al. [1979]

[a] Vaccination.

[b] Insecticide application (based on model).

[c] Anthelmintic treatment.

born animals, providing mineral supplementation,
rotational grazing or controlled breeding are highly
beneficial at relatively low cost and risk. However,
these interventions require a sound understanding of
the livestock system, including some knowledge about
interventions which have been tried earlier, whether
successful or not.

THE HISTORY OF VETERINARY SERVICES

Devastating outbreaks of rinderpest and subsequently of CBPP and trypanosomiasis around the beginning
of the century facilitated not only the introduction of
colonial rule but also the introduction of western
veterinary care. The initial colonial veterinary
service concentrated on the control of major epidemics
and, in retrospect, was fairly successful in containing
rinderpest, CBPP and trypanosomiasis. On the other
hand, there were insufficient time and staff available
to pay much attention to the day-to-day problems of the
livestock owner. In other words, the unthrifty cow was
treated for trypanosomiasis and no further efforts in
diagnosis and treatment were made to salvage the animal
beyond that point.

After independence, the veterinary services continued their effort to control major epidemics, and
this effort was enhanced by the large-scale regional
tsetse eradication or vaccination campaigns such as JP
15 for rinderpest [Plowright 1982], and JP 28 for CBPP
[David-West 1980]. These projects, supported by international agencies, virtually eradicated rinderpest in
the early seventies.

However, change in political priorities and inept
vision of the veterinary authorities have allowed a
decline of the vaccination effort; at present rinderpest and some other major diseases have reached epidemic proportions again, killing thousands of animals.
This debacle, as well as the devastating outbreaks of
tickborne diseases in Zimbabwe during the war [Law-
rence, Foggin, and Norval 1981], raise the question
whether the delivery systems of animal health in Africa
are sufficient under the prevailing circumstances.

Moreover, these debacles occur at a time in which
Africa has more educated veterinarians than ever
before. It is not surprising that the animal health
sector is criticized for its questionable response to
the present problems, and the issue rises whether the
current, educational as well as organizational, system
is indeed suitable of handling Subsaharan Africa's
animal health problems [Halpin 1981]. Our education of
African graduates in MSc or PhD programs in animal
science or health is also to be blamed. The graduate
programs are rarely based on the need of the individual

or his country, and if so they are not always appreciated by the student or his peers. There are exceptions however. ILCA provides practical training for
animal science and veterinary personnel; Michigan State
University, among others, is experimenting with continuing education programs for overseas veterinarians,
which in one case resulted in a Masters degree in
continuing education and communication, a field which
may not at all appeal to a US graduate but is very
useful in rural Africa.

DEVELOPMENTS IN ANIMAL HEALTH DELIVERY

It has to be questioned whether free veterinary
services or veterinary services paid for by a livestock
head tax are optimally serving the livestock industry.
Veterinary services are costly, but it appears from
personal experience that the majority of livestock
owners would rather pay for good services than receive
mediocre services for free. However, there are only
very limited data on delivery systems other than the
traditional government service. Limited field services
are provided by some veterinary schools and institutions for a nominal fee and appear fairly well accepted. Unfortunately most farmers are only exposed to
enforced veterinary services organized without their
consultation or approval.

Although dipping is probably a fairly effective
method to increase survival and production of livestock
in East Africa (the benefits in West Africa are considerably lower), dips are not widely used as farmers
are not in favor of paying for unreliable and often
poor services of the dipping agency. The Veterinary
Cleansing Act in Kenya, for instance, enforces dipping
in certain zones but does not guarantee the quality of
the dipping services. Consequently, the farmers avoid
the government dip and try to solve their problem by
hand picking of ticks or by the purchase and use of
hand sprayers.

Still, a good dipping service could not only be of
benefit to farmers as such but could also be a focus of
other extension services in the area. The most extensive effort was probably made by the Extension Research
Liaison Service and the Faculty of Veterinary Medicine
of the Ahamadu Bello University in Zaria (Nigeria)
which provided for a contact person among the livestock
owners who would coordinate the field trips of veterinary personnel. Moreover, radio and television programs were prepared and broadcasted in various local
languages to instruct and warn livestock owners about
ailments, nutrition and health care of their animals.
These programs are especially relevant if local (traditional) information about disease and husbandry are

incorporated. In West Africa such information is indeed available [Stenning 1959; de St. Croix 1972; Okaiyeto 1980; Maaliki 1981].

In the humid zone the livestock industry developed in a rather different manner. Stimulated by the veterinary departments (which at that time included all livestock services), the local crop farmers with limited experience in raising livestock started to increase their production of chickens and other small animals. Soon their demand for nutritional advice, vaccines and other veterinary services increased beyond the capacity of the government; increasingly these services were provided by private veterinarians and livestock consultants. In many West African countries and in Kenya, this development has been facilitated by the relative ease by which animal feed and animal health products could be imported and obtained. Animal health services and products supplied by government agencies are rarely available in sufficient quality and quantity.

Major changes can be expected in the livestock industry in the humid and subhumid zones of Africa. These changes will be accompanied by new disease problems which require timely interventions provided by government and private sector.

ANIMAL HEALTH PACKAGES

Another major constraint in animal health extension is the lack of sound animal health packages. Recent experiences in the Small Ruminant Collaborative Research Projects demonstrated that increased production of small ruminatns will rarely be achieved without providing a minimal survival package of anthelmintic treatment and, in some areas, vaccination for PPR. Nevertheless many projects are initiated without taking into account that the mortality rate of small ruminants may reach 30-50 percent when not protected against these diseases. Similar remarks can be made of cattle with respect to rinderpest and of poultry and Newcastle disease. All these conditions are relatively easy to control, but require some infrastructure and personnel.

It may be useful then to develop a minimal package which at least ensures survival. Table 19.4 depicts some of these minimal efforts. It is difficult to provide an all-African package and local adjustments have to be made. Tick control for instance is more important in East Africa, where East Coast Fever is a major contraint, than in West Africa; it is also more important for grade cattle than for native livestock. Other diseases also may have a regional difference in distribution or importance. PPR appears to be more common in West Africa; liverfluke disease is more

common in wet areas, especially in the tropical highlands. It is difficult to go into much detail concerning treatment. The treatment schedules for trypanosomiasis and helminthiasis may vary from one region to
another, and are sometimes set by government regulations aimed to prevent the deveopment of drug
resistance. Treatment for East Coast Fever is still
experimental but may become available in the future. A
health package for production is demonstrated in Table
19.5. This package should of course include some
minimal nutritional standards as the interrelation
between disease and nutrition is well known.

There have been instances of introduction of such
packages to livestock owners in the tropics. In the
early fifties the veterinary authorities on the Jose
plateau (Nigeria) provided a package for calves which
included an anthelmintic, a small bag of groundnut cake
and some general advice on calf raising and nutrition.
The demand soon reached levels beyond government's
ability and the program was for some time continued
with local enterprise supplying some of the components.

Beyond these general packages, special recommendations have to be given for exotic animals, young animals, dairy cattle, draft animals, and animals used for

Table 19.4. Minimum veterinary care recommended to ensure survival of livestock

Species	Animal should be vaccinated for	Treatment should be available for	Management practice should include
Cattle	Rinderpest	East Coast Fever	Tick control
	CBPP	Trypanosomiasis	Provision of salt and minerals
	Black leg		
	Anthrax		
Sheep		Haemonchosis	Worm control
		Pneumonia	Provision of salt and minerals
Goats	PPR	Haemonchosis	Worm control
		Coccidiosis	Provision of salt and minerals
		Pneumonia	
Pigs		Pneumonia	Prevention of cysticercosis
Poultry	Newcastle disease	Coccidiosis	Coccidiosis prevention

Source: Unpublished research.

Table 19.5. Veterinary care recommended for ensuring minimum production levels

Species	Animal should be vaccinated for	Treatment should be available for	Management practice should include
Cattle	Heartwater disease	Streptothricosis	Worm and fly control
		Blood parasites	
Sheep	Enterotoxemia	Pneumonia	Tick control
		Mange	Controlled breeding
		Other internal parasites and liver flukes	Shelter
Goats	Enterotoxemia	Mange	Controlled breeding
		Pneumonia	Ectoparasite control
		Other internal parasites	
Pigs	Erysipelas	Babesiosis	Worm control
		Trypanosomiasis	
		Mange	
Poultry	Gumboro disease	Spirochaetosis	Ectoparasite control
	Fowl pox	Parasites	

Source: Unpublished research.

beef production. Draft animals may have to be vaccinated for Foot and Mouth Disease (FMD), a nonfatal
disease of limited importance to the majority of livestock owners. It does cause lameness in adult animals
and an outbreak during the plowing season may be devastating to a mixed farmer without the luxury of replacement animals. It is questionable whether FMD
vaccination is profitable for other cattle [Raikes
1981].

INTEGRATED DISEASE CONTROL

So far the animal health issue has been addressed
as a technical problem to be solved by vaccination and
drug treatment. The original African view of disease
however was more complicated as it involved dominance
of bad over good spirits. Although animal scientists,
including veterinarians, are not used to handling
spiritual methods, it may be useful to discuss briefly
a more holistic approach to disease control.

Disease is not a single problem, but rather pre-
cipitated by other problems. Stress, including social
stress, may have a considerable influence on the function of the immune system making the animal more
susceptible to infection. This issue is especially
important in African livestock which tolerate many
diseases by carrying a low level infection sufficient
to maintain immunity. Any extra challenge or any
stress-induced effect on the immune system may cause a
flare-up of the infection which may lead to clinical
disease. Examples of such occurrences are outbreaks of
babesiosis, anaplasmosis, coccidiosis and to some
extent East Coast Fever and trypanosomiasis in African
cattle after being exposed to some stress situation
such as gathering in vaccination camps.

On the other hand, control methods do also affect
the animal and its environment. This issue is among
others discussed by Ormerod [1979] who questioned the
ecological impact of trypanosomiasis control by tsetse
eradication. Ormerod's discussion more or less coincided with the realization that tsetse eradication in
the humid tropics is technically very difficult if not
impossible. It also coincided with (and enhanced) the
increased interest in the use of trypanotolerant livestock. The use of trypanotolerant livestock has been
discussed [ILCA 1979; Murray, Morrison, and Whitelaw
1982] mainly focusing on the use of the trypanotolerant
N'dama. In recent years, the interest has extended
into genetically determined disease-resistance in
general, including resistance to other diseases and
parasites such as tick infestation, fly worry and
helminthiasis [Utech and Wharton 1982; Preston and
Allonby 1972]. In some cases this disease resistance

can be achieved, but probably at the cost of losing some other genetic trait [Schillhorn van Veen and Folaramni 1978]. In the meantime research is being done to identify these traits by using genetic markers such as blood types, lymphocyte antigen types and other methods. The use of genetic markers to determine disease resistance as well as production potential, combined with modern breeding techniques including multiple ovulation and embryo transfer, could considerably accelerate the relatively slow development of breeds adapted to specific tropical environments.

The most feasible approach to disease control is by adopting good management [Schillhorn van Veen 1982]. Husbandry methods are closely related to the occurrence of disease. Diseases such as tuberculosis, brucellosis, coccidiosis are more common in settled than in migratory livestock. On the other hand, isolation of livestock may prevent spread of epidemic disease. In the highlands of northern Cameroon, for instance, cattle are fattened by keeping them in small huts in the mountains and are hand-fed with grain, grass and farm surplus [Holtzman 1982]. This isolated husbandry system prevents contact with other animals and as such reduces the risk of epizootic diseases and parasite infections, although it may require some extra attention in the area of nutrition.

Other approaches in the reduction of disease challenge in the tropics include the use of tick repellent pasture [Thompson, Roa, and Romero 1978], strategic antiparasitic treatments [Tacher 1982] and the use of slow release devices for continuous drug delivery. Unfortuantely, these devices such as fly and tick repellent eartags, intraruminal bolusses for slow release of drugs, or long-acting drugs are often priced beyond the financial ability of African farmers.

EPIDEMIOLOGICAL INFORMATION

All the above methods will only be successful when sufficient epidemiological knowledge is available. It is unfortunate at present that a trend is developing in the area of veterinary epidemiology emphasizing impact studies using computerized technology without first obtaining sound technical data on the transmission, life cycle and diagnosis in the field. Too few African epidemiologists appreciate the need for sound field information, and would rather punch unreliable data into a computer which in itself is a waste of time and funds despite the fact that occasionally "statistically significant" results may be obtained. This is especially relevant today as many workers, for instance agricultural economists without experience with live-

stock, are increasingly using poorly collected data to
support project analysis and planning.

REFERENCES

Akerejola, A.A., T.W. Schillhorn van Veen, and C.O.
Njoku. "Ovine and Caprine Diseases in Nigeria, a
Review of Economic Losses." Bull. Anim. Hlth
Prod. Afr. 27 (1969):65-70.

David-West, K.B. "Planning and Implementation of
Contagious Bovine Pleuropneumonia Control Program
in Nigeria." Veterinary Epidemiology and Eco-
nomics, Proc 2nd Intern. Symp. ed. W.A. Geering,
R.T. Toe and L.A. Chapman. Canberra, Australia:
Govt. Publ. Service, 1980.

de St. Croix, F.W. The Fulani of Northern Nigeria.
Westmead, England: Gregg International Publ.,
1972.

Felton, M.R. and P.R. Ellis. Studies on the Control of
Rinderpest in Nigeria. Reading: University of
Reading Study No. 23, 1978.

Habtemariam, T.R., E. Howitt, R. Ruppaner and H.P.
Riemann. "The Benefit-Cost Analysis of Alternative Strategies for the Control of Bovine Trypanosomiasis in Ethiopia." Prev Vet Med 1(1983):
157-168.

Halpin, B. "Vets-Barefoot and Otherwise." London:
Overseas Development Institute, Pastoral Network
Paper 11C, 1981.

Holtzman, J.S. "A Socioeconomic Analysis of Stall-Fed
Cattle Production and Marketing in the Mandara
Mountain Region of N. Cameroon." PhD Thesis,
Michigan State University, 1982.

ILCA - International Livestock Centre for Africa.
Trypanotolerant Livestock in West and Central
Africa. Vol. 2. Country Studies. Addis Ababa,
1979.

Ilemobade, A.A. and T.F. Balogun. "Pig Trypanosomia-
sis: Effects of Infection on Feed Intake, Live
Weight Gain and Carcass Traits." Trop Anim Hlth
Prod 13(1981):128-136.

Lawrence, J.A., C.M. Foggin and R.A.I. Norval. "The
Effects of War on the Control of Disease of Livestock in Rhodesia (Zimbabwe)." Vet Rec 107
(1980) 82-85.

Maaliki, A.B. Ngaynaaka: l'Elevage selon les Woodabe
de Niger. Tahoua, Niger: Discussion Paper. Min.
Rural Development, 1981.

Murray, M., W.I. Morrison and D.D. Whitelaw. "Host
Susceptibility to African Trypanosomiasis:
Trypanotolerance." Adv Parasit 21(1982):1-68.

Okaiyeto, P.O. "A Descriptive Study of the Major
Economic Activities of the Settled Pastoral Fulani

in three Zaria Villages of Kaduna State." M.Sc. Thesis, Ahmadu Bello University, 1980.

Ormerod, W.E. "Environmental and Social Consequences of Tsetse Eradication." London: Overseas Development Institute, Pastoral Network Paper 8b, 1979.

Plowright, W. "The Effects of Rinderpest and Rinderpest Control on Wildlife in Africa." Symp Zool Soc (London) 50 (1982):1-28.

Preston, J.M. and E.W. Allonby. "The Influence of Breed on the Susceptibility of Sheep to Haemonchus contortus infection in Kenya." Res Vet Sci 26((1979):134-139.

Raikes, P.L. Livestock Development and Policy in East Africa. Uppsala: Scand Inst Afric Studies, 1981.

Schillhorn van Veen, T.W. "Role of Parasitism in Goat Management." Proc 3rd Internat Conf Goat Prod Disease, pp. 85-89. Scottsdale, Arizona: The Dairy Goat Journal, 1982.

Schillhorn van Veen, T.W. and D.O.B. Folaramni. "The Haemoglobin Types of Northern Nigerian Sheep." Res Vet Sci 25(1978): 397-398.

Schillhorn van Veen, T.W. and B.J. Buntjer. "Some Marketing Aspects of Slaughter Animals in Rural Slaughterslabs in Kaduna State of Nigeria." Rev Elev Med Vet Pays Trop 36(1983): in press.

Stenning, D.J. Savannah Nomads. A Study of the Wodaabe Pastoral Fulani of Western Bornu Province, Northern Region, Nigeria. London: Oxford University Press, 1959.

Tacher, G. "The Use of Drugs in the Development of Livestock Production in Tsetse Infected Areas." World Anim Rev 44(1982):30-35.

Thompson, K.C., E.J. Roa and N.T. Romero. "Antitick Grasses as the Basis for Developing Practical Tropical Tick Control Packages." Trop An Hlth Prod 10(1978):182-197.

Utech, K.B.W. and R.H. Wharton. Breeding for resistance to Boophilus microplus in Australian Illawarra Shorthorn and Brahman and Australian Illawarra Shorthorn cattle. Austr Vet J 58(1982): 41-49.

20
Cattle as Capital, Consumables and Cash: Modelling Age-of-Sale Decisions in African Pastoral Production

Edgar Ariza-Nino and Kenneth H. Shapiro

This chapter presents a capital theory model of herders' decisions concerning the age at which they sell animals. While the model is general, the focus here is primarily on cattle raised under extensive conditions as, for example, in the drier parts of the Sahel. The underlying reason for presenting such a model is to provide a common ground for comparing the great variety of hypotheses, theories, and conclusions offered by the various social and biological sciences involved in livestock development in Africa.

The model is described in the first section and is then applied to a pure beef operation in the following section. In the third section the model is applied to a more realistic beef, milk, and calf operation, and Sahelian conditions are simulated. The results of the model and simulations are then summarized.

A MODEL CONCERNING AGE OF SALE

This section presents a model that derives (or predicts) the age of sale that would be chosen by a pastoralist who wishes to maximize the benefits from his or her herd. Age of sale is a key variable for determining herd size, age/sex composition, animal productivity, and offtake. Value is taken as including cash receipts, the implicit value of animal products consumed by the pastoral family, the value of progeny, liquidity, security, prestige, power, and so forth. The model shows the <u>direction</u> of changes (in age of sale) that are likely to result from interventions that affect prices of meat[1] and milk, costs, mortality, fertility, rate of gain, and the interest rate. A

This chapter is drawn from a larger report entitled "Market Forces and Livestock Development in Africa," 1983, prepared by the authors for the World Bank's Agriculture and Rural Development Department.

simulation of the model with average parameter values
of the Sahel gives estimates of the likely magnitudes
of these changes in age of sale.

The early applications of such models were to wine
storage and tree growing. The optimal "age" of sale is
shown to occur when the rate of increase in the value
of the wine or trees falls to just equal the rate of
interest. At earlier ages the commodity gains in value
faster than money would in a bank or alternate investment. Hence it pays to let the wine age or the tree
grow. At later ages, due to biological factors, the
commodities have a slower rate of gain in value and
eventually it pays to have sold them and have invested
the proceeds elsewhere at the going rate of interest.
Jarvis [1974] applied this model to cattle in Argentina
and more recently [1980] indicated directions for its
applications in Africa. The discussion below follows
this lead.

The Decision Rule

In each period t a pastoralist has the option of
selling any animal or keeping it until the next period
t+1.[2] If the animal is kept, then in the next period
the same decision is faced regarding sale in t+1 or
t+2, and so on. Thus, the long-run decision of when to
sell an animal may be analyzed piecewise as the decision of whether or not to keep an animal one more
period.

The decision to keep or sell an animal may be
characterized as depending on a comparison of the gains
from keeping the animal one more period versus the
costs of doing so. By keeping the animal one more
period, a pastoralist may benefit[3] from (a) the
increase of the animal's sale value due to its
increased weight, and (b) the additional value flowing
from the animal as a living resource. Such flow values
may include milk, calves, security, power, liquidity,
prestige, aesthetic pleasure, and so forth. The cost
of keeping an animal one more period includes (a) the
cost of herding, feeding, watering, maintaining good
health, and the risk of mortality, and (b) the one-
period gains foregone by not selling the animal and
investing the proceeds in another (presumably younger)
animal or some other asset or a bank.

The animal will be sold if the cost of keeping it
one more period outweighs the benefits of keeping it
that additional period. This decision rule may be
written as follows:

Do not sell if
Expected Gains > Expected Costs

More fully, the pastoralist may be considered to make the following comparison:

$$p \frac{\partial w}{\partial \theta} + f \gtrless rpw + c$$

where
- p = the price of meat or the price per kilo of liveweight;
- $\partial w/\partial \theta$ = weight gained by the animal in one period;
- θ = age of sale;
- f = flow benefits derived from a live animal;
- r = interest rate or, more generally, the rate of return available from an alternative investment;
- w = weight of the animal;
- c = direct costs of keeping the animal one more period.

Thus, the two terms on the left hand side (LHS) represent the two types of benefits from keeping an animal one more year, while the terms on the right hand side (RHS) represent the cost of doing so. The first term on the LHS indicates that the gain in sales value is the weight gain, $(\partial w/\partial \theta)$, times the price (p) with price and weight defined in comparable terms.[4] The term f is a summation (and simplification) of the several flow benefits from a live animal. The first term on the RHS indicates that one cost of not selling is the return a pastoralist could have realized from investment of the sale revenue (pw). The term c summarizes the various costs of keeping an animal one more period.[5]

The Variables

Each of these terms bears closer examination. Price is assumed to be determined exogenously in that no single pastoralist can influence prices. The rate of weight gain is a function of the animal's age, and is high at early ages, but slows down when the animal is older. (A roughly sigmoid curve is assumed for weight over time.) This abstracts from dry season weight loss and subsequent compensatory gain in the next rainy season.

The flow of benefits from a live animal (f) is a function, in part, of price and age. To the extent that the animal is valued as currency, or for insurance, or even for prestige and power, these values are enhanced if the potential sales value of the animal is increased by a price increase or by its greater size (a function of age). Also, the value of progeny, part of a female's flow of benefits, is increased if the progeny's potential sales value rises because of a price increase.

On the RHS, pw represents revenue received from a sale. Weight, w, is a function of age of slaughter (θ). The term rpw indicates the possible earnings, for one period, if the sales revenues were invested in another asset. The cost term, c, summarizes a variety of types of cost. Some are opportunity costs of using land and labor for the animal in question rather than for another animal or for completely different purposes. Other costs are cash expenditures for health care, taxes, and maybe grazing or watering fees. A third cost derives from the possibility that the animal may die and provide zero (or greatly diminished) sales revenues and flow benefits the next period.[6]

The pastoralist decision rule elaborated here can be related to a wide range of models and assumptions that typically focus on just one or two of the components discussed above. Placing these different viewpoints in a common context may help draw together some of the diffuse literature on African livestock development.

BEEF PRODUCTION

A simplified form of Jarvis' [1974] model can be used to bring out important features of a system in which production is primarily for beef. In essence, this means that the flow benefits, f, in our model are very low or zero. Thus, the situation describes decisions concerning steers more than cows.[7]

Effect of Low Flow Benefits

First, the implication of a low or zero f should be noted. The pastoralist keeps an animal so long as the benefits to be drived by doing so (the LHS of the equation) exceed the costs of keeping it (the RHS). Initially, due to high weight gains in early ages, the LHS is quite high, but gradually declines to equal the RHS, at which point the animal is sold. If f is low or zero, the LHS (or the increase in benefits in one more period) starts out at a lower level and hence falls to equal the RHS (costs) sooner. Thus, the model predicts that steers, with a low f, will be marketed younger than cows with a high f. And this, of course, is observed in Africa.

Effect of a Pricing Change

Next, the impact of an increase in beef prices can be examined. Intuitively, one may consider a steer as a beef producing machine. If the price of beef goes up, it pays to produce more beef. The only way to do this is to keep the animal longer and allow it to gain more weight as it matures. Thus, an increase in price will lead to an older age of slaughter under conditions posited in this model.

Another way to conceptualize the impact of a price change is by focusing on the relation between costs of maintenance, feed, and so forth compared to price. The decision rule with f = 0 can be written as follows:

$$\text{or} \quad \begin{aligned} p\, \partial w/\partial\Theta &= rpw + c \\ \partial w/\partial\Theta &= rw + \frac{c}{p} \end{aligned}$$

If price rises, c/p will decline and then the RHS will be smaller than the LHS or, in other words, benefits of keeping an animal one more period will be larger than the cost of doing so. Pastoralists will therefore delay sales. The age of sale will increase because, in this formulation, the cost of maintenance, feed, etc., relative to the price of beef has fallen.

Effect of Changes in Costs and Interest Rate

This leads naturally to a consideration of how variations in cost, c, affect the age of slaughter. Obviously, an increase of c increases the RHS and means that the gain from keeping the animal one more year will have fallen to reach the costs of doing so at an earlier age. Thus, an increase of costs tends to lower the sales ages. Similarly, an increase of the interest rate, r, (or the potential rate of return from alternate investments) will increase the RHS and lower the sales age.

Effect of a Change in Production Conditions

An improvement of production conditions that uniformly increases the rate of weight gain at all ages will tend to increase the age of sale since animals will have higher rates of gain at all ages including at older ages. However, an improvement that allows animals to reach their mature weight at younger ages means that the rate of gain slows down sooner. This will lead to a younger age of sale. (This is equivalent to a leftward shift of the growth curve.) A change of conditions that decreases mortality would, in effect, lower the cost of holding an animal one more period. The risk component of c would decline, which would lead to an older age of sale. A change of conditions that increased fertility would mean an increase of flow benefits, f, for females and hence an increase in their age of sale.

BEEF, MILK, AND CALF PRODUCTION

A simulation model is used to calculate the expected present value for cattle in the Sahel using realistic assumptions about weight gain, milk yield, fertility, mortality, costs, prices, and interest rates. The present value calculations, along with the assumption of an optimizing pastoralist, yield predicted age of sale and herd dynamics. The following is a conceptual summary of the simulation and a selection of major predictions.

The Simulation Model

The simulation model is conceptually similar to the general model presented above except that some variables are defined in a slightly different fashion. For convenience we repeat here the profit function form of the general model, which is also shown in note 5.

$$\Pi (\Theta) = pw (\Theta)e^{-r\Theta} + \int_0^\Theta f (p,t)e^{-rt}dt - c \int_0^\Theta e^{-rt}dt$$

The following differences in variable definition are of note: first, the returns from selling the animal, $pw(\Theta)e^{-r\Theta}$, and the flow benefit, f, are the "expected value" of those returns, where the full value is reduced by the probability of mortality. Thus, the cost of mortality is transferred from c, but this has no effect on the outcome. A more important difference is that flow benefits, f, are defined only as milk and calves. No values are assigned to security, liquidity, prestige, power, and so forth. Finally, c is first given a zero value to represent, in only a slightly extreme form, the notion of some writers that there are "no costs" if the range is a common property. Another set of simulation runs assigns to c positive values for the cost (including opportunity cost) of labor and for cash outlays for health care, taxes, and so forth. The actual values of these variables as well as for mortality, fertility and the growth curve are shown in Table 20.1.

The Optimum Age of Sale

The simulation model predicts that males would be sold for slaughter at age 6 and females at age 10 or 11. This accords quite well with the dominant ages observed in West African abattoirs. Thus, a fairly good prediction has been derived using a maximization model that includes values only for beef, milk, and calves, and does not consider other values such as prestige or power.

Table 20.1. Baseline parameter values for the simulations

Age	Liveweight Males[a]	Liveweight Females[b]	Mortality Rates Both Sexes	Fertility Rates Females
	----------kg----------		--percent--	----percent----
0	25.0	25.0	—	0
1	51.4	46.7	.35	0
2	88.8	110.1	.10	0
3	134.6	110.1	.05	0
4	184.6	147.2	.03	.30
5	234.7 ·	184.3	.03	.60
6	281.7	219.4	.03	.70
7	323.6	251.2	.05	.70
8	359.6	278.9	.07	.70
9	389.6	302.5	.09	.65
10	414.1	322.2	.11	.65
11	433.7	338.3	.13	.60
12	449.2	351.3	.15	.55
13	461.4	361.8	.17	.50
14	470.8	370.1	.19	.45
15	478.2	376.6	.21	.40

Source: Authors.

[a]Computed from the following equation: $W_t = 25e^{3(1-.76^t)}$

[b]Computed from the following equation: $W_t = 25e^{2.773(1-.775^t)}$

Note: Other Parameters. Costs are set at zero and at $10 per head in different runs. The interest rate is at 12 percent and 8 percent in different runs. Prices are set at $2.00 per kg liveweight and $.30 per liter milk. Probability of lactation is (.80)x(fertility rate). Each lactation yields 200 liters of milk.

Table 20.2. Expected present value of a steer

Age	Weight	Value	Mortality Rate	Survival Rate	Life Expectancy Rate	Expected Value	Discount[a] Factor	Present Expected Value
-years-	-kg-	-dollars-	-percent-	-percent-	-percent-	-dollars-	-percent-	-dollars-
1	54.4	42.8	.35	.65	.650	27.8	.887	24.7
2	88.5	177.6	.10	.90	.585	103.9	.787	81.8
3	134.6	269.2	.05	.85	.556	149.7	.698	104.5
4	184.6	369.2	.03	.97	.539	199.0	.619	123.2
5	234.7	469.4	.03	.97	.523	245.5	.549	134.8
6	281.7	563.4	.03	.97	.507	285.6	.487	139.1
7	323.6	647.2	.05	.95	.482	312.0	.432	134.8
8	359.6	719.2	.07	.93	.448	322.2	.383	123.4
9	389.6	779.2	.09	.91	.408	317.9	.340	108.1
10	414.1	828.2	.11	.89	.363	300.6	.301	90.5
11	433.7	867.4	.13	.87	.316	274.1	.267	73.2
12	449.2	898.4	.15	.85	.268	240.8	.237	57.1
13	461.4	922.8	.17	.83	.223	205.8	.210	43.2
14	470.8	941.6	.19	.81	.180	169.5	.186	31.5
15	478.2	956.4	.21	.79	.143	136.8	.165	22.6

Source: Authors.

[a]Calculated using a 12 percent rate of interest compounded annually, and costs = 0.

Tables 20.2 and 20.3 show the evolution of the various component values derived from raising steers and cows. These coincide with the profit function formulation and hence, unless otherwise noted, represent the present value of sales revenue and the sum of present values of milk and progeny.[8] The animal is sold when its present value is at its maximum. Tables 20.2 and 20.3 present the case of c = 0. When c ≠ 0 the animal is sold when net present value is a maximum. This is shown in Figures 20.1 and 20.2. The two solutions are not meaningfully different for the Sahelian values used.

The Effects of Changes in Price, Cost, Interest, Mortality and Fertility

Figure 20.3 presents a summary picture of how age of sale changes when key parameters vary from 70 percent to 140 percent of their baseline values.[9] Perhaps the two most striking features of the figure are the insensitivity of the sales age of males and females to variations in price and cost, and the insensitivity of males to variations in mortality and interest rates as well. The results strongly challenge the prospect of altering offtake rates or herd size through manipulation of economic variables.

Table 20.3. Present value of a cow

Age	Current Meat Value	Expected Present Value Components[a]				Expected Present Value of a Female
		Meat	Milk	Male Progeny	Female Progeny	
-Years-				Dollars		
4	294.3	98.2	4.8	7.0	5.8	115.7
5	368.6	105.8	13.1	18.9	21.7	159.5
6	438.9	108.4	21.4	31.0	46.1	206.7
7	502.4	104.5	28.4	41.1	72.8	246.8
8	557.9	95.7	34.1	49.4	98.7	278.1
9	605.1	83.8	38.4	55.7	118.7	296.7
10	664.4	70.4	41.8	60.6	133.7	306.6
11	676.6	57.1	44.3	64.2	141.6	307.2
12	702.6	44.7	46.0	66.6	144.5	301.6
13	723.5	33.9	47.1	68.2	143.4	292.7
14	470.1	24.9	47.8	69.3	140.8	282.8
15	753.3	17.7	48.3	69.9	137.5	273.3

Source: Authors.

[a]12 percent interest rate and costs = 0.

The ineffectiveness of price changes is seen most easily in the case of steers when costs, c, are assumed to be insignificantly different from zero. The decision rule is then to sell when

$$p \; \partial w/\partial \theta = rpw$$

Price cancels out on both sides of the equation and so pastoralists are concerned only with the rate of weight gain compared to the rate of interest times weight, w, at any sales age under consideration. Price does not enter.

If costs are positive we have

$$p \; \partial w/\partial \theta = rpw + c$$

or

$$\partial w/\partial \theta = rw + c/p$$

Here the optimal sales age must change. However, if c is very small relative to p and rw, then that change will be very small, and in the case of the Sahel it is not enough to force sales age up or down even one year.

Finally, for females, with positive costs, we have

$$p \; \partial w/\partial \theta + f = rpw + c$$

or

$$\partial w/\partial \theta + f/p = rw + c/p$$

or

$$\partial w/\partial \theta = rw + \frac{c-f}{p}$$

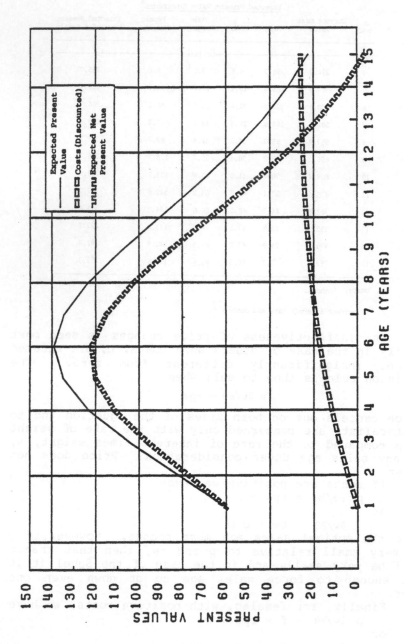

Figure 20.1. Expected net present value of a steer

327

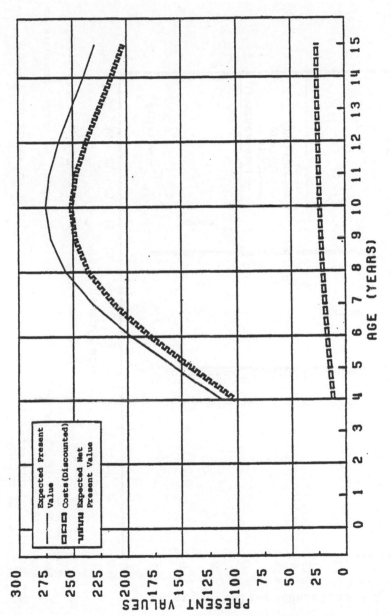

Figure 20.2. Expected net present value of a cow

328

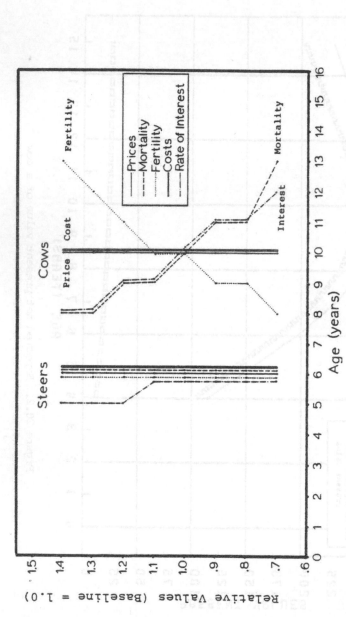

Figure 20.3. Age of sale of cows and steers at various levels of price, cost, mortality, fertility, and interest

Notes: 1. Baseline Cost = $10.00 per head. Baseline Interest = 12 percent. All other parameters as shown in Table 20.1.

2. Mortality and interest have identical effects on female sales age except for a slight difference at 70 percent of the baseline.

Again, age of sale must change if p changes (unless c = f), but in the Sahelian case it does not amount to a change of a full year over a $\pm$ 40 percent variation in price. As shown, c and f work against each other to decrease the impact of price changes.[10]   Figures 20.4 and 20.5 graph the effects of price changes on steers and cows for the case of c > 0.

Another interesting result shown in Figure 20.1 is the mutually reinforcing effect of changes in mortality and fertility on female sales age. If range conditions worsen, say under increased stocking pressure, mortality will rise and fertility will fall.   Both these changes tend to lower the age of sale. Thus there is a double self-correcting mechanism guarding against range degradation: as conditions worsen there are more deaths and fewer births and then, as pastoralists come to perceive or expect this, there is an earlier sale of females.   This latter also has a double effect: it removes animals from the range more quickly and it also lowers births because the females have fewer years for reproduction.   All these mechanisms work in reverse when range conditions improve. This allows for more rapid herd reconstitution than might otherwise be expected.   This may help account for the very rapid herd recovery observed by Meadows and White in Kajiado, Kenya following the drought of 1960 and 1961:
"Herd numbers doubled in three years and had recovered to predisaster numbers in seven years" [1979, p.5]. This phenomenon is contrary to Dahl and Hjort's [1976] doubts that a herd can double itself in even 4 to 5 years. The change in the sales age for females may help explain the very rapid increase.

CONCLUSIONS

Some of the conclusions emerging from this model are:
a. Age of sale is likely to be highly insensitive to changes in economic variables.
b. Herds are likely to adjust very rapidly to changing range conditions as biological changes (in mortality, fertility, milk yield) trigger reinforcing economic decisions (concerning age of sale).
c. The higher the costs, the younger the age of sale. (The simulations show no significant change.)
d. The higher the interest rate, the younger the age of sale. (The simulations show a large change for females.)
e. The higher the rate of weight gain, the older the age of sale.
f. The lower the mortality, the older the age at sale. (The simulations show a large change for females but not for males.)

330

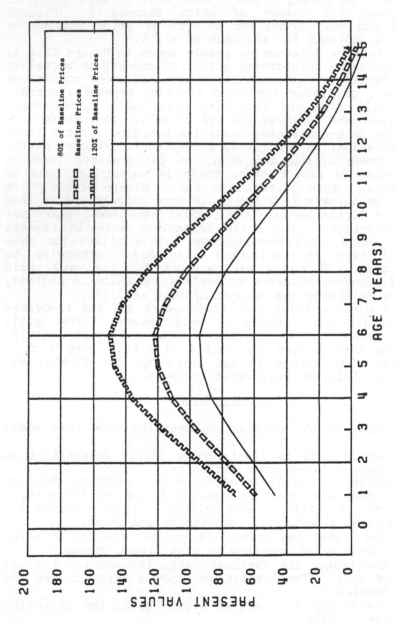

Figure 20.4. Price changes and expected net present value of a steer

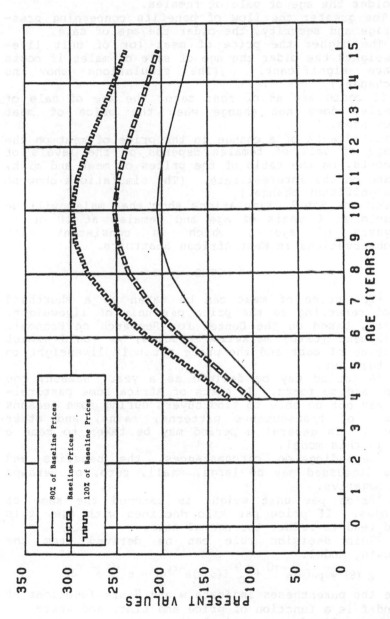

Figure 20.5. Price changes and expected net present value of a cow

g. The higher the fertility, the the older the sale
    age of females.
h. The higher the value (P x Q) of milk produced, the
    older the age of sale of females.
i. The greater the flow of benefits concerning pres-
    tige and security, the older the age of sale.
j. The higher the price of meat (or of unit live-
    weight) the older the age of sale of males if costs
    are significant. (The simulations show no
    change.)
k. If costs are at or near zero, the age of sale of
    males does not change when the price of meat
    changes.
l. The effects of a change in the price of meat on the
    age of sale of females depends on the levels of
    costs, on the ratio of the prices of meat and milk,
    and on the interest rate. (The simulations show no
    significant change.)
m. For the Sahel, simulations show that males will be
    sold at 6 years of age and females at 10 or 11
    years of age , which is consistent with
    observations in West African abattoirs.

NOTES

[1]The price of meat can be taken as a shorthand
way of referring to the price per unit of liveweight.
In fact, based on the Center for Research on Economic
Development (CRED) marketing studies, we would expect
the price of meat and the price per unit liveweight to
move together.
[2]A period may be defined as a year, season, one
month, and so forth. In parts of Africa some pastoralists may not be able to find buyers during some seasons
because of transhumance patterns, rains, and other
factors. In general a period may be taken to mean a
year in this model.
[3]Depending on circumstances, the benefits and
costs discussed may be large, small, zero, or perhaps
even negative.
[4]Price per unit weight is assumed the same at
all ages. If price per kilo declined with age, this
would tend to reduce the age of sale.
[5]This decision rule can be derived from the
following model:

$$\Pi (\Theta) = pw(\Theta)e^{-r\Theta} + \int_0^\Theta f(p,t)e^{-rt}dt - c \int_0^\Theta e^{-rt}dt$$

where the parentheses indicate w and Π are functions of
Θ, and f is a function of price and time, and where
  Π = profit from an animal, i.e., sum of benefits
      over cost,
  e^{-rt} = the discount factor that brings all terms to
      present values,

$\int_0^\Theta$ = summation of the respective present values over the animal's life,

t = time,

and all other terms are as above. Taking the first derivative of Π with respect to Θ , and setting the result equal to zero gives the condition for determining the profit maximizing age of sale. That condition yields the above decision rule for the pastoralist. This follows Jarvis [1974] very closely except for specification of the flow benefits.

[6]While it is conceptually clearer to consider mortality as an element in c, it is computationally easier to consider mortality as a discount factor applied to $\partial w/\partial \Theta$ and w. This is because mortality is a function of age, and our formulation is simplified by having cost independent of age.

[7]Although steers do yield some flow benefits such as security and power.

[8]Note that these are not the one-year differences from holding an animal an additional year.

[9]Each parameter was varied individually.

[10]Indeed, if f > c, an increase in price will increase $\frac{c-f}{p}$ (make it less negative), and hence increase the RHS. This will tend to decrease age of sale for females, the opposite of the effect for males.

REFERENCES

Dahl, Gudrun and Anders Hjort. Having Herds: Pastoral Herd Growth and Household Economy. Stockholm: Department of Sociology, University of Stockholm, 1976.

Jarvis, Lovell S. "Cattle as Capital Goods and Ranchers as Portfolio Managers: An Application to the Argentine Cattle Sector." Journal of Political Economy 82(1974):489-520.

Jarvis, Lovell S. "Cattle as a Store of Wealth: Comment." American Journal of Agricultural Economics 62(1980):606-613.

Meadows, S.J. and J.M. White. "Structure of the Herd and Determinants of Offtake rates in Kajiado District in Kenya 1962-1977." Pastoral Network Paper 7d. London: Overseas Development Institute, Agricultural Administration Unit, March 1979.

21
Production of Meat and Milk from Goats in Mixed Farming Systems in the High Potential Tropics

A.J. De Boer, H.A. Fitzhugh, R.D. Hart,
M.W. Sands, M.O. Job and S. Chema

INTRODUCTION

Many rural development projects are based on objectives related to improving human nutrition and family incomes among target groups. Small farms in the tropics often produce inadequate amounts of high quality protein foodstuffs needed for good health, especially for children and pregnant women. Milk is a good source of this needed protein. Although many farms do not have the land resource base to provide sufficient feed for a lactating cow, they may support several lactating goats. Several goats (3 to 5) in place of a cow would have several advantages: loss of an individual animal has less impact on family welfare than loss of a cow; initial investment per animal is low; mating does to kid in different seasons allows a small, but consistent daily supply of milk throughout the year; litters of two or three kids at 7 to 8 month intervals markedly increases offtake of slaughter stock for family consumption or sale to provide needed income.

A principal constraint to expanded goat enterprises is the amount of feed produced from the limited land area available on smallscale farms, which currently emphasize food crop and cash crop production. Arable land is in such short supply that very little, if any, can be devoted to continuous pasture for livestock. Thus, emphasis must be on use of crop residues and byproducts (e.g., sweet potato vines, maize stover, molasses), on complementary use of cut-and-carry forages and pasture grown as rotation, and on catch crops produced in field interstices or grown on nearby nonarable lands. To be acceptable to small-scale farmers, the dual-purpose goat production system must be based on low-cost, low-risk technology and minimally competitive with cropping activities for land, labor and capital resources. The research problem is, therefore, to develop an animal of the appropriate genotype, to develop a feed resource-nutrition-health management package appropriate to the small farm resource base and to ensure that this component is economically and sociologically acceptable.

The research reported in this paper represents an attempt to develop a new nontraditional technology designed to fit within the constraints under which small mixed farms operate in Western Kenya. The overall objectives of the research project were to develop, test, and modify a technology package which already focused on a dual-purpose (meat and milk) goat production system. The project has been formulated and executed as a cooperative project between the Kenya Ministry of Livestock Development and the U.S. AID-funded Title XII Small Ruminants Collaborative Research Support Program (SR-CRSP). The latter project is a consortium of U.S. institutions working on a worldwide program of small ruminant research, one of the major research sites being Kenya.

RESEARCH FRAMEWORK

The SR-CRSP in Kenya was conceived using the farming systems research approach. The components included sub-projects covering a Production Systems project (focused on goat nutrition and management), Feed Production and Agricultural Economics (all under Winrock International), Rural Sociology (University of Missouri), Animal Health (Washington State University), Animal Breeding (University of California, Davis, 1980-1982, and Texas A&M, 1982-present) plus Systems Analysis and Modeling (Texas A&M, 1980-1983). Throughout the project design stage, researchers recognized that the small farmer target group was attempting to satisfy a number of simultaneous production objectives including family staple food security, maintenance of soil fertility, capital accumulation, risk avoidance, and financial liquidity. These objectives are often conflicting, and progress towards achieving any one objective is often a complex process requiring detailed understanding of farmers' goals, decision-making processes, and production practices. To accomplish this process, direct contact and discussion between researchers and farmers must occur within the farmers' environment. The research team thus designed a program of activities following the farming systems research approach which followed these sequential steps:
1. Gain a basic understanding of the target farming systems.
2. Understand the knowledge base and constraints upon which the farm family base their decisions.

3. Identify possible interventions to improve family welfare.

4. Implement on-farm testing of the interventions to screen their suitability and carry out modifications based on the results.

5. Develop a package appropriate for extension based on results gained from steps 1 through 4 above.

The remainder of this chapter will summarize results gained to date from steps 1 through 4. The program has not yet progressed to the stage where a package can be extended to even a limited recommendation domain of farmers in Kenya. The general conceptual relationship followed by the overall research program is shown in Figure 21.1.

FARMING SYSTEMS DESCRIPTION

The farming systems described in this chapter are located in the Lake Victoria Basin of Western Kenya. These systems share many characteristics with other crop-animal systems in the humid tropics of Subsaharan Africa, so this research has potential for broader application.

Characterization of the target farming systems in Kenya was accomplished through the Small Farm Systems Survey, a collaborative research activity of the Sociology, Economics, and Production Systems projects. Results from this survey have been instrumental in the design of research activities by all projects collaborating in the Kenya program.

The survey was conducted from October 1980 to December 1982 on 40 farms in Kakamega district of Western Province and 40 farms in Siaya district of Nyanza Province (Figure 21.2). Farms were visited at least monthly by enumerators fluent in the local language. Details of the survey procedure are presented by Sands et al. [1982b].

Kakamega district is considered to be agriculturally a high potential area. Rainfall is typically bimodal with an annual rainfall of between 1,750 and 2,100 mm. Altitude ranges from 1,400 to 1,800 m. The most important food crops are maize and beans, usually intercropped. Other food crops include sorghum, cassava, sweet potatoes, and bananas. The principal cash crops of the area are trees, tea, and coffee. Principal livestock species are grade and Zebu cattle, hairsheep, East African goats, and poultry. The dominant tribe in the rural areas of Kakamega are the Abaluhya. The 1979 census estimated that the district population was about 1,033,000. There is an extremely high land pressure of 328 people/km^2. In some parts

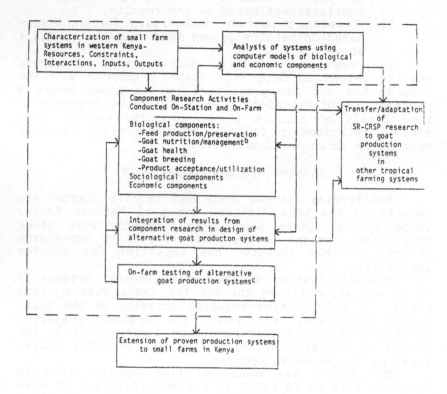

Figure 21.1. Implementation strategy for SR-CRSP research activities involving dual-purpose goat production systems for small farms in Kenya

Note:
SR-CRSP research activities are shown inside dashed line; extension activities to be conducted by MLD and other agencies are shown outside dashed lines.

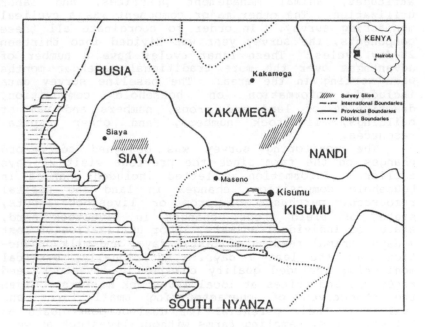

Figure 21.2. Map of western Kenya

of the district, pressure is as high as 880 people/km².

Siaya district, by comparison, is considered a medium potential zone. Rainfall is bimodally distributed with an annual mean of 1,000 to 1,300 mm. Altitude ranges from 1,140 to 1,500 m. Natural vegetation is dominated by invasive shrubs (e.g. Albizia coriana and Lantana ssp.) and poor quality grasses including Hyparhenia ssp.; Cymbopogon ssp.; Arsitida ssp., Panicum ssp.; and Digitaria ssp. Principal food crops include maize and beans, cassava, sorghum, sweet potatoes, bananas, and minor pulses. Cash crops include trees, cotton, and sugarcane. Grade cattle are rare in Siaya, but there are large numbers of Zebu cattle, hairsheep, East African goats, and poultry. The 1979 census indicated that the district population is 470,000. The means there are approximately 185 people/km². The Luos are the dominant tribe of Siaya.

Survey Methodology

The survey was comprised of several distinct components. Single point questionnaires included a baseline module as well as modules on sociological

attitudes, animal management practices, and labor
utilization. The other major component was a cyclical
monitoring survey. In order to coordinate all these
components, the survey year was divided into thirteen
28-day cycles. These lunar cycles have a number of
advantages over the more traditional calendar months
when working in the area. The baseline survey data
included information on household composition,
description of land and crops, numbers and use of
buildings, livestock numbers, and other capital
resources.

The monitoring survey was designed to record
changes on the farm since the previous visit 28 days
earlier. Information collected included changes in
household composition, changes in land and capital
resources, purchases of crop or livestock inputs,
status of field crops, changes in livestock herd,
status of individual animals, labor utilization of last
7 days, animal feeding of last 7 days, and milk production over the last 7 days. Additionally, cyclical
monitoring included quality changes in livestock feed
resources and prices at local livestock markets. Given
the objectives of characterizing small farms and
evaluating their potential for dual-purpose goats on
typical farms, sampling farms without livestock as well
as those that had livestock was important.

Results and Discussion

Table 21.1 shows mean farm size by class and by
district. Siaya farmers tended to have slightly larger
farms (1.09 ha) than Kakamega farmers (0.98 ha). The
scarcity of land is an important production constraint
since 54 percent of the farms are less than 0.8 ha in
size.

The household average in Kakamega and Siaya
district was 7.95 and 4.65 persons, respectively. Of
this total, 3.68 and 2.23 were males and 4.27 and 2.42
were females. Farm family size increased with farm size
from 4.91 members on farms less than 0.4 ha, to 7.82
members on farms of over 1.6 ha. However, of even more
interest is the dramatic drop in people per ha of farm
with increasing farm size. There were 28.47 family
members per ha on farms of less than 0.4 ha. This
number declined to a level of 3.5 people per ha on
farms over 1.6 ha. The constraint of scarce land
resources is emphasized by the high land pressure in
both Kakamega (15.33 members/ha) and Siaya (9.90
members/ha).

In order to better visualize various interactions
among components of farming systems, schematic models
are used. Figures 21.3 and 21.4 are schematic repre-
sentations of "typical" small farms in Western Kenya.

Table 21.1. Average farm sizes of surveyed farms

Farm Size Class	Mean	Std. Dev.	Number of Farms	Percentage of Sample
	——Ha——			
Less than .4 ha	.26	.12	22	28
.4 ha - .8 ha	.64	.13	21	26
.8 ha - 1.2 ha	1.00	.12	11	14
1.2 ha - 1.6 ha	1.36	.01	9	11
Over 1.6 ha	2.39	.77	17	21
Siaya	1.09	.89	40	50
Kakamega	.98	.84	40	50

Source: Sands et al. [1982a], p. 18.

Boxes labeled "Market" represent off-farm activities involving purchased inputs (except land) such as supplies, animals, and household items plus products and labor sold off the farm. Commodities stored on the farm are represented by tank symbols. Crop and livestock subsystems are represented by rectangles. The "Household" is the core of the family farm. Labor use, human food, and animal feed are identified as are interactions between the crop and livestock subsystems. Solid lines (___) depict strong flows while the broken lines (----) intermittent or weak interchanges.

Farmers sold maize, beans, livestock, and labor. They bought maize, salt, sugar, and tea. Major cropping activities were inter-cropped maize and beans, maize alone, sorghum, cassava, sweet potatoes and vegetables (kale, tomatoes, etc.). Wild vegetables were also collected and consumed. Seed for improved maize varieties was the major crop input from the market.

Livestock included cattle, sheep and goats, as well as chickens. Animals were generally grazed on fallow land or cropped fields. Crop residues or cut forages were rarely brought to tethered livestock. Other crop-livestock interactions included occasional use of manure for fertilizer or oxen for plowing.

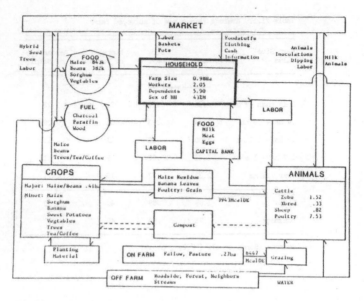

Figure 21.3. Model of a Kakamega small farm (long rains, 1981)

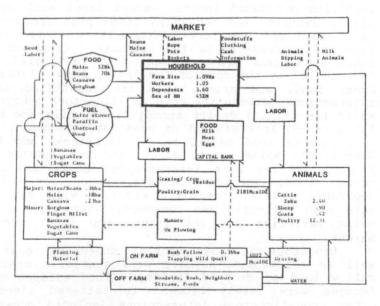

Figure 21.4. Model of a Siaya small farm (long rains, 1981)

Animals and milk were sold, while animals and dip were principal purchased inputs. Labor for herding was the primary household input to the livestock subsystem.

Typical Kakamega farm. The "typical" small farm in Kakamega district is shown in Figure 21.3. The farm had 0.98 ha of land worked by 2.05 workers and supported 5.90 dependents. The head of household was female on 58 percent of the farms because the male had left to work in the urban areas.

The major crop was maize/beans intercropped on 0.41 ha during the long rains (March-August). Minor crops included maize, sorghum, bananas, sweet potatoes and vegetables planted on 0.3 ha. Improved maize seed was purchased for the long rains. Some purchased seed, as well as native varieties, were used during the short rains (September-December). Native varieties were used for planting materials for other crops. Most labor for the maize/beans intercrop (71 days) came from the available household supply (642) days). Family labor was also used for tending minor crops and performing household activities such as fetching water, collecting firewood, and cooking.

The majority of maize harvested (843 kg) was consumed on the farm as ugali, a maize meal mush. Relishes included sukuma wiki (kale), wild vegetables, and fish or meat. Harvested beans (382 kg) were eaten and sold in the market. Half of the families had one or two cows and one sheep; most families had 7 to 10 chickens. When cows were milked, a small proportion (20-30 percent) of milk was sold and the rest consumed on the farm, primarily with tea. No milk was processed as cheese but some was soured. Calves and lambs were usually sold and served as a major source of income to pay school fees and to purchase clothing and other items.

Interactions between crop and animal components were not particularly strong. The most important direct interaction was feeding crop residue to livestock. Maize stover and thinnings provided about 3,943 Mega calories (Mcal) of digestable energy (DE) available for livestock consumption. Banana leaves were also fed. Spoiled maize grain was occasionally fed to poultry.

Grazing on unimproved land provided the majority of energy and protein for the ruminants. On-farm non-cropped land provided approximately 6,447 Mcal DE annually. Off-farm grazing along roadsides and other common land was used extensively. Forage quality could be quite high at certain times of the year. Grasses such as Cyndon dactylon and Digitaria scalarum retained high levels of protein and digestibility when grazed frequently. In addition, local legumes (Glycine weightii, Trifolium ssp.) were found in many heavily grazed areas.

Manure was not utilized in a systematic fashion. Composted household scraps, manure, and crop residues were used by some farmers on their long rains hybrid maize. Oxen plowing was usually not utilized in this area of Kakamega district because of small plot sizes, so most cultivation was by human labor.

The principal constraints to crop production are water availability and soil fertility. Though total rainfall is generally adequate, yields would probably benefit from improved timeliness of planting and weeding. Also, hybrid maize would respond to nutrient additions (manure, purchased fertilizers) to the humic nitosols in the area. Anthracnose was an important constraint to crop production in Kakamega.

The principal constraint to livestock productivity was inadequate nutrition. Because of small farm size, devoting crop land to forage crops is not a viable option; therefore, increased forage production will depend on changing the cropping system. Residue yields should increase with crop productivity. Cover crops or relay planting a forage source after harvest of beans and maize are options.

Typical Siaya farm. The Siaya small farm is modeled in Figure 21.4. The farm had 1.09 ha; the family had 1.05 workers and 3.60 dependents. Because of outmigration of males in search of wage employment, 55 percent of the farms were managed by a female head of household. The main crops were maize/beans intercropped (0.36 ha), maize alone (0.18 ha) and cassava (0.23 ha). Minor crops included sorghum, finger millet, sweet potatoes, bananas and vegetables. Local varieties were used for all crops. Labor required for the maize/beans intercrop (86 days) and other crops was drawn from the pool of family labor (328 days). Household activities such as fetching water from streams, collecting firewood and cooking required considerable family labor.

The maize harvested (528 kg) was primarily consumed on the farm. This was supplemented by purchased maize, cassava, and sorghum. Harvested beans (70 kg) were consumed and sold in the local market.

The typical farm had 2 to 3 cows, and several sheep and goats. Milk was consumed on the farm primarily in tea. Calves, kids, and lambs were usually sold to meet major cash needs. Family labor, usually a male child or older male, was used to herd the livestock 7 to 8 hours per day. Livestock required little labor because feed was only rarely cut and carried to them.

Poultry included native chickens (10-15 birds) and ducks. Eggs were usually hatched and only rarely eaten or sold. Birds were sold when cash was required. An important visitor was often occasion for slaughter of a chicken. Poultry meat was supplemented by trapping wild quail on fallow bush land.

Interactions between crop and livestock components were limited. Some crop residues were grazed after grain harvest. Though some 2,081 Mcal of DE was potentially available from maize residues, the actual amount consumed by livestock was probably much less due to trampling and the use of stalks for fuel. Farmers rarely allowed livestock to graze cassava leaves because of potential toxic effects.

Grazing on unimproved land provided the vast majority of feed. On-farm grazing land provided some 4,022 Mcal DE. Off-farm grazing provided a minimum of 15 percent and probably more than 35 percent of the maintenance requirements for production. The primary grasses were poor quality Cymbopogon ssp. and Brachiaria ssp. growing under bushes (Lantana camara, Albizia coriana, and Markhamia platycalyx).

Manure was not systematically utilized as fertilizer; however, some manure from the night boma may have been used when planting maize.

Intensification of the system is primarily dependent on increasing the low yields of maize (900-1,050 kg/ha). Water availability is probably the major limiting factor. More timely planting and weeding are important to take advantage of the water that is available. More use of manure and purchased fertilizer would increase nutrient availability. The parasitic weed, Striga hermontheca, is a major problem in maize and sorghum; it is best controlled by rotation of maize and sorghum with resistant crops (e.g., cassava, millet, or forages).

Improving crop productivity would increase residue yields. These increases should make feasible the collection and storage of residues for later use. Increased use of higher quality forages would require additional management to discourage poor quality forages such as Lantana and Cymbopogon. Relay cropping of forages into mature cassava stands or fallow maize fields may provide more animal feed.

Livestock Component

Livestock represent an important component of small farms in Western Kenya. Some type of livestock are seen on almost all farms as, for example, poultry were found on 83 percent of all farms. Ruminant livestock were found on 61 percent of all farms. Cattle are the most valued livestock and in Kakamega, there has been some influence of the national artificial insemination program. Crossbred cattle were found on 15 percent of the Kakamega farms. No farms in Siaya kept crossbred cattle. Crossbred cattle numbers increased with farm size.

Sheep and goats play a lesser role on the farms in both districts. All sheep were of the native hair type, with either fat tails or fat rumps. Goats were of the small East African breed but were only found on one farm in Kakamega as they are considered difficult to control and keep out of crops. The one farmer who kept goats grazed them in a nearby national forest.

Animal resources are presented in Table 21.2. The means were computed over all farms in each class and were, therefore, influenced both by number of farms having a specific breed, as well as animals the farmer had. Livestock units were calculated from the following:

Bulls/Oxen, 2 years or younger	0.5	Zebu Heifers	0.4
Bulls/Oxen, 2 years or older	1.0	Crossbred Heifer	0.2
Cows	0.6	Calves	0.2
Crossbred Cows	0.8	Sheep/Goats	0.2

Numbers of all types of livestock increased with increasing farm size. Total Livestock Units (TLU) increased from 0.99 to 3.07 on farms between 1.2 and 1.6 ha and then dropped down to 1.41 units per farm for farms over 1.6 ha. TLU per ha tended to decrease with increasing farm size. On farms less than 0.4 ha, farmers have to feed their livestock from resources gathered off the farm as 3.7 animals per ha exceeds the capacity of the farm feed base.

Within all species, the herds typically had between 60 and 74 pecent females. Females of reproductive age represented 39 percent to 45 percent of the herd or flock. Only Siaya had any castrated animals. The oxen were kept for plowing while the goats were kept for later slaughter. Animals were usually castrated at older ages (3 to 4 years of age). Older male animals were typically sold to raise cash for the purchase of food grains or for payment of school fees.

The live weights of the animals were low. Mature weights and other production coefficients are presented in Table 21.3. Cattle weights were calculated from heart girth measurement using equations for East African Zebu derived by Semenye [1979]. Mature cattle had a mean weight of 174 kg and 214 kg in Siaya and Kakamega, respectively. Sheep weighed 21.2 and 26.0 kg in Siaya and Kakamega while goats weighed 26.4 and 25.1 kg, respectively. Weights are for mature animals of both sexes.

Parturition intervals were calculated for all animals that had at least two parturitions the farmer remembered. This biases the means towards shorter intervals as animals that had very long intervals often

Table 21.2. Animal resources

Farm Size Class	Zebu Cattle	Crossbred Cattle	Cattle	Sheep	Goats	Poultry	TLU	TLU/ha
Less than .4 ha	1.45	-	1.45	.41	.14	7.40	.99	3.75
.4 ha - .8 ha	1.04	.10	1.14	.52	.14	8.85	.65	1.04
.8 ha - 1.2 ha	2.36	-	2.36	1.82	.91	10.33	1.57	1.58
1.2 ha - 1.6 ha	2.80	.22	5.00	1.22	.67	7.14	3.07	2.25
Over 1.6 ha	2.00	.53	2.53	1.06	.06	14.60	1.41	.59
Siaya	2.40	-	2.40	.90	.42	12.31	1.58	1.46
Kakamega	1.52	.33	1.85	.82	.15	7.53	1.04	1.06

Source: Sands et al. [1982a], p. 24.

Table 21.3. Livestock production coefficients

Variable	Unit	Siaya	Kakamega
Cattle			
Mature weight, cows	kg	174.0	214.0
Parturition interval[a]	months	21.3	19.6
Lactation milk yield[b]	kg	525.0	610.0
Lactation length	months	9.8	10.9
Sheep			
Growth rate[c]	g/day	43.0	90.0
Mature weight			
Rams	kg	23.8	27.5
Ewes	kg	20.9	26.1
Parturition interval[a]	months	9.8	9.1
Lambs/parturition	number	1.0	1.0
Goats			
Growth rate[c]	g/day	70.0	--
Mature weight			
Bucks	kg	31.3	28.2
Does	kg	23.6	23.4
Parturition interval[a]	months	11.0	10.0
Kids/parturition	number	1.1	1.0

Source: Sands et al. [1982a], p. 29.

[a] Parturition interval is calculated only for animals having at least two parturitions.

[b] Over and above that taken by calf.

[c] From birth to 180 days.

had to be dropped from the analysis. For cattle, the district means for parturition interval were 20.1 and 18.9 months for Siaya and Kakamega. Sheep had means of 9.0 and 7.8 months for Siaya and Kakamega. These are extremely short intervals and are probably not realistic. Goat kidding intervals were 11.0 and 10.0 months, respectively, for Siaya and Kakamega districts. Cattle and sheep had only one young born per parturition. In two of 15 recorded goat parturtions, twins were born.

The milk yield of cattle above that used by the calf was calculated by extrapolating yield over a 7-day test period to a monthly yield and summing across months of lactation. The milk yields were 525 kg and 610 kg for cows in Siaya and Kakamega, respectively.

The principal constraint on livestock production on the small farms in Western Kenya is undoubtedly inadequate nutrition. Using DM yields from small farm unimproved pastures in Western Kenya reported by Goldson, the estimated DM production from farm grazing areas was computed (Table 21.4). Then, using quality analyses on samples taken over the year, the Mcal of Digestible Energy (DE) produced for every farm was computed. This was supplemented by the DE from maize stover. The total annual DE production per farm is shown in Table 21.4.

Maize stover DE provides 45 percent of the total in Kakamega, but only 33 percent of the total in Siaya (Table 21.4). The average total Mcal of DE produced on the farm was not sufficient to cover maintenance requirements of the average livestock holdings. For instance, the maintenance requirements of the livestock on a Kakamega farm are 6,466 Mcal of DE. With the farm only producing 6,348, the farm has a deficit of 118 Mcal. This deficit will be much larger if the cost of milk production and growth are added to the requirements. This deficit suggests farmers utilize feed from off the farm to meet the nutrient requirements of their livestock.

No farmer in the survey reported ever buying commercial feed for ruminants. However, all farmers took their cattle, sheep, and goats off the farm to graze at some time. Usually animals would be released from the night boma in the morning and tethered on the farm for the remainder of the morning. Around noon the animals would be herded off the farm to be watered and fed along public ways until about five or six in the evening when they would be returned to the night boma.

With these scarce feed resources, dual-purpose milk goats have been proposed as an alternative for subsistence milk production on these small farms. Under tropical conditions, cattle rarely are able to consume more than 2 multiples of maintenance. Given its ability to be more selective in its feeding, the goat might

Table 21.4. Annual fodder production

Farm Size class	Farm Area	Ratio	DM	Grazing DE	Maize DE	Stover DE	Total DE
			-MT-	----------Mcal----------			
Less than .4 ha	.26	.28	.32	713	333	783	1496
.4 ha – .8 ha	.64	.26	.73	1629	626	1471	3100
.8 ha – 1.2 ha	1.00	.36	1.58	3525	927	2178	5703
1.2 ha – 1.6 ha	1.36	.28	1.68	3749	1113	2616	6365
Over 1.6 ha	2.39	.25	2.65	5914	1792	4211	10125
Siaya	1.09	.30	1.23	2804	583	1370	4174
Kakamega	.98	.26	1.56	3481	1220	2867	6348

Source: Sands et al. [1982a], p. 24.

be able to consume 2.2 to 2.3 multiples of maintenance. However, in order to support this consumption, the animal must be presented with enough feed to allow for the selectivity. Typically, goats in the tropics would have to be presented with enough feed to allow 50 percent to 60 percent refusals. A 1/4 crossbred doe producing 272 kg of milk per year would require 1,340 Mcal DE to allow for 50 percent refusals during the high demand period of lactation (180 days). Thus, a farm of between 0.4 and 0.8 ha could theoretically produce enough feed to support two 35 kg does producing a total of 544 kg of milk. This assumes that the goats can be managed to utilize the seasonality of feed production. In trying to design production systems appropriate to the small farmer, particular attention should be paid to difference between farms and between locations. Dual-purpose goat systems in Siaya would have to be based on a herding/grazing management system as feed production is principally from low quality grasses. Stover production is low, and not sufficient to warrant the extra labor required to cut and transport it to a confined goat herd. With only 2.3 workers per ha, there is not enough extra labor to support the large amount of time required for a cut and carry system in an area with such low productivity.

In explaining why so few goats are kept in Kakamega, farmers stated that they are difficult to herd and threaten the growing crops. Thus, a goat production system in this area would require at least partial confinement. With maize stover now producing about 45 percent of the farm DE and enough labor (4.3 workers/ha) to harvest it, crop residues could potentially be utilized in a semiconfinement system. More efficient storage and use of crop residues and wet season fodder production will be necessary if dual-purpose production systems are to be successful.

PRELIMINARY RESEARCH RESULTS

Feed Resources

The Small Farm System Survey identified the key role that must be played by improving animal nutrition through increased on-farm production of high-quality forages during specific seasons. The goal of the feed production research is to identify feed production options that can be integrated into a dual-purpose goat production system. These feed resource options must provide sufficient quantity and quality of feed at the appropriate times during the production cycle so that the dual-purpose goat component can efficiently produce meat and milk within existing farm systems and satisfy family objectives.

Interventions to increase feed from existing cropping systems which have been evaluated included planting at higher than optimum densities to allow thinning; stripping, topping and ratooning; and intercropping and relay cropping forage crops in the food cropping system. Interventions to increase feed from existing noncropped areas include traditional cut-and-carry forage grasses, such as Napier (<u>Penisetum purpureum</u>), and grazing legumes such as <u>leucaena</u>; and nontraditional cut-and-carry crops, such as pigeon pea. Interventions to increase feed from off-farm grazing areas include evaluation of native legumes found in communal grazing areas and testing of introduced legumes such as <u>Stylosanthes spp</u>.

The most important research results have been identification of feed production strategies with highest potential for meeting needs of an integrated dual-purpose goat production system. As these strategies have been identified, research has focused on testing specific feed-increasing interventions.

Figure 21.5 summarizes the process followed to analyze different strategies to increase feed production from existing cropping systems, and to identify interventions with the highest potential. Without forage-producing interventions, animal feed production is concentrated at the two time periods when crops are harvested (July-August and November-December). The different feed-producing interventions that were tested had the potential of making feed available during most of the year; unfortunately, many of the interventions also have the potential of decreasing grain production. Experiments were conducted to measure forage and grain production and interactions among interventions.

Treatments such as stripping and topping maize plants at different intervals before grain harvest had a significant impact on total forage production producing up to 1.0 MT/ha of feed, but both of these treatments produced a significant reduction of grain yield (25 percent reduction from stripping and a 15 percent reduction from topping). Treatments such as planting at higher-than-optimum densities and thinning at the first weeding had no negative effect on grain yields and produced high quality feed (15 percent crude protein), but produced less than 200 kg/ha of feed.

Preliminary results from the analysis of different strategies to increase feed from existing cropping systems suggest that, at this stage in the research, the feed production strategies with the most potential are intercropping sorghum and intercropping and relay cropping pigeon pea. Sorghum is intercropped with maize 4 to 6 weeks later and allowed to ratoon. A high quality feed (12 percent crude protein) is produced at the first cutting (up to 1.0 MT/ha) and the stover from

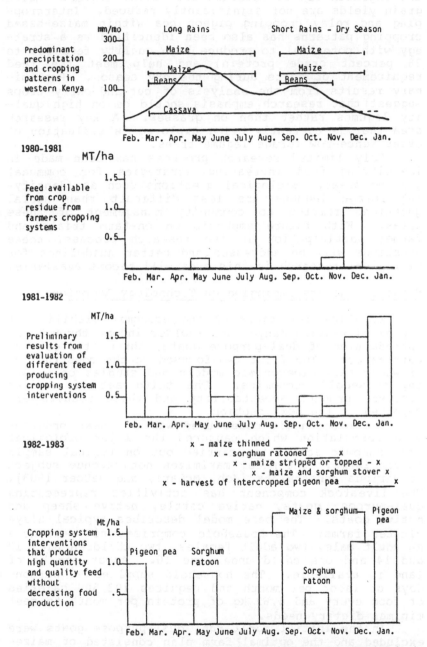

Figure 21.5. **General research strategy followed by the feed resources subproject in the analysis and identification of feed resource interventions to increase feed from cropped areas. Predominant cropping patterns were analyzed, potential interventions were evaluated and interventions with most potential were selected for further research.**

the regrowth can be combined with maize stover. Maize grain yields are not significantly reduced. Intercropping and relay cropping pigeon pea within maize-based cropping patterns has also been identified as a strategy with potential to produce high quality feed (up to 28 percent crude protein) and help meet the feed requirement of goats during the dry season. Preliminary results from the analysis of cut-and-carry crops suggest that research emphasis should be on high quality legumes rather than on grasses. A key research area that is currently being explored is evaluation of other fence-row forage legume trees.

Only limited research progress has been made in identifying feed increasing strategies for communal grazing areas. Biological questions such as identifying forage legumes are less difficult than social questions related to community management of these areas. With future emphasis on on-farm trials and farmer participation in the research process, these questions will be addressed and better guidelines for biological research in this area will become available.

Economic Factors Impinging on Technology Adoption

To allow assessment of the economic viability of improving Western Kenyan smallholder income through the introduction of dual-purpose goats, three studies were carried out. The first two focused on the construction of whole-farm bioeconomic models and studies on existing household economics. The third set of studies assessed product acceptability and these results are reported in the next section.

The bioeconomic model used was a linear programming formulation which captured the major biological and economic activities carried out on typical sample survey farms. The model maximizes net revenue subject to various constraints [Job, McCarl, and DeBoer 1983]. The livestock component has activities representing dual-purpose goats, native cattle, native sheep and native goats. The base model describes typical Siaya cluster farms. The household comprises six persons-- an adult male, two adult females, two children ages 10 and 14 and one child under age 10. About one ha of land is available. The household supplies 90 person-days of labor per month and requires 403,500 calories of food every and 4.62 kg of protein per month to meet minimal dietary needs.

In the initial model run, dual-purpose goats were excluded and the optimal farm plan consisted of maize-beans intercropped. Some maize and beans are sold and net returns are Ksh 930. Next, dual-purpose goats are allowed. For the one ha model farm, this activity does not enter the optimal farm plan. When farm size is increases to 1.6 ha, dual-purpose goats do enter and net farm returns rise to Ksh 3,195. Experiments were also carried out to examine the impact of subsidized capital and the use of small-scale silage making. If the capital cost of goats, housing and fodder planting material is subsidized by 50 percent, dual-purpose goats are very profitable and net returns on even the small farm site (0.8 ha) rise to Ksh 1,917. The silage making option also proved highly profitable with net returns on the one ha model farm rising to Ksh 3,780.

It was also necessary to examine the impact that the introduction of dual-purpose goats would have on household food consumption and purchasing patterns [Nyaribo 1983]. More specifically, we were interested in knowing whether survey farmers were able to produce adequate maize and beans for family needs, whether animal products were regularly purchased and consumed, household cash needs, and the potential impact that dual-purpose goats would have on these patterns. Food purchases were much more important in terms of both volume and regularity than we had anticipated. Surveys were conducted in April and June of 1981 and in October 1982. A surprisingly high proportion of households purchased the staple good, maize meal, during all three survey periods. Between 30 to 70 percent of households relied exclusively on purchased maize meal for at least part of the year. The major protein sources were beans, beef, tripe, poultry, goat meat, milk and fish. Almost all households consume meat at least once a week. Milk consumption was also higher than expected and this indicates a good potential demand for locally produced goat milk. Households also had high cash requirements for clothes and school fees. These results indicate that farmers in the survey are not self-sufficient in food, participate regularly and actively in local markets, regularly purchase milk from a variety of sources, and purchase a considerable amount of staple foods. This market orientation implies that the introduction of dual-purpose goats will probably be treated as a market opportunity and that low-cost, locally produced goat milk could allow farmers to marginally reduce expenditures on milk. The occasional sale of milk and surplus kids could also provide an important source of cash for food purchases and other household needs.

Product Acceptability

This section focuses on goat milk and goat milk products. No difficulties are anticipated in disposing of surplus breeding stock, at least in the initial stages of the project, since a strong demand exists for dairy goat breeding stock in Kenya.

356

Although goat milk is consumed by pastoralists in
other parts of Kenya, the Small Farm Systems Survey
revealed no consumption of goat milk by families in the
survey region and some hesitancy to try goat milk.
Reasons for this hesitancy were only vaguely stated.
However, there have been taboos in the region against
consumption of goat meat and milk by women. Previous
disease outbreaks such as brucellosis may have been
linked with goat milk consumption, and goat milk can
have a distinctive smell and flavor. Fresh milk has
excellent nutritional value for humans but also serves
as an excellent medium for disease organisms.

Results of research include [Boor, Fitzhugh, and
Ochieng 1983; Fitzhugh 1983]:

1. Goat milk--fresh or soured--was readily accepted in a series of taste panel tests using
   school children and members of farm families
   of different age and sex. These tests involved over 390 persons. Comparisons between goat
   and cow milk products showed no difference in
   average acceptance.
2. Milk is commonly boiled in tea, cooked with
   food and/or soured; thereby minimizing health
   hazard to consumers.
3. Quantities of milk currently or potentially
   available from dual-purpose goats are limited
   and are generally consumed on the day of milking so longer term preservation (e.g., cheese)
   of larger quantities for later use is not an
   immediate priority for research.

Results of research to date have largely put to
rest early concerns about acceptability and hygienic
safety of goat milk products. Additional research will
be conducted on the effects on nutritional value of
goat milk as most commonly consumed (e.g., the possible
binding of protein by tannins in tea).

Other Research Inputs

This chapter has reported results obtained primarily from the Production Systems, Feed Resources, and
Agricultural Economics subprojects of the Small
Ruminant CRSP. The other subprojects have made major
contributions to our understanding of the constraints
facing the introduction of a dual-purpose goat technology and to the conduct of research to overcome some
of the major biological problems that will inhibit
performances and survivability of dual-purpose goats
within the small farm environment of Western Kenya. A
capsule summary of these components is given by De Boer
et al. [1983].

SUMMARY AND CONCLUSIONS

The characterization of small farm systems has been instrumental in the identification of constraints, design of appropriate research projects, and the provision of a baseline set of information against which future interventions may be assessed. The on-farm testing and evaluation phase is in progress and has built upon experience gained through working with farmers during the survey. It is realized that our on-farm testing stage may be somewhat premature relative to an ideal FSR program where the more promising research results obtained from international and national research centers are tested on carefully supervised farm plots. In a project as complicated as the SR-CRSP, it has been impossible for all the research components to maintain the time schedule needed for formulation of production "packages." For example, genetic improvement and the formation of new breeds of dual-purpose goats is intrinsically a long-term process whereas the farming system characterization phase can be accomplished in a much shorter time frame. Also, the program did not have a substantial knowledge base to work from in Kenya on most components of goat production, and consequently much of the research had to commence without benefit of the substantial knowledge base that exists for the major cereal crops or for cattle.

The major lessons learned from the first four years of work in the SR-CRSP in Kenya can be summarized as follows:

Long-term commitment. There is no question that animal improvement programs are expensive, time consuming, and long-term in nature. Our experience has re-emphasized this fact. A minimal time horizon of ten years seems essential to obtain adequate understanding of the complex farm-household and crop-livestock interactions which characterize much of the high potential areas of Africa and to design, test, and modify the component technologies required to make substantial improvements in the traditional farming system.

Flexibility. Many of the operational and methodological details were worked out and revised as the research progressed. We consistently emphasized the FSR approach rather than any specific methodology. Revisions in research workplans, budgets, staffing, and recommendations to be tested on farms all evolved over time in response to cultural norms, institutional arrangements, and changes in the project funding base. A "cookbook" approach to FSR would not have been appropriate.

Coordination. Strong local support, coordination, and leadership has been essential in moving the research to its current level of maturity. The project has been successful in assisting the Kenyan counterpart agency, the Ministry of Livestock Development, in utilizing the FSR approach and in understanding how it can be applied to livestock improvement programs. This has led, in turn, to a long-term commitment for support to the SR-CRSP. Local needs, rather than U.S. institution needs, have been pushed to the forefront, and operational problems can be freely discussed by both groups.

In conclusion, the research reported here differs from most earlier applications of the FSR approach to improving small farm productivity. First, the focus is on livestock, and second, the approach is to develop an entirely new enterprise (dual-purpose goats) rather than to attempt to improve existing farm enterprises. It is much too early to say whether Western Kenya will witness the widespread adoption and utilization of dual-purpose goats, but we are convinced that the approach adopted for the generation and testing of technology for small farm improvement in Western Kenya is sound.

REFERENCES

Boor, K.J., H.A. Fitzhugh, and E. Ochieng. "Consumer Acceptability of Goat Milk Products in Western Kenya." Proceedings of the Small Ruminant CRSP Workshop, p. 6. Kenya, February 1983.

De Boer, A.J., H.A. Fitzhugh, R.D. Hart, M.W. Sands, M.O. Job, and S. Chema. "Production of Meat and Milk from Goats in Mixed Farming Systems in the High Potential Tropics." Small Ruminant CRSP Working Paper, Winrock International, Morrilton, Arkansas, July, 1983.

Fitzhugh, H.A. "Dual-Purpose Goat Production Systems for Smallholder Agriculturalists." Report submitted to Small Ruminant CRSP Management Entity, University of California-Davis, April 1983.

Goldson, J.R., "Factors Limiting Milk Yield of Cattle on Smallholder Farms." Ministry of Agriculture, National Animal Husbandry Research Stations Pasture Research Project Technical Report No. 21, Kitale, 1977.

Hart, R.D. "Synthesis and Evaluation of Dual-Purpose Goat Pruduction for Small Mixed Farms in Western Kenya." Proceedings of the Small Ruminant CRSP Workshop, pp. 35-40.

Job, Morgan. "A Study of Technology Adoption by Small Farmers with a Case Study Involving Dairy Goats in Kenya." Ph.D. thesis, Purdue University, 1982.

Job, Morgan, Bruce McCarl, and John De Boer. "An Economic Analysis of Dual-Purpose Goat Adoption on

Small Farms in Western Kenya." Draft manuscript, Winrock International, Morrilton, Arkansas, March 1983.

Nyaribo, F. "Analysis of Household Consumption in Western Kenya during the Short and Long Rains with a Comparison of On-farm Produced Food and Off-farm Food Purchases." Paper presented at Small Ruminant CRSP Scientific Meeting, Naivasha, Kenya, February 22, 1983.

Sands, Michael W. "Role of Livestock on Smallholder Farms in Western Kenya." Ph.D. thesis, Cornell University, 1983.

Sands, M.W., H.A. Fitzhugh, J. Kekovole, and P. Gachuki. "Results of Small Farm Systems Survey with Implications to the Potential for Dual-Purpose Goats on Small Farms in Kenya." Proceedings of the Small Ruminant CRSP Workshop, pp. 12-34. Kenya, March 1982a.

Sands, M.W., H.A. Fitzhugh, R.E. McDowell and S. Chema. "Mixed Crop-Animal Systems on Small Farms in Western Kenya." Morrilton, Arkansas: Winrock International-Small. Ruminant Collaborative Research Support Program Technical Report Series No.17, December 1982b.

Semenye, P. "The Estimation of Live Weight of Zebu Cattle from Heart Girth Measurements." Working Document, International Livestock Centre for Africa, Nairobi, 1979.

Appendixes

Appendices

Appendix 1 Population and associated data
for Subsaharan Africa, 1970-2000

Region and country	1963	Total population 1970	1980	2000	Hypothetical site of stationary population	Population growth rate 1970-1979
	---------Millions---------				---------Percent---------	
Low income, semi-arid						
Chad	3.2	3.6	4.4	7	19	2.0
Gambia	0.4	0.4	0.6	1	3	3.0
Mali	4.3	5.4	6.9	12	35	2.6
Mauritania	1.1	1.2	1.6	3	9	2.7
Niger	3.3	4.0	5.3	10	29	2.8
Somalia	2.4	2.8	4.6	6	17	2.3
Upper Volta	4.2	5.4	6.9	10	28	1.6
Subtotal	18.9	22.8	30.3	49	140	2.9
Low income, other						
Benin	2.2	2.6	3.5	6	19	2.9
Burundi	3.0	3.5	4.2	7	17	2.0
Centr. Afr. Rep.	1.4	1.9	2.3	3	9	2.2
Ethiopia	20.8	25.5	31.5	53	162	2.1
Guinea	· 3.3	3.9	5.0	9	23	2.9
Guinea-Bissau	0.4	0.5	0.6	1	3	1.6
Lesotho	0.9	1.1	1.3	2	5	2.3
Madagascar	5.7	6.8	8.7	15	45	2.5
Malawi	3.7	4.5	6.2	11	36	2.8
Mozambique	7.0	8.1	10.5	20	51	2.5
Rwanda	3.0	3.6	4.8	9	29	2.8
Sierra Leone	2.1	2.7	3.5	6	17	2.5
Sudan	10.6	14.1	18.4	31	86	2.6
Tanzania	10.9	13.3	18.0	35	97	3.4
Togo	1.6	2.0	2.6	4	13	2.4
Uganda	8.1	9.8	13.2	24	67	3.0
Zaire	17.7	21.6	28.3	49	139	2.7
Subtotal	102.4	125.5	162.6	285	818	2.7
Middle income, oil exporters						
Bostwana	0.5	0.6	0.8	2	6	2.2
Cameroon	5.0	6.8	8.4	14	37	2.2
Ghana	7.3	8.6	11.7	21	52	3.0
Ivory Coast	4.9	5.3	8.0	15	45	5.5
Kenya	9.0	11.3	16.5	34	109	3.4
Liberia	1.2	1.4	2.0	4	11	3.3
Mauritius	0.7	0.8	1.0	1	2	1.4
Senegal	3.7	4.3	5.7	10	30	2.6
Swaziland	0.3	0.4	0.6	1	3	2.6
Zambia	3.5	4.2	5.8	11	31	3.0
Zimbabwe	4.1	5.3	7.4	15	42	3.3
Subtotal	40.2	49.0	67.9	128	368	3.4
Oil exporters						
Angola	4.8	5.6	7.1	12	35	2.3
Congo	1.0	1.2	1.5	3	7	2.5
Gabon	0.4	0.5	0.6	1	2	1.2
Nigeria	45.9	56.3	77.1	161	459	2.5
Subtotal	52.1	63.6	86.3	177	503	3.1
Total 39 countries	213.6	260.9	347.1	639	1,829	2.9
Others						
Cape Verde	0.2	0.3	0.3	NA	NA	2.0
Comoros	0.2	0.3	0.4	NA	NA	2.8
Djibouti	0.2	0.2	0.3	NA	NA	6.9
Equatorial Guinea	0.2	0.3	0.4	NA	NA	2.3
Sao Tome and Principe	NA	0.1	0.1	NA	NA	1.5
Seychelles	NA	NA	NA	NA	NA	NA

Appendix 2 Land associated data and
GNP for Subsaharan Africa, 1980

Region and country	Land area	Land area as a proportion of		Population density	Population in agriculture	GNP per capita
		Arable and permanent cropland	permanent pasture[a]			
	-1,000 HA-	-------Percent-------		--km² ---	----Pct----	--$US--
Low income, semi-arid						
Chad	125,920	3	36	3	84	110
Gambia	1,000	27	16	52	78	250
Mali	122,000	2	25	5	87	140
Mauritania	103,040	--	38	2	83	320
Niger	126,670	3	8	4	88	270
Somalia	62,734	2	46	6	80	—
Upper Volta	27,380	9	37	25	81	180
Subtotal	568,744	7	29	5	—	—
Low income, other						
Benin	11,062	16	4	30	46	250
Burundi	2,565	51	35	157	83	180
Centr. Afr. Rep.	62,298	3	5	3	87	290
Ethiopia	110,100	13	41	26	79	130
Guinea	24,586	6	12	20	80	280
Guinea-Bissau	2,800	10	46	16	82	170
Lesotho	3,035	10	66	43	84	340
Madagascar	58,154	5	58	14	83	290
Malawi	9,408	25	20	50	84	200
Mozambique	78,409	4	56	13	64	250
Rwanda	2,495	39	19	176	90	200
Sierra Leone	7,162	25	31	47	65	250
Sudan	237,600	9	24	7	77	370
Tanzania	88,604	6	40	18	81	260
Togo	5,439	26	4	46	68	250
Uganda	19,971	28	25	54	81	290
Zaire	226,760	3	4	12	74	260
Subtotal	950,448	8	26	17	--	--
Middle income, oil importers						
Botswana	58,537	2	75	1	80	720
Cameroon	46,944	15	18	17	81	560
Ghana	23,002	12	15	47	51	400
Ivory Coast	31,800	12	9	20	79	1,040
Kenya	56,925	4	7	27	78	380
Liberia	9,632	4	2	16	69	500
Mauritius	185	58	4	525	28	1,030
Senegal	19,200	27	30	28	74	430
Swaziland	1,720	12	73	31	73	650
Zambia	74,072	7	47	7	67	500
Zimbabwe	38,667	7	13	18	59	470
Subtotal	360,684	9	30	19	—	—
Oil exporters						
Angola	124,670	3	23	6	58	440
Congo	34,150	2	29	4	34	630
Gabon	25,767	2	18	2	76	3,280
Nigeria	91,077	33	23	81	53	670
Subtotal	275,664	13	23	31	—	—
Total 39 countries	2,155,540	8	27	16	--	--

Source: FAO Production Yearbook, 1981.

[a]Refers to land used permanently (5 years or more) for herbacious forage crops, either
cultivated or growing wild (wild prarie or grazing land).

Appendix 3 Inventory, indigenous production and production per head of inventory, cattle, sheep and goats, Subsaharan Africa, 1961-65

Region and country	Inventory		Indigenous production		Indigenous production per head of inventory	
	Cattle	Sheep and goats	Beef and veal	Sheep and goats	Cattle	Sheep and goats
	----1,000 Hd----		----1,000 MT----		----------Kg----------	
Low income, semi-arid						
Chad	4,250	4,000	47	11	11.1	2.8
Gambia	175	166	2	—	11.4	—
Mali	4,040	9,072	38	28	9.4	3.1
Mauritania	2,244	5,888	24	13	10.7	2.2
Niger	3,700	7,166	41	24	11.1	3.3
Somalia	2,410	8,276	25	28	10.4	3.4
Upper Volta	1,956	3,060	21	11	10.7	3.6
Subtotal	18,775	37,628	198	115	10.6	3.1
Low income, other						
Benin	376	812	4	2	10.6	2.5
Burundi	467	551	8	2	17.1	3.6
Centr. Afr. Rep.	420	508	7	1	16.7	2.0
Ethiopia	25,159	22,254	222	86	8.8	3.9
Guinea	1,581	805	16	2	10.1	2.5
Guinea-Bissau	230	203	4	0	17.4	0
Lesotho	365	2,142	5	3	13.7	1.4
Madagascar	9,282	575	108	2	11.6	3.5
Malawi	399	558	5	2	12.5	3.6
Mozambique	1,775	874	27	2	15.2	2.3
Rwanda	492	1,123	7	4	14.2	3.6
Sierra Leone	148	158	1	—	6.8	—
Sudan	8,052	14,834	99	56	12.3	3.8
Tanzania	8,745	7,293	85	30	9.7	4.1
Togo	156	968	2	2	12.8	2.1
Uganda	3,485	3,109	54	11	15.5	3.5
Zaire	913	2,225	13	7	14.2	3.1
Subtotal	62,045	58,992	667	212	10.8	3.6
Middle income, oil importers						
Botswana	1,514	451	31	1	20.5	2.2
Cameroon	1,738	2,716	24	7	13.8	2.6
Ghana	486	554	7	1	14.4	1.8
Ivory Coast	301	1,138	5	3	16.6	2.6
Kenya	7,253	9,096	101	26	13.9	2.9
Liberia	21	230	—	1	—	4.3
Mauritius	44	61	1	—	22.7	—
Senegal	1,760	2,148	21	6	11.9	2.8
Swaziland	533	267	19	3	35.6	11.2
Zambia	1,279	192	20	1	15.6	5.2
Zimbabwe	3,586	881	87	3	24.3	3.4
Subtotal	18,515	17,734	316	52	17.1	2.9
Oil exporters						
Angola	1,541	666	21	2	13.6	3.0
Congo	26	91	--	—	--	—
Gabon	3	96	--	—	—	—
Nigeria	10,860	28,348	131	90	12.1	3.2
Subtotal	12,430	29,201	152	92	12.2	3.2
Total 39 Countries	112,065	143,555	1,333	471	11.9	3.3

Source: Derived from FAO, Production Yearbook, 1973.

Appendix 4 Cattle inventory and indigenous beef and veal
production in Subsaharan Africa, 1970

Region and country	Inventory	Slaughter	Inventory per person	Indigenous beef and veal production		
				Total	Per head of inventory	Per person
	————1,000 Hd————		——Hd——	1,000MT	————Kg————	
Low income, semi arid						
Chad	4,500	245	1.3	54	12.0	15.0
Gambia	249	27	0.6	3	12.0	7.5
Mali	5,350	405	1.0	50	9.3	9.3
Mauritania	2,660	130	2.2	27	10.2	22.5
Niger	4,300	235	1.1	49	11.4	12.3
Somalia	2,800	125	1.0	20	7.1	7.1
Upper Volta	2,800	200	0.5	30	10.7	5.6
Subtotal	22,659	1,367	1.0	233	10.3	10.2
Low income, other						
Benin	570		0.2	7	12.3	2.7
Burundi	685	69	0.2	10	14.6	2.9
Centr. Afr. Rep.	470	77	0.2	10	21.3	5.3
Ethiopia	26,232	2,322	1.0	249	9.5	9.8
Guinea	1,800	134	0.5	14	7.8	3.6
Guinea-Bissau	260	26	0.5	1	3.8	2.0
Lesotho	400	55	0.4	13	32.5	11.8
Madagascar	9,881	879	1.5	115	11.6	16.9
Malawi	491	56	0.1	8	16.3	1.8
Mozambique	2,050	205	0.3	12[a]	5.9[a]	1.5[a]
Rwanda	710	73	0.2	9	12.7	2.5
Sierra Leone	230	48	0.1	2	8.7	0.7
Sudan	13,500	1,020	1.0	139	10.3	9.9
Tanzania	13,206	1,208	1.0	127	9.6	9.5
Togo	180	30	0.1	4	22.2	2.0
Uganda	4,145	509	0.4	55	13.3	5.6
Zaire	900	124	—	15	16.7	0.7
Subtotal	75,710	6,835	0.6	790	10.4	6.3
Middle income, oil importers						
Botswana	1,481	150	2.5	23	15.5	38.3
Cameroon	2,100	167	0.3	30	14.3	4.4
Ghana	606	170	0.1	9	14.9	1.0
Ivory Coast	396	272	0.1	5	12.6	0.9
Kenya	8,600	905	0.8	30[a]	3.5[a]	2.7[a]
Liberia	28	10	—	2	71.4	1.4
Maurituis	48	17	0.1	—	—	—
Senegal	2,600	315	0.6	36	13.8	8.4
Swaziland	568	53	1.4	13	22.9	32.5
Zambia	1,550	173	0.4	27	17.4	6.4
Zimbabwe	3,900	555	0.7	70[b]	17.9[b]	13.2[b]
Subtotal	21,877	2,777	0.5	245	11.2	5.0
Oil exporters						
Angola	2,514	251[a]	0.4	19[a]	7.6[a]	3.4[a]
Congo	31	6	—	—	—	—
Gabon	4	—	—	—	—	—
Nigeria	11,550	1,642	0.2	139	12.0	2.5
Subtotal	14,099	1,899	0.2	158	11.2	2.5
39 countries	134,345	12,878	0.5	1,426	10.6	5.5

Source: Derived from FAO Production Yearbook, 1971
  a) Commercial only
  b) Inspected only
  c) Not specified

Appendix 5 Cattle inventory and indigenous beef and veal
production in Subsaharan Africa, 1980

Region and country	Cattle			Indigenous beef production		
	Inventory	Slaughter	Inventory per person	Total	Per head of inventory	Per person
	——1,000 Hd——		——Hd——	-1,000MT-	————Kg————	
Low income, semi-arid						
Chad	4,000	200	0.9	56	14.0	12.7
Gambia	321	38	0.5	5	15.6	8.3
Mali	4,960	306	0.7	64	12.9	9.3
Mauritania	1,200	145	1.0	29	24.2	18.1
Niger	3,206	321	0.6	46	14.3	8.7
Somalia	3,900	410	0.9	61	15.6	13.3
Upper Volta	2,760	240	0.4	36	13.0	5.2
Subtotal	20,347	1,660	0.7	297	14.6	9.8
Low income, other						
Benin	771	102	0.2	10	13.0	2.9
Burundi	846	90	0.2	14	16.5	3.3
Centr. Afr. Rep.	1,236 .	161	0.5	17	13.8	7.4
Ethiopia	26,000	1,950	0.8	215	8.3	6.8
Guinea	1,760	176	0.3	21	11.9	4.2
Guinea-Bissau	200	27	0.3	3	15.0	5.0
Lesotho	590	72	0.5	8	13.6	6.2
Madagascar	10,000	990	1.2	126	12.6	14.5
Malawi	823	78	0.2	12	14.6	2.3
Mozambique	1,400	240	0.1	36	25.7	3.4
Rwanda	640	105	0.1	14	21.9	2.9
Sierra Leone	343	60	0.1	3	8.7	0.9
Sudan	18,354	1,261	1.0	209	11.4	11.4
Tanzania	12,578	1,240	0.7	124	9.9	6.9
Togo	230	32	0.1	4	17.4	1.5
Uganda	4,800	530	0.4	80	16.7	6.1
Zaire	1,186	152	0.0	22	18.5	0.8
Subtotal	81,757	7,266	0.5	918	11.2	5.7
Middle income, oil importers						
Botswana	2,911	141	3.6	28	9.6	35.0
Cameroon	3,200	330	0.4	48	15.0	5.7
Ghana	950	114	0.1	13	13.7	1.1
Ivory Coast	700	340	0.1	12	17.1	1.5
Kenya	11,000	1,540	0.7	193	17.5	11.7
Liberia	39	30	0.0	1	25.6	0.5
Maurituis	56	10	0.1	—	—	—
Senegal	2,238	246	0.1	29	13.0	5.1
Swaziland	665	80	0.1	15	22.6	25.0
Zambia	2,151	172	0.4	22	12.6	4.7
Zimbabwe	5,279	710	0.7	107	20.3	14.5
Subtotal	29,189	3,713	0.4	473	16.2	7.0
Oil exporters						
Angola	3,150	310	0.4	49	15.6	6.9
Congo	74	14	0.1	1	13.5	0.7
Gabon	3	2	0.0	—	—	—
Nigeria	12,300	1,960	0.2	205	16.7	2.7
Subtotal	15,527	2,286	0.2	255	16.4	3.0
Total 39 countries	146,820	14,925	0.4	1,943	13.2	5.6

Source: Derived from FAO, Production Yearbook, 1981.

368

Appendix 6   Inventory and indigenous production of sheep
and goats, Subsaharan Africa, 1970

Region and country	Inventory				Indigenous production mutton and goat meat	
	Sheep	Goats	Total	Per person	Total	Per person
	------1,000 Hd------				1,000MT	----Kg----
Low income, semi-arid						
Chad	1,800	2,300	4,100	1.1	12	3.3
Gambia	80	91	171	0.4	1	2.5
Mali	5,750	5,500	11,250	2.1	37	6.9
Mauritania	3,800	3,050	6,850	5.7	15	12.5
Niger	2,700	6,000	8,700	2.2	28	7.0
Somalia	3,900	4,900	8,800	3.1	31	11.1
Upper Volta	1,500	2,500	4,000	0.7	13	2.4
Subtotal	19,530	24,341	43,871	1.9	177	7.8
Low income, other						
Benin	580	610	1,190	0.5	3	1.2
Burundi	227	472	699	0.2	3	0.9
Centr. Afr. Rep.	64	520	584	0.3	1	0.5
Ethiopia	12,679	11,263	23,942	0.9	53	2.1
Guinea	480	495	975	0.2	3	0.8
Guinea-Bissau	64	174	238	0.5	—	—
Lesotho	1,655	897	2,552	2.3	5	4.6
Madagascar	492	876	1,368	0.2	4	0.6
Malawi	118	636	754	0.2	3	0.7
Mozambique	210	880	1,090	0.1	3	0.4
Rwanda	228	600	828	0.2	3	0.8
Sierra Leone	55	156	211	0.1	1	0.4
Sudan	13,500	10,050	23,550	1.7	96	6.8
Tanzania	2,823	4,456	7,279	0.6	28	2.1
Togo	619	547	1,166	0.6	2	1.0
Uganda	855	1,911	2,766	0.3	9	0.9
Zaire	646	1,804	2,450	0.1	6	0.3
Subtotal	35,295	36,347	71,642	0.6	223	1.8
Middle income, oil importers						
Botswana	400	1,082	1,482	2.5	4	6.7
Cameroon	1,500	2,050	3,550	0.5	9	1.3
Ghana	1,331	1,412	2,743	0.3	8	0.9
Ivory Coast	833	833	1,666	0.3	4	0.8
Kenya	3,700	4,000	7,700	0.7	22	2.0
Liberia	150	145	295	0.2	1	0.7
Mauritius	3	67	70	0.1	—	—
Senegal	1,380	1,400	2,780	0.7	8	1.9
Swaziland	40	259	299	0.8	4	10.0
Zambia	28	180	208	0.1	1	0.2
Zimbabwe	450	690	1,140	0.2	4	0.3
Subtotal	9,815	12,118	21,933	0.5	65	1.3
Oil exporters						
Angola	159	770	929	0.2	2	0.4
Congo	34	48	82	0.1	—	—
Gabon	53	60	113	0.2	—	—
Nigeria	8,100	23,400	31,500	0.6	100	1.8
Subtotal	8,346	24,278	32,624	0.5	102	1.6
Total 39 Countries	72,986	97,084	170,070	0.7	567	2.2

Source: Derived from FAO, Production Yearbook, 1973.

Appendix 7 Inventory and indigenous production of sheep and goats, and lamb and mutton, Subsaharan Africa, 1980

Region and country	Inventory				Indigenous production, lamb, mutton and goat meat	
	Sheep	Goats	Total	Per person	Total	Per person
	——————1,000 Hd——————				1,000MT	——Kg——
Low income, semi-arid						
Chad	2,400	2,300	4,700	1.07	21	4.8
Gambia	158	170	328	0.55	1	1.7
Mali	6,250	6,750	13,000	1.88	49	7.1
Mauritania	5,200	2,600	7,800	4.88	19	11.9
Niger	2,800	7,000	9,800	1.85	42	7.9
Somalia	10,100	16,300	26,400	5.74	85	18.5
Upper Volta	1,855	2,782	4,637	0.67	10	1.4
Subtotal	28,763	37,902	66,665	2.20	227	7.9
Low income, other						
Benin	964	926	1,890	0.54	5	1.4
Burundi	316	657	973	0.23	3	0.7
Centr. Afr. Rep.	85	925	1,010	0.44	3	1.3
Ethiopia	23,250	17,180	40,430	1.28	132	4.2
Guinea	437	405	842	0.17	2	0.4
Guinea-Bissau	50	120	170	0.28	--	—
Lesotho	1,168	767	1,935	1.49	5	3.8
Madagascar	600	1,300	1,900	0.22	5	0.6
Malawi	75	630	705	0.11	3	0.5
Mozambique	106	335	441	0.04	2	0.2
Rwanda	290	880	1,170	0.24	4	0.8
Sierra Leone	260	144	404	0.12	1	0.3
Sudan	17,800	12,570	30,370	1.65	140	7.6
Tanzania	3,775	5,662	9,437	0.52	26	1.4
Togo	810	720	1,530	0.59	3	1.2
Uganda	1,072	2,155	3,227	0.24	14	1.1
Zaire	734	2,732	3,466	0.12	9	0.3
Subtotal	51,792	48,108	99,900	0.61	357	2.2
Middle income, oil importers						
Botswana	149	637	786	0.98	3	3.8
Cameroon	2,160	2,340	4,500	0.54	15	1.8
Ghana	1,700	2,100	3,800	0.32	11	0.9
Ivory Coast	1,200	1,250	2,450	0.31	7	0.9
Kenya	4,300	4,530	8,830	0.54	38	2.3
Liberia	200	200	400	0.20	1	0.5
Mauritius	4	70	74	0.07	--	—
Senegal	2,070	1,100	3,170	0.56	8	1.4
Swaziland	45	262	307	0.51	3	5.0
Zambia	49	310	359	0.06	1-	0.2
Zimbabwe	387	982	1,369	0.19	5	0.7
Subtotal	12,264	13,781	26,045	0.38	92	1.4
Oil Exporters						
Angola	225	935	1,160	0.16	3	0.4
Congo	68	128	196	0.13	--	—
Gabon	100	90	190	0.32	1	1.7
Nigeria	11,700	24,500	36,200	0.47	163	2.1
Subtotal	12,093	25,653	37,746	0.44	167	1.9
Total 39 countries	104,912	125,444	230,356	0.66	843	2.4

Source: Derived from FAO, Production Yearbook, 1981.

Appendix 8 Cattle, sheep and goat inventory, animal unit
basis, Subsaharan Africa, 1980

Region and country	Total Animal Units (AU) [a/]			Cattle area proportion of total	Total AU per person
	Cattle	goats	Total		
	————1,000 AU————			-percent-	-AU-
Low income, semi-arid					
Chad	4,000	940	4,940	81	1.12
Gambia	321	66	387	83	0.65
Mali	4,960	2,600	7,560	66	1.10
Mauritania	1,200	1,560	2,760	43	1.73
Niger	3,206	1,960	5,166	62	0.97
Somalia	3,900	5,280	9,180	42	2.00
Upper Volta	2,760	927	3,687	75	0.53
Subtotal	20,347	13,333	33,680	60	1.11
Low income, other					
Benin	771	378	1,149	67	0.33
Burundi	846	195	1,041	81	0.25
Centr. Afr. Rep.	1,236	202	1,438	86	0.63
Ethiopia	26,000	8,086	34,086	76	1.08
Guinea	1,760	168	1,928	91	0.39
Guinea-Bissau	200	34	234	85	0.39
Lesotho	590	387	977	60	0.75
Madagascar	10,000	380	10,380	96	1.19
Malawi	823	141	964	85	0.16
Mozambique	1,400	88	1,488	94	0.14
Rwanda	640	234	874	73	0.18
Sierra Leone	343	81	424	81	0.12
Sudan	18,354	6,074	24,428	75	1.33
Tanzania	12,578	1,887	14,465	87	0.80
Togo	230	306	536	43	0.21
Uganda	4,800	645	5,445	88	0.41
Zaire	1,186	693	1,879	63	0.07
Subtotal	81,757	19,979	101,736	79	0.63
Middle income, oil importers					
Botswana	2,911	157	3,068	95	3.84
Cameroon	3,200	900	4,100	78	0.49
Ghana	950	760	1,710	56	0.15
Ivory Coast	700	490	1,190	59	7.15
Kenya	11,000	1,766	12,766	86	0.77
Liberia	39	80	119	33	0.06
Mauritius	56	15	71	79	0.07
Senegal	2,238	634	2,872	78	0.50
Swaziland	665	61	726	92	1.21
Zambia	2,151	72	2,223	97	0.38
Zimbabwe	5,279	274	5,553	95	0.75
Subtotal	29,189	5,209	34,398	85	0.51
Oil exporters					
Angola	3,150	232	3,382	93	0.14
Congo	74	39	113	65	0.08
Gabon	3	38	41	7	0.07
Nigeria	12,300	7,240	19,540	63	0.25
Subtotal	15,527	7,549	23,076	67	0.27
Total 39 Countries	146,820	46,070	192,890	76	0.56

Source: Converted to AU from data in Appendix's 5 and 7.

[a/] Cattle = 1 AU
Sheep and goats = 0.2 AU

Appendix 9 Trade in sheep, goats and cattle,
Subsaharan Africa, 1980

Region and Country	Goats and Sheep			Cattle		
	Imports	Exports	Net Exports	Imports	Exports	Net Exports
			Head			
Low income, semi-arid						
Chad	0	159,800	159,800	0	255,000	255,000
Gambia	0	0	0	0	0	0
Mali	0	307,000	307,000	0	225,000	225,000
Mauritania	0	610,000	610,000	0	93,000	93,000
Niger	0	200,000	200,000	0	175,000	175,000
Somalia	0	1,481,000	1,481,000	0	143,000	143,000
Upper Volta	1,000	152,000	152,000	0	57,000	57,000
Subtotal	1,000	2,909,800	2,908,800	0	948,000	948,000
Low income, other						
Benin	12,000	0	-12,000	9,486	0	-9,486
Burundi	0	6,700	6,700	0	0	0
Centr. Afr. Rep.	30,800	0	-30,800	42,000	4,100	-37,900
Ethiopia	249	29,796	29,547	200	13,647	13,447
Guinea	15,000	52,500	37,500	0	35,000	35,000
Guinea - Bissau	83	0	-83	11	0	-11
Lesotho	45,000	5,000	-40,000	50,000	28,000	-22,000
Madagascar	0	316	316	0	0	0
Malawi	0	0	0	0	0	0
Mozambique	0	0	0	0	0	0
Rwanda	0	0	0	0	0	0
Sierra Leone	46,000	0	-46,000	26,000	0	-26,000
Sudan	0	422,271	422,271	1,953	8,659	6,706
Tanzania	0	0	0	100	1,000	900
Togo	8,000	0	-8,000	3,000	0	-3,000
Uganda	0	0	0	0	0	0
Zaire	6,000	0	-6,000	90	0	-90
Subtotal	163,132	516,583	353,451	132,840	90,406	-42,434
Middle Income, oil importers						
Botswana	200	0	-200	2,000	0	-2,000
Cameroon	1,600	1,700	100	11,535	1,950	-9,585
Ghana	11,000	0	-11,000	1,400	0	-1,400
Ivory Coast	580,000	0	-580,000	224,991	0	-224,991
Kenya	3,300	11,550	8,250	42	1,847	1,805
Liberia	8,000	0	-8,000	22,000	0	-22,000
Mauritius	1,200	0	-1,200	7,207	0	-7,207
Senegal	200,000	10,000	-190,000	25,000	140	-24,860
Swaziland	19,000	20,000	1,000	9,000	560	-8,440
Zambia	213	0	-213	279	0	-279
Zimbabwe	0	169	169	0	1,807	1,807
Subtotal	824,513	43,419	-781,094	303,454	6,304	-297,150
Oil Exporters						
Angola	0	0	0	8,000	0	-8,000
Congo	0	0	0	7,000	0	-7,000
Gabon	0	0	0	2,000	0	-2,000
Nigeria	285,000	0	-285,000	356,000	0	-356,000
Subtotal	285,000	0	-285,000	373,000	0	-373,000
Total 39 Countries	1,273,645	3,469,802	2,196,157	809,294	1,044,710	235,416

Source: Derived from FAO, Trade Yearbook, 1981.

Appendix 10 Beef and veal production, trade and estimated
consumption, Subsaharan Africa, 1980

Region and country	Production	Imports	Exports	Net or total consumption	Consumption per capita
	------------------1,000 MT------------------				----Kg----
Low income, semi-arid					
Chad	25	0	0	25.0	5.7
Gambia	5	0.1	0	5.1	8.5
Mali	37	0	0	37.0	5.4
Mauritania	17	0	0	17.0	10.6
Niger	37	3.0	0	40.0	7.5
Somalia	45	0	0	45.0	9.8
Upper Volta	29	0	0.4	28.6	4.1
Subtotal	195	3.1	0.4	197.7	6.5
Low income, other					
Benin	11	0.2	0	11.2	3.2
Burundi	12	—	0	12.0	2.9
Centr. Afr. Rep.	23	—	0	23.0	10.0
Ethiopia	214	0	1.2	212.8	6.8
Guinea	18	0	0	18.0	3.6
Guinea-Bissau	3	0	0	3.0	5.0
Lesotho	11	0	0	11.0	8.5
Madagascar	126	0	6.3	119.7	13.8
Malawi	12	0	0	12.0	2.3
Mozambique	36	2.0	0	38.0	3.6
Rwanda	11	0	0	11.0	2.3
Sierra Leone	5	3.0	0	8.0	2.3
Sudan	208	0	—	208.0	11.3
Tanzania	124	0	—	124.0	6.9
Togo	4	0.7	0	4.7	1.8
Uganda	80	0	0	80.0	6.1
Zaire	22	1.0	0	23.0	0.8
Subtotal	920	6.9	7.5	919.4	5.6
Middle income, oil importer					
Botswana	29	0.2	16.1	13.1	16.4
Cameroon	50	0.1	1.2	48.9	5.8
Ghana	13	2.8	0	15.8	1.4
Ivory Coast	47	9.0	0	56.0	7.0
Kenya	193	0	0.1	192.9	11.7
Liberia	4	0.3	--	4.3	2.2
Mauritius	2	2.1	0	4.1	4.1
Senegal	32	—	0.1	31.9	5.6
Swaziland	17	2.0	3.5	15.5	25.8
Zambia	27	—	0	27.0	4.7
Zimbabwe	107	2.2	12.1	97.1	13.1
Subtotal	521	18.7	33.1	506.6	7.5
Oil exporters					
Angola	50	2.0	0	52.0	7.3
Congo	2	3.1	0	5.1	3.4
Gabon	—	4.5	0	4.5	7.5
Nigeria	251	20.0	0	271.0	3.5
Subtotal	303	29.6	0	332.6	4.9
Total 39 Countries	1,939	58.3	41.0	1,956.3	5.6

Source: Derived from FAO Production Yearbook, 1981.

Appendix 11 Lamb, mutton and goat meat production trade
and consumption, Subsaharan Africa, 1980

Region and country	Meat production	Lamb, mutton production	Lamb and Mutton		Net production, consumption	Consumption per capita
			imports	exports		
	------------------------------1,000 MT------------------------------					----kg----
Low income, semi arid						
Chad	8	10	0	0	18.0	4.0
Gambia	1	1	0	0	2.0	3.3
Mali	21	24	0	0	45.0	6.5
Mauritania	5	7	0	0	12.0	7.5
Niger	29	11	0	0	40.0	7.6
Somalia	53	12	0	0	65.0	14.1
Upper Volta	0	3	0	0	3.0	0.4
Subtotal	117	68	0	0	185.0	6.1
Low income, other						
Benin	3	3	0	0	6.0	1.7
Burundi	2	1	0	0	3.0	0.7
Centr. Afr. Rep.	3	1	--	0	4.0	1.7
Ethiopia	55	77	0	0.2	131.8	4.2
Guinea	1	1	0	0	2.0	0.4
Guinea-Bissau	--	--	0	0	0.0	--
Lesotho	2	4	0	0	6.0	4.6
Madagascar	3	2	0	0.2	4.8	0.6
Malawi	2	--	--	0	2.0	0.3
Mozambique	1	1	0	0	2.0	0.2
Rwanda	3	1	0	0	4.0	0.8
Sierra Leone	0	1	0	0	1.0	0.3
Sudan	44	82	0	--	126.0	6.9
Tanzania	15	10	0	--	25.0	1.4
Togo	2	2	0	0	4.0	1.5
Uganda	6	5	0	0	11.0	0.8
Zaire	7	2	--	0	9.0	0.3
Subtotal	149	193	0	0.4	341.1	2.1
Middle income, oil importer						
Botswana	3	1	0.1	0	4.1	5.1
Cameroon	7	8	0	0	15.0	1.8
Ghana	6	5	0.1	0	11.1	1.0
Ivory Coast	7	6	0.4	0	13.4	1.7
Kenya	17	21	0	--	38.0	2.3
Liberia	1	1	--	0	2.0	1.0
Mauritius	0	--	1.7	0	1.7	1.7
Senegal	3	7	0	--	10.0	1.8
Swaziland	3	--	0	0	3.0	5.0
Zambia	1	--	--	0	1.0	0.2
Zimbabwe	4	1	0.1	0	5.1	0.7
Subtotal	52	50	2.4	0	104.4	1.5
Oil exporters						
Angola	3	1	0	0	4.0	0.6
Congo	0	--	--	0	0.0	--
Gabon	0	--	0.1	0	0.1	0.2
Nigeria	126	40	0.2	0	166.2	2.2
Subtotal	129	41	0.3	0	170.3	2.0
Total 39 Countries	447	352	2.7	0.4	801.3	2.3

Source: Derived from FAO, Production Yearbook, 1981.

Appendix 12

AGENCY FOR INTERNATIONAL DEVELOPMENT

AFRICA BUREAU
LIVESTOCK DEVELOPMENT ASSISTANCE STRATEGY PAPER

Contents

I. Introduction
   A. Purpose of the Strategy Paper
   B. Scope of the Strategy Paper
   C. Main Components of the Strategy

II. The Africa Livestock Subsector
   A. Importance of Livestock
   B. Livestock Production Systems in Africa

III. Constraints on Livestock Development and
        Lessons Learned
   A. Constraints
   B. Lessons Learned from Previous Donor-Assisted
        Livestock Projects in Africa

IV. Elaboration of the Strategy
   A. Integrating Livestock into the Overall
        Agricultural Sector
   B. Developing Research and Other Institutions
        and Delivery Systems
   C. Support for Market Specific Initiatives
        and Opportunities
   D. Support for Private Sector Development

V. Implementation Considerations
   A. Tactical Orientations
   B. Institutional Resources
   C. Long-Term Program Assistance

Bibliography

Africa Bureau
Office of Technical Resources
Agriculture & Rural Development Division
December 22, 1982

Preface

This paper has been in preparation since mid-1981 when
AID decided to sponsor a second workshop on African
pastoral livestock production. It includes ideas from
(1) an issues paper prepared for that conference by
Peter D. Little, in October 1981; (2) the conference
report prepared by the Institute for Development
Anthropology in March 1982; (3) "Guidelines for a Livestock Sub-Sector Strategy in Africa" prepared by the
Institute for Development Anthropology; (4) a paper on
"Livestock Program Strategies and Priorities" prepared
by Winrock International in October, 1981; (5) an
original draft of this paper and subsequent comments by
James Dickey, Sahel Regional Livestock Advisor; (6) the
Africa Bureau Senior Agricultural and Rural Development
Officers workshop held at Ibadan, Nigeria in May, 1982;
(7) revised drafts of this report by Tjaart W. Shill-
horn and Tom Zalla, Michigan State University; and (8)
comments and editing by AFR/TR/ARD staff, AFR/TR and
PPC. The paper incorporates comments from AFR/DP,
PPC/PDPR, S&T/AGR and the Deputy Assistant Administrators for Africa provided in response to requests for
clearance of an earlier draft.

AFRICA BUREAU
LIVESTOCK DEVELOPMENT ASSISTANCE STRATEGY PAPER

1. Introduction

A. Purpose of the Strategy Paper

The purpose of this Strategy Paper is to provide a framework within which AID Missions in Africa should prepare their country-specific strategies for assisting the livestock production and marketing components of their agricultural development programs. It also will serve as a guide in the selection and design of specific livestock projects and programs intended to implement these strategies.

B. Scope of the Strategy Paper

The Strategy Paper has been developed within the framework of the approved Africa Bureau Food Sector Assistance Strategy Paper, dated October 1981, and takes into consideration the complimentarity of livestock and crop production systems in Africa. The Strategy aims at increasing the efficiency of livestock production and the level of farm employment and income arising from livestock. It emphasizes approaches with a history of success and points to the need to continue to monitor and develop production systems in environments where success to date has been more difficult to achieve.

C. Main Components of the Strategy

This livestock sub-sector strategy has four principal components:

1. Assist in formulating and implementing regional, national and institutional policies and programs which integrate livestock and agricultural systems at all levels.
2. Support the building and improvement of research and other institutions and delivery systems which promote livestock development.
3. Support localized, market specific initiatives which will enhance livestock production.
4. Promote increased involvement of the local private sector in the supply of inputs and in the production, preservation or utilization of livestock and livestock products.

II. The African Livestock Subsector

A. Importance of Livestock

In many African countries livestock account for as
much as one-third of the agricultural GDP, providing
income, employment and food for a substantial part of
the population. Moreover, national accounts often
understate livestock's contribution to the gross domestic product because they ignore the role of livestock
and livestock products in crop production, and frequently, illegal livestock exports. Africa has by no
means reached its potential in livestock production.
The area possesses approximately 200 million livestock
units, generally characterized by low productivity from
the viewpoint of livestock production in developed
countries. However, slight improvements in management
systems can sometimes produce considerable pay-off.
Ecologically, Sub-Saharan Africa can be divided
into desert, savanna, highlands and humid forest zones.
Numerically, livestock production is concentrated in
the savannas while a substantial part of the human
population lives in or close to the highland and forest
zones. Animal diseases have limited the full
development of ruminant production in the humid areas
of Africa and an important part of ruminant populations
elsewhere migrate seasonally in order to avoid either
disease or malnutrition.
The majority of African households own some livestock. They are owned by individual family members and
a contribute in a substantial way to the quality of the
human diet and the household economy. Most milk and a
considerable amount of the meat production in Africa,
especially poultry and small ruminants, are consumed
directly by producing households. However, meat and
milk imports have increased 242 and 102 percent,
respectively, between 1975 and 1980, attesting to a
very high commercial demand and a relatively inelastic
supply response for livestock products in Africa. In
addition, small ruminants and poultry often provide
independent income for women, while livestock in
general continue to play an important role in the
culture of many African people. Farm animals are used
for savings, capital accumulation and as a hedge
against a wide range of future uncertainties. By
better utilizing available manure and animal power,
African farmers could considerably improve their crop
production. Thus, appropriate improvements in the
livestock sector will not only improve nutrition and
well-being of the rural poor, but may also relieve
farmers and governments from increased energy costs
related to agricultural growth.

B. Livestock Production Systems in Africa

The traditional livestock production systems in Africa are diverse and complex and have evolved over centuries. Although they may appear primitive in certain ways, they are, in many cases, the optimal systems under climates, pastures and range conditions, diseases, markets and political and social circumstances.

For purposes of this strategy, the production systems in Africa will be categorized as (1) commercial systems, (2) pastoral and semi-pastoral systems, and (3) mixed farming systems.

1. Commercial Systems

These are mainly concentrated in poultry and commercial ranching. The large poultry enterprises, generally located near urban centers, use production systems similar to large poultry establishments in Europe and the United States. Commercial ranches use modern ranching technology, including disease control, pasture and water-management, and produce for relatively certain markets. These privately or commercially owned ranches generally operate with low labor input under relatively good management.

2. Pastoral and Semi-Pastoral Systems

The majority of African cattle and camels, and many small ruminants, are held in pastoral and semi-pastoral herds. Many of the animals are owned by sedentary farmers and non-farm households who entrust them to herders in return for various forms of payment. These herders move their herds along regular patterns, often around one or more permanent or semi-permanent homesteads or temporary encampments. The migration patterns may vary from year to year according to seasonal constraints and opportunities, such as the availability of pasture and water, and the avoidance of disease. It is generally agreed that this system is an excellent adaptation to uncertain environmental conditions and probably promotes the most efficient use of forage and water in the vast marginal rainfall areas.

The animals of pastoralists are usually an integral part of their culture. They provide milk, meat, fibre, transport, fuel and other products, such as manure, and, as such, are important for the maintenance of social relationships. Livestock products are consumed and sold or exchanged with settled farmers for grain or for grazing rights.

Migratory movement of livestock has become increasingly difficult in recent years due to cropping
pressure on land, restrictions on movement across
borders, and other governmental regulations. In addition, the long drought decimated many herds, forcing
some herders to shift toward mixed agriculture. These
factors appear to be inducing a shift toward mixed
farming and a more sedentary lifestyle by some pastoral
groups. Though still relatively limited in some ways,
this trend promises to accelerate in the future. In
the meantime, there will be a need for continuing
selective support for pastoralists via measures which
encourage the economic use of rangeland.

3. Mixed Farming Systems

Settled crop/livestock farming systems are generally found in the semi-humid savanna and in highland
areas, and, to a lesser extent, in the forest zone
where cattle do not thrive. In all those areas, small
ruminants and poultry are more numerous than cattle,
with swine assuming generally greater importance as
rainfall increases. In these ecological zones, local
grazing, fodder harvest, and crop residues are sufficient to sustain a limited number of livestock under more
sedentary conditions throughout the year. The greater
labor input required to keep livestock in mixed farming, as opposed to pastoralist and semi-pastoralist
systems, favors the accumulation of the smaller animals
by smaller sized, lower resource households. For this
reason, support of small stock production in mixed
farming systems deserves special attention in AID
development assistance programs.

Mixed farmers are usually permanently settled, and
livestock production is only one part of an integrated
agricultural system that is often highly dependent on
crop production. Like their semi-nomadic counterparts,
mixed farmers keep livestock for a variety of purposes,
but with more emphasis given to commercial sales,
religious festivities, manure production and animal
traction. A system of pooled herds or entrusted care
may still exist for local grazing of ruminants or where
extended family members need to move animals to distant
pastures at critical times of the year. More attention
is given to cutting and preserving forage than in the
semi-nomadic systems. Small scale feeding schemes and
intensive production of animals and animal products for
sale are common and provide at least a part of the
family cash income. Mixed farming systems offer potential for a multitude of interventions, though more applied research is needed in non-health areas before AID
can support livestock production in the mixed farming
context on a broad scale.

III. Constraints on Livestock Development and Lessons Learned

A. Constraints

1. Policy Constraints

In many African countries policy constraints are proving to be particularly intractable impediments to livestock development. Trade policy and pricing policy for meat and foreign exchange, for example, often favor domestic urban consumers and reduce returns to live- stock owners both through lower prices and through higher marketing costs for those who run the risk of moving animals through illegal markets. Similar policies for food grains reduce the profitability of animal traction and manure, and thereby discourage the intensification of production necessary to feed a grow- ing, increasingly urbanized population. Commercial policies that favor the export of high-protein agricul- tural by-products raise feed supplement costs and discourage fattening schemes, animal traction and intensive milk production.

Another important national policy area for the animal-based systems of Africa's semi-arid areas is land tenure. Most African governments do not have clear land tenure policies for the pastoral areas in general, and there are often gross discrepancies between formal (i.e., governmental) and locally recognized and accept- ed land tenure systems. For the livestock producer, this creates uncertainties regarding access to water and dry season grazing. Often tenure relationships encourage overstocking which, in the long run, negates much of the gain from otherwise sound interventions. Other government policies restrict the availability of vaccines, medicines, feedstuffs and equipment and impede the efficient functioning of markets.

2. Institutional Constraints

Livestock development in African countries is severely constrained by the lack of an adequate insti- tutional capacity to conduct the kind of research needed to provide improved livestock production techno- logies and to guide livestock development programs. This includes the capacity for undertaking pre-design research, monitoring project implementation and evalu- ating project impact. As a result, learning from experience and past mistakes is very slow. Periodic turnover of both national and expatriate personnel aggravates this problem by inhibiting the development of an institutional memory. Livestock sub-sector activities need an "institutional home" in order to maintain continuity in addressing livestock issues and to develop the necessary political base for forcing policy reforms.

The lack of indigenous technical and managerial resources and the poor use of available public sector manpower further constrains the livestock sub-sector. The common practice of placing scarce, highly trained scientists in administrative positions befitting their level of education rather than in technical positions which require their skills is due, in part, to an institutional incentive structure that favors administration and discourages application, teaching and research. Excessive formal training in relation to the needs of the job often inhibits the supply of lower level technical personnel such as artificial inseminators and vaccinators. For such manpower, a shorter period of practical training would often be a more effective and less costly way of meeting required skill levels.

3. Technical Constraints

African livestock production is fundamentally constrained by poor animal nutrition related to the scarcity and quality of dry season feed supplies and, at least partly related to this, poor animal health. In some areas, lack of stock watering facilities and poor animal housing also limit production. These factors contribute to high calf mortality, prolonged physical and sexual maturation cycles and long calving intervals. It is not likely that the productivity of the livestock sector can be increased without addressing these fundamental problems.

4. Social Constraints

Inadequate understanding of the multiple economic and social objectives of present livestock producers, and their interrelationships, has especially plagued projects aimed at pastoralists. Risk management, dry season milk and grain supplies, the demands of family members for available cash, the need for certain portions of animals which are slaughtered as opposed to those which die, reciprocal social obligations, the need for cash for school fees and unexpected events such as funerals, all can influence producers' attitudes toward livestock management, production and marketing as much as price and more direct economic variables.

B. Lessons Learned from Previous Donor-Assisted
   Livestock Projects In Africa

USAID, the IBRD and other donors have financed a
large number of livestock projects in Africa over the
past two decades. This experience yields several
lessons that will prove helpful for designing and
implementing new projects over the next decade.

1. Overcoming technical constraints relating to
dry season feed and water supplies, animal health and
marketing infrastructure is proving more difficult than
anticipated. Range management systems as practiced in
the western United States have yet to be proven economically, socially and politically feasible in Africa, or
necessarily ecologically relevant.
2. Project design has paid insufficient attention
to the ecological, socio-economic and technical milieu
in which projects are implemented.

   a. There is need for more pre-design research;
      more experimentation in project design and
      implementation; and more flexible implementa-
      tion agendas that can easily incorporate
      lessons learned as projects unfold.
   b. Careful monitoring of interventions should be
      an integral component of project design and
      implementation.
   c. Project design should relate holistically to
      the diverse elements of livestock/cropping/
      household systems on the one hand and deal
      with discrete activities which AID has shown
      it can sucessfully assist on the other.

3. Not enough attention has been given to building up coalitions of interests to support projects or
to obtaining qualified host-country staff and establishing effective project implementation management
systems.
4. The active participation of pastoralists,
livestock farmers and lower level livestock agents can
provide information needed for good project design and
monitoring, and the means to reduce recurrent costs and
improve project implementation.
5. Traditional livestock trade, including movement over long distances from producing areas to market
centers, appears to be fairly efficient and adaptable,
under normal circumstances. For this and other reasons
that are not always clear, donor projects aimed at
increasing off-take rates have had limited success.
Off-take rates continue low in spite of heavy investments in marketing infrastructure. Indeed, marketing

investments which centralize control over marketing
activities many actually reduce off-take.

6. Successful proven technical interventions in
pastoral Africa in the near future will probably remain
in the veterinary field. In some cases, closely
monitored pilot projects which integrate water development with pasture management merit consideration. The
range of promising interventions in mixed farming
situations is wider, but confirmatory applied research
is necessary before they are implemented on a large
scale. The technology for successful artificial
insemination and disease control aimed at small farmers
is known, but effective delivery systems are still the
exception rather than the rule.

7. Improvements in the livestock sub-sector will
be very difficult to realize in the absence of coherent
supportive host country policies concerning access to
water, land tenure, external trade, pricing, marketing,
taxation and institution development, or without the
active participation of livestock producers in project
design and implementation.

8. Due to AID's own lack of knowledge as well as
the nature of the biological cycle of livestock under
African conditions, successful livestock projects and
programs require a long-term commitment. American
institutions which are involved need to work in close
collaboration with host country institutions on a long-
term basis.

IV. Elaboration of the Strategy

A. Integrating Livestock into the Overall
Agricultural Sector

1. Improving the National Policy and Planning
Environment

Although AID's ability to influence host government policy is limited, in some cases policy changes to
the livestock sub-sector may be the only means of
improving herder income and livestock production or
bringing about more effective integration of livestock
and cropping activities. This is especially true where
livestock and arable agriculture are served by different ministries.

AID assistance to policy and planning could take
the form of assisting with frequent short-term assessments of the livestock/agriculture sector to formulate
policy alternatives for discussions with host governments. If governments prove unwilling to undertake
needed policy reforms, such actions can be made conditions precedent to continued project development or
disbursement of program funds. In some cases, the

formation of an <u>ad</u> <u>hoc</u> Livestock Secretariat, including technical, policy analysis and plannning entities of those ministries concerned with livestock/cropping interactions, might be appropriate. AID could provide a senior level advisor to the unit while training host country analysts. This unit should have the capability to obtain and analyze data. The ultimate goal would be to institutionalize a host country capability to plan, formulate and evaluate its livestock development policies and programs.

2. Formulating and Implementing Projects and Programs

Social scientists and agricultural technicians are only beginning to understand and appreciate linkages between livestock and cropping activities both within the rural household and between specialized crop and livestock producers. Before those linkages can be fully exploited we need a better understanding of them. This can be obtained both by focusing on such interactions during project preparation, design, implementation and evaluation as well as through well defined farm level diagnostic surveys conducted by animal and crop scientists, agricultural economists, and anthropologists working together in a farming systems context. Special attention will have to be given to the role of women in exploring these interactions since women often have substantial economic interests in household/livestock/crop interactions.

B. Developing Research and Other Institutions and Delivery Systems

1. Institution Building

Decades of donor-supported livestock projects have generated little data that can be used in planning future livestock programs. AID should increase the capacity of both public and private African institutions to conduct their own livestock-related research, strengthen researcher-producer links, and develop data storage and exchange systems with neighboring nations. AID should also ensure that research in animal health and husbandry be integrated in specific farming systems research and development projects across the continent.

The similarity of livestock health and production problems across wide areas of Africa, and the overall scarcity of resources and manpower available to address them points to the need to approach such problems through regional networks of scientists working on common problems. This should be done in the context of the Africa Bureau's Agricultural Research Strategy

Paper and Cooperative Development in Africa (CDA), efforts to strengthen African research capabilities through broad based support to selected countries within each of five agro-climatic zones. The international research centers, U.S. Title XII universities, USDA and other U.S. institutions can assist the development of this capacity. Such institutions can play an important role in training the human resources needed for staffing national institutions. AID could also support host country institutions by contracting directly with them for studies which provide location specific information needed to make livestock production successful.

2. Improving Delivery Systems

There is an urgent need to analyze and improve animal health delivery systems, especially for small stock in mixed farming systems. Originally, the colonial governments imposed a livestock tax which in part paid for public sector extension efforts to control epidemic diseases. The taxes, however, were unpopular and occasionally jeopardized other efforts at improvement. Some countries (Nigeria for example) have since abolished the cattle tax but are still providing officially free veterinary and range management services to livestock owners.

Ultimately, livestock owners themselves should be paying for these services. There is considerable evidence that producers would agree to making at least partial payment for veterinary services, especially if reliable supplies are established. Producers, through their purchasing power, would then have an opportunity to participate in the decision making process concerning the quality and the direction of the services provided them while the role of the private sector in this area could expand. Village veterinary volunteers and private herder associations are examples of promising innovations in this area. The former offer special promise for addressing the health problems of livestock on small mixed farms.

Animal husbandry, artificial insemination, and crop/forage extension are other areas where delivery innovations are badly needed. Closer links between research, training, extension and the farmers need to be forged. Supporting projects which give cooperatives or other local institutions the responsibility for paying part of an extension agent's salary or salary supplement should provide accountability for performance, enhance local institutions and provide participation in managing development. Appropriately designed water management activities can provide similar benefits in addition to controlling grazing.

C. Support for Market Specific Initiatives and Opportunities

Many types of livestock development activities make sense only in localized areas where special conditions or special markets, either local or export, prevail. These include such things as cattle and small ruminant fattening targeted at specific markets; intensive milk production in highland areas or near urban markets; or other local opportunities. In addition support could be given to new "high" technologies, such as genetic engineering in vaccine production or other interventions which potentially have far reaching effects but which may require some applied research, or need testing and monitoring on a pilot scale, before they can be extended with confidence. This component of the strategy endorses AID support of applied research and pilot projects which, through careful monitoring, will provide information which can be used to design better, perhaps larger projects, which will strengthen the overall livestock sub-sector.

D. Support for Private Sector Development

The fourth component encourages the participation of the local private sector in livestock development. The encouragement of these activities may be in the form of policy reforms or financial, socio-economic or technical assistance. It should stimulate or improve production, processing or marketing of livestock and livestock products, either locally or for export, by the private sector. It could also include assistance to groups of private entrepreneurs or groups of famers for such things as producing vaccines and distributing inputs.

African livestock sub-sector programs should support producers' participation and eventual control wherever possible. Local producer organizations can often manage project infrastructure such as water points or veterinary posts, and may also be able to manage cost recovery schemes. Such interventions will require training of local producers in management, marketing and para-veterinary care.

V. Implementation Considerations

A. Tactical Orientations

Several types of interventions in the livestock subsector are consistent with this strategy. In addition to traditional interventions in the animal health area, these might include, but not be limited to, some of the following activities and projects where environmental and social factors are favorable. In the arid and semiarid zones, AID can begin very carefully to experiment with pilot projects involving seasonal livestock water supplies designed to optimize the stocking rate and reduce overgrazing near permanent water points, increase grazing in overgrown fire risk areas, improve animal nutrition and reduce animal stress. Water points might be established on trekking routes to facilitate normal marketing in the dry season, with maximum possible beneficiary participation and control. Credit, livestock inputs, commodities to enhance food security, animal health and monitoring services and training may also be selectively supplied through traditionally organized herder associations where these are demonstrated to be effective.

In the higher rainfall agricultural zones, AID can promote increased animal production in mixed farming projects via intensified use of natural forage, crop residue and agricultural and other by-products. Small farmer animal health delivery systems also merit attention. However, the agency's limited experience with such activities to date suggests a cautious approach to them initially, emphasizing action research and development rather than full scale implementation until benefits are confirmed. Intercropping and rotational production of legumes for their residual feed value as well as for cash income, and livestock feeding programs aimed at special markets and/or occasions should be a particular focus of this kind of activity. AID could sponsor on-farm monitored experiments in forage/silage production, storage and use to identify the most cost-effective practices at the farm level. In addition, research to identify other cost effective ways of intensifying forage production should be supported.

Aid should also support trypanosomiasis and tick-borne disease research in the higher rainfall areas with emphasis on determining the relative efficiency of trypanotolerant breeds, chemotherapy maintenance and vector eradication-control programs. Technical interventions aimed at vector-borne disease control should be linked to planning for the use of land areas opened up by such programs.

In other areas, AID should support pre-design studies of the ecological, socio-economic and technical context of potential projects. More general market and herd composition and/or management studies should also be supported where these promise to resolve perplexing marketing questions or generate data that will help derive full benefits from livestock sub-sector assistance. With respect to infrastructure, cold storage for medicines and vaccines, dips and on-farm feed storage may be appropriate in selected circumstances.

B. Institutional Resources

The financial and human resources to implement this livestock development assistance strategy must come from private and public U.S. institutions working in close collaboration with host country and African regional organizations such as the Inter African Bureau of Animal Resources (IBAR) under the OAU. U.S. resources will come from Title XII and other agricultural universities; USDA; other federal agencies; International Agricultural Research Centers; private voluntary organizations; direct hire AID staff; agribusiness firms; foundations and other "not-for-profit" entities; national associations; and consulting firms. As much as possible, host country resources should come from existing organizations. Where these are weak, first consideration should be given to strengthening them or forming ad hoc working relationships with other existing services rather than creating new institutions. Existing African regional institutions should be strengthened appropriately.

C. Long-Term Program Assistance

AID is aware that livestock development is a relatively slow process and encourages long-term program commitment without a rapid turnover of personnel and resources. Cooperative linkages between established U.S. institutions and existing or emerging host country institutions should be encouraged as a means of supporting the formulation and implementation of long term livestock development strategies which are integrated with crop production and other rural household activities.

BIBLIOGRAPHY

Eicher, Carl K. and Doyle C. Baker, 1982. "Research on Agricultural Development in Sub-Saharan Africa: A Critical Survey." MSU International Development Paper No. 1. Department of Agricultural Economics, Michigan State University, East Lansing, Contract No. AID/AFR-G-1261.

Ferguson, D.S., Circa 1977. " A conceptual Framework for the Evaluation of Livestock Production Development Projects and Programs in Sub-Saharan West Africa." Center for Research on Economic Development, University of Michigan, Ann Arbor.

Hoben, Allen, 1979. "Lessons from a Critical Examination of Livestock Projects in Africa." AID Program Evaluation Working Paper No. 26. USAID, Washington, D.C.

Horowitz, Michael M., 1979. "The Sociology of Pas-
    toralism and African Livestock Projects." AID
    Program Evaluation Discussion Paper No. 6. USAID,
    Washington, D.C.
Institute for Development Anthropology, 1980. "The
    Workshop on Pastoralism and African Livestock
    Development." USAID Program Evaluation Report No.
    4. USAID, Washington, D.C. Contract No. AID/OTR-
    G-1741.
International Livestock Centre for Africa, 1980. Pas-
    toral Development Project: Report on a Workshop
    on the Design and Implementation of Pastoral
    Development Projects for Tropical Africa." ILCA
    Bulletin No. 8, Nairobi.
Little, Peter D., 1981. "Issues Paper: For the Work-
    shop on African Pastoral Production Systems."
    Institute for Development Anthropology, Bingham-
    ton, New York. Contract No. AID/AFR-0085-C-
    00-1033.
_____, 1982. "The Workshop on Development and
    African Pastoral Livestock Production." Institute
    for Development Anthropology, Binghamton, New
    York. Contract No. AFR-0085-C-00-1033.
Little, Peter D. and Michael M. Horowitz, 1982.
    "Guidelines for a Livestock Sub-Sector Strategy in
    Africa." Institute for Development Anthropology,
    Binghamton, New York. Contract No. AFR-0085-C-00-
    1033.
McDowell, R.F. and P.E. Hildebrand, 1980. "Integrated
    Crop and Animal Production: Making the Most of
    Resources Available to Small Farms in Developing
    Countries." Rockefeller Foundation Conference
    Paper, Bellagio.
Norman, D.W., 1982. "Institutionalizing the Farming
    System Approach to Research." Paper Prepared for
    the Africa Bureau Agriculture and Rural Develop-
    ment Officers Workshop, Ibadan, Nigeria, May
    10-13, 1982.
Sandford, Stephen, 1981. "Organizing Governments' Role
    in Pastoral Sector." John Galaty, P. Salzman and
    D. Aronson, eds: The Future of Pastoral Peoples:
    Research Priorities for the 1980s. International
    Development Research Center, Ottawa.
_____, 1981. "Review of World Bank Livestock
    Activities in Dry Tropical Africa." IBRD, Washing-
    ton, D.C.
Sperling, Louise, 1980. "African Livestock Projects:
    A Documentary Review." Agency for International
    Development, Washington, D.C.
Winrock International Livestock Research and Training
    Center. "Livestock Program Strategy and Priori-
    ties." Winrock International. Morrilton, Arkansas.

List of Contributors

EDGAR J. ARIZA-NINO is assistant research scientist, Center for Research on Economic Development, the University of Michigan, Ann Arbor, MI 48109. He has been involved in numerous livestock production-marketing studies in Africa, including the Bakel Range and Livestock Project in Senegal (Project Director), Livestock and Meat Marketing in West Africa (Project Director), and Livestock and Agroforestry Integration in Kenya (with ICRAF). Publications include: Livestock and Meat Marketing in West Africa (with others); Consumption Effects of Agriculture Policies (with others).

JOAN ATHERTON is social scientist analyst, Program and Policy Coordination Bureau, USAID (PPC/PDPR/RD, Room 2675 NS, AID, Washington, D.C. 20523). She is involved in reviewing AID's livestock project designs for conformance with rural development policies and with related land tenure and participatory development strategies. She has worked on a short-term basis in Botswana, Uganda, Malawi, and Kenya on project evaluations and agricultural sector assessments.

ROY H. BEHNKE, JR. is an anthropologist currently residing at Cosme Road, Box 12A, Odessa, FL 33556. His research in Africa has included the study of the political organization of a pastoral society in Libya, the role of rural out-migration in creating and sustaining an indigenous form of Libyan small stock ranching, and in Botswana on the use of cattle in arable agriculture and other cattle management practices. Publications include The Herders of Cyrenaica, University of Illinois Press, 1980.

JOHN W. BENNETT is professor of anthropology at Washington University, St. Louis, Missouri, and the Land Tenure Center, University of Wisconsin, Madison, WI 53705. He has worked in Sudan and Egypt on problems of desertification and in Sudan, Kenya, and Tanzania involving pastoralism and livestock. Publications include: The Ecological Transition, Pergamon Press, 1976; Of Time and the Enterprise, University of Minnesota Press, 1982.

SAMSON CHEMA is Deputy Director for Livestock Research, Ministry of Livestock Development, Dept. of Veterinary Service, Veterinary Research Lab, P.O. Kabete, Kenya. Publications include: "Mixed Crop-Animal Systems on Small Farms in Western Kenya." SR-CRSP Technical Report #17, with others; "Seasonal variation in Feed Resources on Small Farms in Western Kenya." Paper presented at 74th Annual Meeting American Society Animal Science, Guelph, Ontario, August 1982, with others.

A. JOHN DE BOER is agricultural economist and program officer for Asia, Winrock International, Petit Jean Mountain, Route 3, Morrilton, AR 72110. Since 1978 he has been Principal Investigator (Agricultural Economics) for the Kenya Small Ruminant Collaborative Research Support Program. Publications include: "An Economic Analysis of Dual-Purpose Goat Adoption on Small Farms in Western Kenya," Winrock International; "Role of Small Ruminants in Agricultural Development Projects," Agriculture and Rural Development Department, World Bank.

ALEX DICKIE, IV is research assistant in the Range Science Department, Utah State University, Logan, UT 84322, where he is completing his doctorate. He has been a range management advisor for the USAID Maasai Range Livestock Development Project in Tanzania and the USAID Bakel Range Livestock Development Project in Senegal, and a range management representative in the multi-donor assessment of drought conditions in Mali for AID/Bamako.

PHYLO EVANGELOU recently completed his doctoral studies in the Food and Resource Economics Department, University of Florida, Gainesville, FL 32611. He has worked with ILCA's Kenya Country Programme in the study of pastoral livestock systems, and has also taught school in Botswana.

H.A. "HANK" FITZHUGH is Program Officer, Latin America and Caribbean, Winrock International, Route 3,

Petit Jean, Morrilton, AR 72110. He has been Principal Investigator, Title XII, SR-CRSP Dual Purpose Goat Production Systems Project in Maseno, Kenya, and has performed short-term consultancies in Nigeria, Liberia, Cameroon, and Ethiopia. Publications include: "The Role of Ruminants in Support of Man." Winrock International, Morrilton, AK, 1978, with others; Hair Sheep of Western Africa and the Americas. A Winrock International Study. Westview Press, Boulder, CO, 1983, co-editor.

KEITH E. GREGORY is a research geneticist at the U.S. Meat Animal Research Center, ARS, USDA, P.O. Box 166, Clay Center, NE 68933. He has conducted livestock development programs in Uganda and has been a consultant on cattle research and development programs elsewhere in Subsaharan Africa for ILCA. Publications include: "Comparison of Crossbreeding Systems and Breeding Stock Used in Suckling Herds in Continental and Temperate Areas," Proceedings, 2nd World Congress on Genetics Applied to Livestock Production, V (1982):482; "Heterosis, Crossbreeding and Composite Breed Utilization in the Tropics," as above, IV (1982):279.

R.D. HART is an agronomist with Winrock International, Route 3, Petit Jean, Morrilton, AR 72110, who has worked in Kenya. Publications include: "Integrative Agricultural Systems Research," in: Servant, J. and A. Pinchinat, Caribbean Seminar on Farming Systems Research Methodology, 555-565, IICA San Jose, Costa Rica, 1982, co-author; "One Farm System in Honduras; A Case Study in Farm System Research." in: Shaner, W.W.; P.F. Philipp, and W.R. Schmehl. Readings in Farming Systems Research and Development. 59-73. Westview Press, Boulder, CO, 1982.

MORGAN O. JOB is an agricultural economist living at Oropouche Road, Sangre Grande, Trinidad & Tobago, West Indies. He was a Research Assistant in Kenya, supported by Winrock International and the Economics portion of the title XII USAID Small Ruminant CRSP Project. Publications include: "Economic Factors Influencing the Simulated Adoption of Dual-Purpose Goats on Small Farms in Western Kenya, in: Proceedings of the First Small Ruminant CRSP, Kenya, Workshop. Nairobi, Kenya. pp. 40-48, 1983, with others.

JAMES A. KNIGHT is completing his doctoral dissertation on pastoral systems in central Niger at the University of Chicago. His research focuses on social and symbolic determinants of economic behavior and their mobilization in the course of economic change in pastoral systems.

LING-JUNG "KELVIN" KOONG is research leader, production systems unit, U.S. Meat Animal Research Center, ARS, USDA, P.O. Box 166, Clay Center, NE 68933. His primary interest is the application of mathematical modeling and systems analysis to animal production. He is known for his work in the quantitative description of nutritional energetics. Publications include: "The Application of Systems Analysis of Mathematical Modeling Techniques to Animal Science Research," Symposium on Use of the Computer in Animal Science Teaching, Research and Extension. Amer. Soc. Anim. Sci. (1978) 9-19; "A Quantitative Model of Energy Intake and Partition in Grazing Sheep in Various Physiological States," Anim. Prod. 25 (1977):133, co-author.

STEVEN W. LAWRY is a research assistant at the Land Tenure Center and is completing his doctoral studies in the Land Resources Program, Institute for Environmental Studies, University of Wisconsin, Madison, WI 53705. He has been engaged in land use planning and livestock development in Botswana and Sudan. Publications include: "The Matsberg Land Use Plan: A Case Study in Settlement and Agricultural Development Planning in Kgalagadi District," Settlement in Botswana, Heinemann Books, 1981; "Land Tenure, Land Policy and Smallholder Livestock Development in Botswana," Land Tenure Center Research Paper No. 78, March, 1983.

PETER D. LITTLE is research associate, Institute for Development Anthropology and adjunct assistant professor, SUNY-Binghamton, NY 13901. He has engaged in agricultural production and marketing research in several areas of Kenya, including a sociological study for the World Bank's Baringo Pilot Semi-Arid Area Project. Publications include: Food Production, Marketing and Consumption in the Semi-Arid Area of Baringo District, Kenya, Rome, FAO, 1982; "Evolution of Policy Toward the Development of Pastoral Areas in Kenya," co-authored with S.E. Migot-Adholla, The Future of Pastoral Peoples, ed. J. Galaty, D.

Aronson, P. Salzman, and A. Chouinard. Ottawa: International Development Research Centre, 1981.

ROBERT E. MCDOWELL is professor of international animal science, Cornell University, Ithaca, NY 14850. His work in Africa has included livestock research in Egypt, the study of trypantolerant cattle in the Gambia, program reviews in Mali, Ivory Coast, and Nigeria, analysis of game ranching in Kenya, and serving on the Agriculture Faculty, University of Zimbabwe. He is chairman of the board of ILCA. Publications include: Improvement of Livestock Production in Warm Climates; "Game or Cattle for Meat Production on Kenya Rangelands."

WILLIAM M. MOULTON is professor of International Veterinary Medicine at Tufts University. He has worked on livestock health problems in Kenya, Uganda, Tanzania, Mali, and Sudan, including the development of the animal health component of the USAID supported Mali livestock program and a review of the Small Ruminant CRSP in Kenya. Publications include various articles on animal health.

JAMES T. O'ROURKE is assistant professor in the Range Science Department, UMC 52, Utah State University, Logan, UT 84322. As a range management specialist, he has worked extensively in Tanzania, Senegal, and Morocco. Publications include: "Grazing Rate and System Trial over Five Years in a Medium-Height Grassland of Northern Tanzania," Proceedings, First International Range Congress, Denver, Colorado, August 14-18, 1978; "Toward More Effective Range Management Training," Natural Resource Technical Bulletin No. 3, AID/NPS Natural Resources Project, 1982.

DAVID J. PRATT is principal scientific officer, ODA/ LRDC, Tolworth Tower, Surbiton, Surrey KT6 7DY, England. He has worked for the Kenya Ministry of Agriculture, 1956-68, as Pasture Research Officer, Senior Range Research Officer, and Head of the Range Management Division; for UK Overseas Development Administration, 1968-77, including project planning and supervision (Swaziland and Ethiopia) and consultancies for FAO and World Bank (Somalia and Mali); and was Director General of ILCA, 1977-82. Publications include: Rangeland Management and Ecology in East Africa, co-edited with M.D. Gwynne, Stodder & Stoughton, 1977; "BushControl Studies in the Drier Areas of Kenya," Journal of Applied Ecology 3,4, & 8 (1966-67).

JAMES C. RIDDELL is associate professor at the Land
Tenure Center, University of Wisconsin, Madison.
He has participated in projects involving agricultural land use and land management training in
Liberia, Ivory Coast, Guinea, Ghana, Cameroon, and
Mauritania; continuing education in Egypt; and
short-term consultancies in Senegal, Niger, Sudan,
Kenya, Zambia, and Botswana. Publications
include: "Land Tenure Issues in Africa Development," co-authored with Dan Kanel and K.H. Parsons, Land Tenure Center Position Paper No. 12,
1978; "Reinventing the Square Wheel: Land Resources and Land Tenure in West African livestock
and Range Development Projects," in Migratory
Pastoralism and Its Development, Westview Press,
forthcoming.

MICHAEL W. SANDS is an animal scientist at IDIAP-CATIE,
Apartado Postal 178, David, Panama. He was
Research Associate with Title XII Dual-Purpose
Goat Production Systems project, Maseno, Kenya,
for 2 years. Publications include: "Mixed Crop-
Animals Systems on Small Farms in Western Kenya."
SR-CRSP Technical Report #17, 1982, with others;
"Results of Small Farm Systems Survey with
Implication to the Potential for Dual-Purpose
Goats on Small Farms in Western Kenya."
Proceedings of the Small Ruminant CRSP Workshop,
Nairobi, Kenya, pp 12-34, 1982, with others.

TJAART W. SCHILLHORN VAN VEEN is assistant dean for
research, College of Veterinary Medicine, Michigan
State University, East Lansing, MI 48823, and a
private farmer (sheep and cattle). He spent eight
years in northern Nigeria, mainly working on
epidemiology of animal diseases, especially for
small ruminants. He has held various consultancies in West and East Africa, as well as in
Central America.

HAROLD K. SCHNEIDER is professor of anthropology,
Indiana University, Bloomington, Indiana 47401.
He has conducted ethnographies of the Pokot in
Kenya and the Turu in Tanzania, studied the Bots-
wana Tribal Grazing Scheme, and been involved in
projects in Liberia, Nigeria, Uganda, and Ethi-
opia. Publications include: Livestock and
Equality in East Africa, I.U. Press, 1979; The
Africans, Prentice-Hall, 1981.

KENNETH H. SHAPIRO is associate dean and director of
International Agriculture Programs, University of
Wisconsin, Madison, WI 53705. He has conducted

socioeconomic research on farming systems in Tanzania, Niger, Upper Volta, Ivory Coast, and Mali. Publications include: "Sources of Technical Efficiency: The Roles of Modernization and Information," Economic Development and Cultural Change, 1978; "Market Forces and Livestock Development in Africa," World Bank, Agriculture and Rural Development Technical Note, 1983.

AHMED E. SIDAHMED is conducting systems research in western Kenya for Winrock International. He has worked on research and management policies for range and irrigated pasture livestock systems in Sudan, and has held a post-doctoral fellowship to study the plant/animal interface using mathematical modeling at the U.S. Meat Animal Research Center. Publications include: "Contribution of Browse to Protein and Energy Requirements of Spanish Goats," J. Animal Science (1981); "Effect of Fasting on Digestibility, Intake and Composition of Esophageal Samples Collected by Grazing Sheep," J. Animal Science (1977).

JAMES R. SIMPSON is professor in the Food and Resource Economics Department, University of Florida Gainesville, FL 32611. He has conducted cattle production and marketing research in Botswana, Ethiopia, and Kenya. Publications include: The World's Beef Business, Iowa State University Press, 1982; over 100 other publications on development, and livestock and meat industries, worldwide.

ALBERT E. SOLLOD is associate professor of medicine and head of the Section of International Veterinary Medicine, Tufts Veterinary School, Grafton, MA 01536. He has filled long and short-term assignments in Mali, Senegal, Mauritania, Niger, Upper Volta, Togo, Kenya, Burundi, and Ivory Coast, and his current research entails the integration of veterinary epidemiology into pastoral systems studies. Publications include: "Bovine Trypanosomiasis: Effect on the Immune Response of the Infected Host," co-authored with G.H. Frank, American Journal of Veterinary Research 40 (1979):658-64; "Patterns of Disease in Sylvopastoral Herds of Central Niger." USAID, Niamey, 1981.

J. DIRCK STRYKER is associate professor of international economic relations, Fletcher School, Tufts University, and president of Associates for International Resources and Development, 287 Alewife

Brook Parkway, Somerville, MA 01244. He has worked widely in Africa, including studies of comparative costs and incentives in the livestock and other agricultural sectors of Ivory Coast, Ghana, Mali, Senegal, Guinea, Niger, Upper Volta, Senegal, Cameroon, and Madagascar. Publications include: Rice in West Africa: Policy and Economics, coauthor, 1981; "Malian Livestock Production and Distribution," 1973.

GREGORY M. SULLIVAN is assistant professor of agricultural economics, Auburn University, AL 36849. He has conducted livestock production and marketing research in Tanzania and Ghana, as well as a socioeconomic analysis of the Niger River system. Publications include: "Simulation of Production Systems in East Africa by Use of Interfaced Forage and Cattle Models," Agricultural Systems, Vol. 7, pp. 245-265, November 1981, with others; A Simulation Model for Measuring the Impact of an Improved Management Practice in Traditional Livestock Systems in Tanzania," Journal of Range Management, Vol. 33, No. 3, pp. 174-178, May 1980, with others.

THURSTON F. TEELE, an economist by profession, is founder and director of Chemonics International Consulting Division, 2000 M St. N.W., Suite 200, Washington, DC 20036. He has been Chief-of-Party and Study Director on livestock projects in Mali and Kenya, and has prepared proposals, negotiated contracts and supervised technical assistance projects in livestock, agriculture and rural development throughtout Africa, Asia and the Middle East. Publications include: "Kenya Livestock and Meat Development Study, 1977," Chemonics; "Final Report, Mali Livestock II and I," Chemonics and USAID, 1983.

JOHN C.M. TRAIL is principal scientist, International Livestock Centre for Africa (ILCA), Box 46847, Nairobi, Kenya. He has conducted livestock production research in Uganda (1957-69), Botswana (1970-76), Ethiopia (1976-77), and Kenya (1978 to present), and has provided input for a considerable number of projects throughout Africa. Publications include: "Trypanotolerant Livestock in West and Central Africa"; "Aspects of Indigenous and Crossbreed Beef Cattle Production in East and Southern Africa."

R. TREVOR WILSON is coordinator and principal animal scientist for the Small Ruminant and Camel Group

of the International Livestock Centre for Africa,
P.O. Box 5689, Addis Ababa, Ethiopia. He has
spent 20 years in Africa during which time he has
studied commercial ranch management in Tanzania
and Kenya, traditional livestock systems in Kenya,
Sudan, Ethiopia, and Mali, and wildlife ecology in
Sudan, and Ethiopia. In addition, he has conduct-
ed short-term research in Niger, Upper Volta, and
Nigeria. Publications include: "Studies on the
Livestock of Southern Darfur, Sudan," Trop. Anim.
Hlth. Prod. (eight papers, 1975-83); The Camel,
London: Longmans, 1983.

KATHERINE WOLFGANG is currently a student at Tufts
University, School of Veterinary Medicine,
Grafton, MA 01536, where she plans to work with
the Department of International Health. She has
recently returned from Upper Volta where she
completed a study funded by AID on traditional
cattle health care by the Fulani of Southeast
Upper Volta.

Index

T - #0134 - 071024 - C0 - 222/146/24 - PB - 9780367167011 - Gloss Lamination